IoT and Low-Power Wireless

Devices, Circuits, and Systems

Series Editor
Krzysztof Iniewski

Wireless Technologies
Circuits, Systems, and Devices
Krzysztof Iniewski

Circuits at the Nanoscale
Communications, Imaging, and Sensing
Krzysztof Iniewski

Internet Networks
Wired, Wireless, and Optical Technologies
Krzysztof Iniewski

Semiconductor Radiation Detection Systems
Krzysztof Iniewski

Electronics for Radiation Detection
Krzysztof Iniewski

Radiation Effects in Semiconductors
Krzysztof Iniewski

Electrical Solitons
Theory, Design, and Applications
David Ricketts and Donhee Ham

Semiconductors
Integrated Circuit Design for Manufacturability
Artur Balasinski

Integrated Microsystems
Electronics, Photonics, and Biotechnology
Krzysztof Iniewski

Nano-Semiconductors
Devices and Technology
Krzysztof Iniewski

Atomic Nanoscale Technology in the Nuclear Industry
Taeho Woo

Telecommunication Networks
Eugenio Iannone

For more information about this series, please visit: https://www.crcpress.
com/Devices-Circuits-and-Systems/book-series/CRCDEVCIRSYS

IoT and Low-Power Wireless

Circuits, Architectures, and Techniques

Edited by
Christopher Siu

Managing Editor
Krzysztof Iniewski

CRC Press
Taylor & Francis Group
Boca Raton London New York

CRC Press is an imprint of the
Taylor & Francis Group, an **informa** business

CRC Press
Taylor & Francis Group
6000 Broken Sound Parkway NW, Suite 300
Boca Raton, FL 33487-2742

First issued in paperback 2020

© 2018 by Taylor & Francis Group, LLC
CRC Press is an imprint of Taylor & Francis Group, an Informa business

No claim to original U.S. Government works

ISBN 13: 978-0-367-65602-7 (pbk)
ISBN 13: 978-0-8153-6971-4 (hbk)

Library of Congress Cataloging-in-Publication Data

Names: Siu, Christopher, author. | Iniewski, Krzysztof, 1960- author.
Title: IoT and low-power wireless : circuits, architectures, and techniques /
Christopher Siu and Krzysztof Iniewski.
Description: Boca Raton, FL: CRC Press/Taylor & Francis Group, 2018. |
"A CRC title, part of the Taylor & Francis imprint, a member of the Taylor &
Francis Group, the academic division of T&F Informa plc." | Includes
bibliograpical references and index.
Identifiers: LCCN 2018010550 | ISBN 9780815369714 (hardback: acid-free paper) |
ISBN 9781351251662 (ebook)
Subjects: LCSH: Internet of things–Equipment and supplies. | Near-field
communication.
Classification: LCC TK5105.8857 .S57 2018 | DDC 621.39/81–dc23
LC record available at https://lccn.loc.gov/2018010550

Visit the Taylor & Francis Web site at
http://www.taylorandfrancis.com

and the CRC Press Web site at
http://www.crcpress.com

Table of Contents

List of Figures

List of Tables

Preface

Sometime in the future, we may look back and reflect that we are living in times during which a unique confluence of technologies is creating a new paradigm for networked devices, commonly referred to as the internet of things (IoT). The idea behind IoT dates back to the 1990s, when Kevin Ashton was a brand manager at Proctor & Gamble (P&G). In 1997, Ashton and his team were tasked with promoting Oil of Olay lipsticks. When Ashton noticed that some retail stores were not stocked with the product, he realized that human data entry for restocking the lipstick is unreliable. He thus came up with the idea of taking the Radio Frequency Identification (RFID) chip out of a contactless smart card and attaching one to each lipstick to track store inventory. Ashton then extended this idea, and pitched a solution to solve P&G's supply chain problem to the executives. Although the price of RFID tags was still prohibitive at that time, Ashton was convinced that one day the price will drop enough for this idea to be economically feasible. P&G executives funded the research project, and Ashton eventually became the executive director of Massachusetts Institute of Technology (MIT)'s Auto-ID Center, where he was able to further his vision.

Today, roughly 20 years after Ashton's idea, we are able to see his IoT concept coming to fruition. The convergence and advancement of several technologies have made this possible, including

- Sensor and actuator technology

- Wireless technology

- Computational power and network protocol

- Miniaturization of devices, with integrated circuit technology riding Moore's law to the limit

The chapters in this book cover some of the wireless research that will enable the implementation of IoT. The book also looks ahead at advanced wireless techniques that will continue the evolution in ubiquitous wireless communication.

Chapter 1: This chapter provides an overview of IoT, focusing on the technologies deployed for the physical and link layers. Emerging standards for IoT are also outlined.

Chapter 2: Low-power wearables have entered into the mainstream consumer market, with fitness devices that monitor exercise and heart rate being the most prevalent. This chapter explores the usage of wearables in the medical market, and the challenges that come with designing sensors and electronics for such devices.

Chapter 3: The challenge of wearable medical monitoring is further explored in the context of algorithms and firmware. Algorithms that can reliably interpret the physiological and biomechanical signals, derive metrics from them, and predict clinically significant events are one of the keys to success in this market.

Chapter 4: Connecting numerous devices into a wireless sensor network is the focus of this chapter. Distributed versus centralized architectures are discussed, including techniques that can improve the efficiency and robustness of the network.

Chapter 5: A key technique for IoT devices to run years on a single battery is to put the receiver to sleep for as long and as often as possible. This chapter addresses this important issue with the wake-up receiver method to achieve energy-efficient communication.

Chapter 6: As Complementary Metal Oxide Semiconductor (CMOS) process scaling continues and the supply voltage continues to shrink, voltage resolution and dynamic range in analog circuits also deteriorate. Over the past decade, engineers have adjusted their design strategy by taking advantage of the time resolution of CMOS, resulting in the time-to-digital converter (TDC). Various innovations have been developed for TDCs, and this chapter presents an all-digital TDC architecture with delta-sigma noise shaping.

Chapter 7: The power amplifier is one of the power hungry blocks within a Radio Frequency (RF) transmitter. The aim of achieving high efficiency and high linearity is a continual design challenge. In this chapter, a systematic design technique is presented, along with the analysis of a current mode digital RF power amplifier incorporating predistortion.

Chapter 8: Frequency synthesis using a phase locked loop (PLL) is another power hungry function within an RF transceiver. Within the PLL, the voltage-controlled oscillator and frequency divider consumes much of the power. As a result, injection locking has been studied to reduce power consumption, and this chapter provides an analysis of various injection-locked techniques.

Chapter 9: The Cartesian In-Phase and Quadrature (I/Q) modulator driven by a PLL has been a conventional architecture used in RF transmitters, but the need for RF mixers and filters has presented challenges in deep-submicron CMOS. Over the past decade, efficient digital transmitter architectures that avoid the use of mixers and filters have gained traction. In this chapter, the use of powerful digital calibration techniques in a direct modulation PLL has enabled further performance gains.

Chapter 10: As the spectra at 2.4 and 5 GHz have become very crowded, engineers are looking at higher frequencies for future deployment. WiGig is one example of moving WiFi to the 60 GHz band for enabling multi-Gbps wireless communication. Techniques for frequency synthesis at 60 GHz are discussed in this chapter. An injection-locked 60 GHz oscillator is used in conjunction with a subsampling PLL to achieve low-power and low-phase noise. An implementation of these techniques in 65-nm CMOS is presented along with the measured results.

Chapter 11: Fifth generation wireless is presently under definition and development, and one consideration is the integration of IoT into the network. Heterogeneous architectures have been proposed, where Wireless Local Area Network (WLAN) is used in dense small cells. The latest status of IEEE 802.11ad/WiGig in the 60 GHz band is presented in this chapter, including a low-power CMOS transceiver with beamforming capability.

Chapter 12: Battery life has always been a key issue in portable devices, and it has become crucial for IoT as it is impractical to replace the battery in billions of devices regularly. While the earlier chapters focused on circuit techniques and protocol innovations to extend the battery life, this chapter looks at ways that we can recharge the battery without user intervention. While energy scavenging has been considered for IoT nodes, wireless charging has also made its way into the consumer market. This chapter presents an efficient power management structure for inductive power delivery and its applications in markets such as implantable medical devices.

Christopher Siu
Kris Iniewski
Editors
Vancouver, Canada

Series Editor

Krzysztof (Kris) Iniewski is managing R&D at Redlen Technologies Inc., a start-up company in Vancouver, Canada. Redlen's revolutionary production process for advanced semiconductor materials enables a new generation of more accurate, all-digital, radiation-based imaging solutions. Kris is also a President of CMOS Emerging Technologies (www.cmoset.com), an organization of high-tech events covering Communications, Microsystems, Optoelectronics, and Sensors. In his career, Dr. Iniewski held numerous faculty and management positions at University of Toronto, University of Alberta, SFU, and PMC-Sierra Inc. He has published over 100 research papers in international journals and conferences. He holds 18 international patents granted in USA, Canada, France, Germany, and Japan. He is a frequent invited speaker and has consulted for multiple organizations internationally. He has written and edited several books for IEEE Press, Wiley, CRC Press, McGraw Hill, Artech House, and Springer. His personal goal is to contribute to healthy living and sustainability through innovative engineering solutions. In his leisurely time, Kris can be found hiking, sailing, skiing, or biking in beautiful British Columbia. He can be reached at kris.iniewski@gmail.com.

Editor

Christopher Siu is a faculty member at the Department of Electrical and Computer Engineering Technology, British Columbia Institute of Technology (BCIT), located in Burnaby, British Columbia, Canada. Chris is also a founder of Wavelink Electronics Ltd. and Tyche Technologies Inc., consulting companies specializing in the design of analog and radio frequency electronics. He obtained a master's degree from Stanford University, California and a bachelor's degree from Simon Fraser University, British Columbia, both in electrical engineering. Chris is also a licensed professional engineer in the province of British Columbia. During his career, Chris has worked in Silicon Valley and in Canada, for companies such as Hewlett Packard, Philips Semiconductor, and PMC-Sierra. He has designed analog and RF integrated circuits that have been released to production as well as managed engineering teams across multiple sites. When not teaching or practicing engineering, he likes to spend his time skiing, playing tennis, and traveling.

List of Contributors

Su-Shin Ang
Inova Design Solutions Ltd.
London, United Kingdom

Miriam Carlos-Mancilla
Universidad del Valle de Mexico
Tlaquepaque, Jalisco, Mexico

Wei Deng
Apple Inc.
Cupertino, California

Hesam S. Gougheri
Department of Electrical Engineering
School of Electrical Engineering and
    Computer Science
Pennsylvania State University
University Park, Pennsylvania

Miguel Hernandez-Silveira
Sensium® Healthcare Ltd.
Abingdon, United Kingdom

Kris Iniewski
ET CMOS Inc.
Port Moody, British Columbia,
    Canada and
Redlen Technologies Inc.
Saanichton, British Columbia,
    Canada

Masataka Irie
Wireless Technology Department
Platform Development Center
Automotive & Industrial Systems
Company Panasonic Corporation
Yokohama, Japan

Syed K. Islam
Department of Health Sciences
University of Tennessee
Knoxville, Tennessee

Heikki Karvonen
Centre for Wireless
    Communication
University of Oulu
Oulu, Finland

Makihiko Katsuragi
Semiconductor Research and
    Development
Toshiba Corporation
Kawasaki, Japan

Mehdi Kiani
Department of Electrical
    Engineering
School of Electrical Engineering and
    Computer Science
Pennsylvania State University
University Park, Pennsylvania

Kento Kimura
Fujitsu Ltd.
Kawasaki, Japan

Masashi Kobayashi
Wireless Technology Department
Platform Development Center
Automotive & Industrial Systems
Company Panasonic Corporation
Yokohama, Japan

Satoshi Kondo
Corporate Research and
    Development Center
Toshiba Corporation
Kawasaki, Japan

Salvatore Levantino
Dipartimento di elettronica,
    informazione e bioingegneria
    (DEIB)
Politecnico di Milano
Milan, Italy

Hanli Liu
Department of Physical Electronics
Tokyo Institute of Technology
Tokyo, Japan

Ernesto López-Mellado
CINVESTAV Unidad Guadalajara
Zapopan, Jalisco, Mexico

Ifana Mahbub
Department of Electrical Engineering
University of North Texas
Denton, Texas

Akira Matsuzawa
Department of Physical Electronics
Tokyo Institute of Technology
Tokyo, Japan

Hiroyuki Motozuka
Wireless Technology Department
Platform Development Center
Automotive & Industrial Systems
    Company
Panasonic Corporation
Yokohama, Japan

Kenichi Okada
Department of Physical Electronics
Tokyo Institute of Technology
Tokyo, Japan

Juha Petäjäjärvi
Centre for Wireless Communication
University of Oulu
Oulu, Finland

Suraj Prakash
Department of Electrical and
    Computer Engineering
Texas A&M University
College Station, Texas

Salvatore Pullano
Department of Electrical Engineering
    and Computer Science
University of Tennessee
Knoxville, Tennessee
and
Department of Health Sciences
University Magna Graecia of
Catanzaro Catanzaro, Italy

Haoyu Qian
Qualcom Technologies Inc.
San Diego, California

Carlo Samori
Dipartimento di elettronica,
    informazione e bioingegneria
    (DEIB)
Politecnico di Milano
Milan, Italy

Samira Shamsir
Department of Electrical Engineering
    and Computer Science
University of Tennessee
Knoxville, Tennessee

Naganori Shirakata
Wireless Technology Department
Platform Development Center
Automotive & Industrial Systems
Company Panasonic Corporation
Yokohama, Japan

Mario Siller
CINVESTAV Unidad Guadalajara
Zapopan, Jalisco, Mexico

Jose Silva-Martinez
Department of Electrical and
    Computer Engineering
Texas A&M University
College Station, Texas

Teerachot Siriburanon
School of Electrical, Electronic &
    Communications Engineering
University College Dublin
Dublin, Ireland

Christopher Siu
Department of Electrical and
    Computer Engineering Technology
British Columbia Institute of
    Technology
Burnaby, British Columbia,
Canada

Hiroshi Takahashi
Wireless Technology Department
Platform Development Center
Automotive & Industrial Systems
Company Panasonic Corporation
Yokohama, Japan

Kazuaki Takahashi
Wireless Technology Department
Platform Development Center
Automotive & Industrial Systems
Company Panasonic Corporation
Yokohama, Japan

Koji Takinami
Wireless Technology Department
Platform Development Center
Automotive & Industrial Systems
Company Panasonic Corporation
Yokohama, Japan

Tomoya Urushihara
Wireless Technology Department
Platform Development Center
Automotive & Industrial Systems
Company Panasonic Corporation
Yokohama, Japan

Fei Yuan
Department of Electrical and
    Computer Engineering
Ryerson University
Toronto, Ontario, Canada

Yushi Zhou
Department of Electrical Engineering
Lakehead University
Thunder Bay, Ontario, Canada

1

The Internet of Things—Physical and Link Layers Overview

Christopher Siu

British Columbia Institute of Technology (BCIT)

Kris Iniewski

Redlen Technologies Inc.

CONTENTS

1.1 Introduction

The internet of things (IoT) has sometimes been referred to as the digitization of the physical world. It is a confluence of different technologies at low-enough costs that makes this possible. While different definitions of IoT exist, we will use the following description for this book:

> A device embedded with a sensor and/or actuator, connected to the internet, that shares its information with other devices and hosts, with the potential to act on this information based upon some rules and intelligence.

1

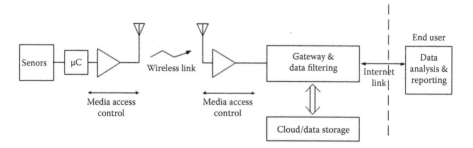

FIGURE 1.1
Simplified IoT system block diagram.

In the simplified block diagram given later, sensors in an end node collect data at specified intervals. The data are framed into packets by the microcontroller, which also contains parts of the protocol stack to perform media access control (MAC). The packets are modulated and transmitted over the wireless link, which is received by a gateway connected to the internet. The gateway may have a rules engine to reduce the amount of data before it is stored. The sensor data may then be transferred to an end user for further analysis and report generation (Figure 1.1).

Note that while the gateway may be mains powered, the sensor nodes will be powered by battery and/or energy scavenging. Since it is not feasible to change the battery regularly on a large number of sensor nodes, there is great motivation to reduce the power consumption of end nodes as much as possible.

1.2   Radio and MAC Technologies for IoT

In conceptualizing computer networks, many of us have seen the 7-layer open systems interconnection (OSI) model. The 7 layers, from the lowest to the highest, are the physical, link, network, transport, session, presentation, and application layers. Over the past two decades, with the exponential growth of the internet running transmission control protocol/internet protocol (TCP/IP), the OSI model has been eclipsed by a 5-layer model, sometimes referred to as the TCP model or the IP stack [1]. Shown later is the TCP model with the corresponding standards and protocols for WiFi (Figure 1.2).

The physical layer defines the hardware aspects of the communication link, such as the modulation method, voltage levels, and physical medium (e.g., copper wire, over-the-air). The link layer provides several services, typically implemented with a combination of hardware and software. If the physical medium is shared by multiple users, such as wireless communication on a certain frequency band, then orderly access to the medium must be controlled

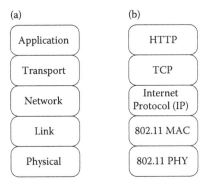

FIGURE 1.2
The 5-layer model in relation to WiFi.

so that users don't interfere with each. The mechanism for this is aptly named MAC, and it is a key function of the link layer. Other services provided by the link layer include framing of higher layer data and delivering the data reliably.

The focus of this book is on the physical and link layer technologies that are in development for IoT. As such, this chapter provides an overview of these technologies, but the higher layers will also be mentioned where it is appropriate.

1.2.1 Physical layer with existing radio frequency (RF) standards

One of the main energy consumers in mobile systems is the wireless transceiver. Hence, research into low-power circuit techniques is ongoing, but there are limits on using this approach alone. Additional innovations in MAC and network architecture have also been necessary to drastically reduce the transceiver power consumption. At the present time, there is no de facto RF standard for IoT; existing standards are repositioning themselves, and new standards are being introduced to support this new market. In the following, we will briefly survey some of these RF technologies and standards.

Bluetooth (IEEE 802.15.1) was conceived to be a wire replacement for computer peripherals, for example, the connection between a PC and a mouse. As such, it falls within the classification of a wireless personal area network (WPAN), for short-range point-to-point connections. Today, a large number of mobile devices like smartphones include Bluetooth capability, and the standard is constantly evolving to create Bluetooth low energy with mesh networking to support new market needs.

IEEE 802.15.4 was created as a lower power, lower data rate alternative to Bluetooth. In 802.15.4–2006, a 2.4 GHz physical layer using spread spectrum is specified at 250 kbps. Over the years, it has been used as the platform for Zigbee, Thread, and other proprietary solutions. It is also part of IPv6 over Low-Power Wireless Personal Area Network (6LoWPAN), which supports IPv6 addressing for network nodes. Although there is no native support for mesh networking in 802.15.4, it has been implemented in the higher layers for various applications. Similar to Bluetooth, however, it is a short-range standard.

WiFi (IEEE 802.11) has become one of the most ubiquitous wireless standards on the planet. The standard is designed to support an ethernet-based wireless local area network (WLAN), and so the range and power consumption are necessarily higher than those of Bluetooth and 802.15.4. Although WiFi radio transceivers are not a popular choice for low-power wireless systems, a new task group called 802.11ba has been formed to address this. In particular, this task group is creating a new standard for low-power wake-up radio (LP-WUR) in WiFi, intended to make WiFi an attractive technology for IoT.

One of the key ideas for LP-WUR is to use an ultra-low power auxiliary receiver to detect a wake-up packet, while keeping the main WiFi radio transceiver in sleep mode most of the time. In fact, the auxiliary receiver itself may be duty-cycled between sleep and wake to further reduce power consumption (Figure 1.3).

To make an ultra-low power receiver, some obvious tradeoffs such as performance, data rate, and modulation scheme need to be considered. For the IEEE 802.11ba initiative, on–off keying is used to allow for a simple demodulation. Furthermore, low-power radio circuits can be used, including techniques such as

- superregenerative receiver (Figure 1.4)

- envelope detection

- injection locking

- subsampling architectures

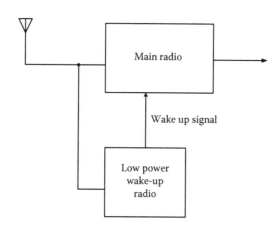

FIGURE 1.3
Wake-up radio concept.

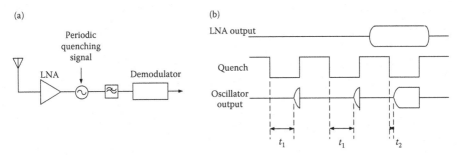

FIGURE 1.4
Superregenerative receiver: (a) block diagram and (b) internal waveforms.

Superregeneration is an idea developed by Edwin Armstrong in the early 1920s, and in its modern implementation, it removes the phase locked loop from a typical radio receiver. In conjunction with on–off keying, the oscillator start-up time depends on whether a signal is received by the low noise amplifier (LNA) or not. By detecting this time difference, the receiver decides whether a logic 0 or 1 was transmitted.

1.2.2 Physical layer with emerging radio frequency (RF) standards

The existing RF standards competing for market share in IoT have tended to be WPAN and WLAN standards, since the strict need for low power consumption favors these short-range applications. The architecture implied here is a large number of sensor nodes connected to gateway(s) either directly or via a mesh network. The short range of these standards also creates potential problems if one node is not in range of any other nodes and/or gateways.

Many of us are accustomed to a wide area cellular coverage; we never think about being near a gateway or base station before communicating on our mobile phones. A wireless wide area network (WWAN) is thus very attractive in terms of network access, but devices connected to a WWAN also have high power consumption. Just as the IEEE 802.x standards are evolving to meet IoT needs, so are the cellular standards. We will survey the following WWAN for IoT:

- Narrowband IoT (NB-IoT)

- Sigfox

- LoRaWAN

The 3rd Generation Partnership Project has been defining cellular standards since the third generation, and this now includes the 4th generation long term evolution (LTE) standard that is in use. LTE has undergone a number of revisions, and one of the latest releases (Rel 13) defines NB-IoT, which is a low-power, low-data rate service at 250 kbps.

Sigfox is a proprietary standard operated by a company of the same name. Sigfox uses a scheme called ultra narrow band modulation, which requires only 100 Hz of bandwidth per message, with a correspondingly low rate of 100–600 bps. At the present time, the coverage and deployment are much more extensive in Western Europe than in the United States.

LoRaWAN and LoRa are open standards for low-power WWAN; LoRaWAN specifies the MAC, and LoRa specifies the physical layer. LoRa uses spread spectrum modulation, and hence has built-in resistance to interference and multipath fading. LoRa also has a low-data rate in the tens of kbps, allowing the present integrated circuit implementations (SEMTECH SX127n series) to receive sensitivity in the −140 to −150 dBm range.

1.2.3   Link layer considerations for WUR

If radio duty cycling is fundamental to low-power wireless, then the MAC layer must be designed to support this need. For example, the 802.11ba LP-WUR uses a new wake-up packet to inform the wake-up receiver that the main radio needs to be taken out of sleep and prepare for data exchange (Figure 1.5).

Since the WUR itself is duty-cycled, this scheme inevitably introduces latency into system. Optimization between power consumption and latency is a topic of research, extending into the format of the wake-up packet. For example, Yoon [2] implemented a WUR with two modes of operation. The WUR starts in monitoring mode used to detect the packet start bits at a very low data rate; this mode is duty-cycled to further reduce the WUR power consumption. After detecting the start bits, the WUR enters identification mode (ID) to receive device addresses at a higher data rate.

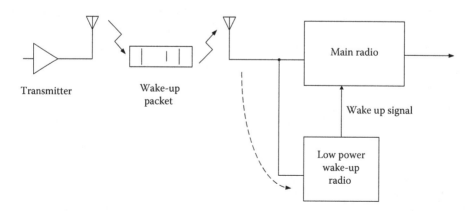

FIGURE 1.5
Wake-up radio MAC layer requirement.

1.2.4 Link layer example—6LoWPAN

A common IoT usage scenario is a network of sensors designed to run for years on battery and/or energy scavenging. We have already illustrated how low-power radio transceivers and duty cycling play a huge role in making this possible. If we look at this usage scenario more closely, typical sensors need to transfer small volumes of data, in contrast to the requirements of PCs, smartphones, etc. Since short messages conserve power and bandwidth, the design of the link layer must compromise between the overhead versus the payload of a frame.

Over the past two decades, the network infrastructure using the TCP/IP suite of protocols has grown exponentially. As a result, efforts to use IP addressing for IoT has resulted in the *Thread* specification [3,4], which uses IEEE 802.15.4 as the physical and MAC layers, and 6LoWPAN as a bridge between the 802.15.4 MAC and the IP (Figure 1.6).

6LoWPAN is an open IoT networking protocol that is specified by the Internet Engineering Task Force (IETF). It creates an adaptation mechanism between IPv6 in layer 3 and the 802.15.4 MAC in layer 2. Since a full unmodified TCP/IP stack may be incompatible with the limited hardware in IoT devices, 6LoWPAN creates a streamlined routing protocol that reduces network overhead and latency.

In using TCP/IP with 802.15.4, the following issues are addressed by 6LoWPAN:

- Adaptation needed for maximum transmission unit (MTU) size

- Reduction of overhead

- User datagram protocol (UDP) instead of TCP to reduce latency

Adaptation is needed to accommodate the different MTU sizes between IPv6 and IEEE 802.15.4. MTU is the size of the largest network layer protocol data

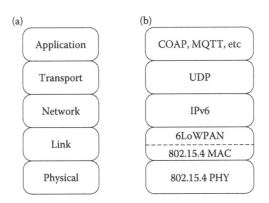

FIGURE 1.6
Thread specification in the 5-layer model.

unit that can be communicated in a single network transaction [5]. IPv6 has a packet size of 1280 bytes, while IEEE 802.15.4 allows for an MTU of only 127 bytes. 6LoWPAN introduces a fragmentation scheme to allow IPv6 to operate over an 802.15.4 network, using a 11-bit fragmentation header that allows for 2048 bytes packet size [6]. However, fragmentation can still lead to bad performance over a lossy network, so it is best to avoid big packet sizes.

Of the 127 bytes allowed by the 802.15.4 MTU, the upper layers like the IPv6 and UDP headers can consume significant amount of the MTU, leaving only 33 bytes for the actual payload [7]. To reduce this overhead, header compression is used to create more room for the payload [8].

IEEE 802.15.4 does not include mesh routing in the MAC specification; it uses simple addressing that supports star and peer-to-peer topologies. Mesh networks are thus outside of the 802.15.4 standard, and mesh support is implemented between the MAC and network layers. 6LoWPAN has a field for mesh headers and allows fast forwarding of packets in a mesh without traveling through the IP stack; this is referred to as mesh-under (layer 2) forwarding and route-over (layer 3) forwarding.

1.2.5   Application layer protocols

Given all the potential data collected by sensors and other "things," certain questions arise. How we should store all this data, if we should store it at all? Are there any time constraints on analyzing this data and acting on it? The answers to these questions depend on the application, and it may be useful to put IoT devices into some of these frameworks:

- **D2D**—Device-to-device communication: intelligent machines collect data and coordinate some action together. Also referred to as machine-to-machine communication.

- **D2S**—Device-to-server communication: device data are collected and sent to a server and the IT infrastructure.

- **S2S**—Server-to-server communication: the servers share data to command some action back to the devices, to analyze the data, and/or to generate a report for humans to view.

In all these frameworks, the ability to connect thousands of devices and interact with them in real time is crucial. Real time, however, varies depending on the application; the tolerable latency is different between car accident avoidance and farmland irrigation. As a result, different IoT protocols have emerged that have strengths in different frameworks. Some of the protocols available today include

- MQTT

- XMPP

- AMQP

- CoAP

- DDS

Message queue telemetry transport (MQTT) is targeted for D2S, collecting data from large number of devices and transferring that to the server infrastructure. The telemetry in its name stems from using this data for remote monitoring and control. Many sensors may connect to a data concentrator such as IBM's MessageSight appliance. Because the latency of several seconds is tolerable and the data transfer must be reliable, TCP instead of UDP is used in the transport layer. By design, MQTT is meant for data transfer from the end node to the IT infrastructure, and not useful for D2D communication. MQTT has been defined as a standard under the Organization for the Advancement of Structured Information Standards [9].

Extensible messaging and presence protocol (XMPP) is designed for connecting devices to humans; it uses XML text and name@domain.com addressing. As a result, it is useful in consumer IoT applications, such as connecting home appliances to a web server that a person can access using a smartphone.

Advanced message queuing protocol (AMQP) is a queuing system designed to connect servers together (S2S). Transactional messages are exchanged between servers, buffered using a queue. Since reliability is of great importance here, TCP is used in this protocol. AMQP has been defined as an Organization for the Advancement of Structured Information Standards standard [10] since 2012.

Constrained application protocol (CoAP) uses HTTP commands like GET and PUT for D2D communication. It is a web transfer protocol designed for constrained nodes and constrained networks that are low power but lossy. The end nodes are typically constrained in computing power and memory, while IP networks using 6LoWPAN can experience high error rates. CoAP has been defined under IETF RFC 7252 [11].

Finally, the main purpose of data distribution service (DDS) is for D2D communication. DDS can deliver millions of messages per second to many devices and offers ways to filter data and select multiple destinations for this data, in effect, implementing a multicast operation. D2D requires low latency that can vary depending on the device, and hence instead of TCP, DDS uses a quality-of-service (QoS) control scheme. The application of DDS includes the hospital environment and military systems.

1.2.6 Future directions

The allocation of the industrial, scientific, and medical (ISM) frequency bands for unlicensed use has been a resounding success. The 2.4 GHz band, available worldwide, is already overcrowded by the proliferation of wireless devices such as WiFi, Bluetooth, and so forth. Efforts have been made to migrate devices to other bands, with one notable example being the definition of 802.11a WLAN for operation in the 5.7 GHz band.

Although IoT deployment is still in its infancy, it is not too early to think how this exponential growth in wireless nodes will coexist with other devices in the ISM band. By design, wireless standards for the ISM bands are resistant to some amount of interference, using techniques such as spread spectrum. However, one can also look at the other ISM bands, such as 24 and 60 GHz, to accommodate new sensors and "things." Circuit technique at 60 GHz is an ongoing area of research, and while the state of the art is still too power hungry for IoT, performance will continue to improve. This improvement will accelerate once 60 GHz ICs are used in the consumer space, such as the IEEE 802.11ad standard to provide multi-Gbps wireless capability in this band. Although there are fundamental limits such as increased path loss at 60 GHz, the small wavelengths in the millimeter range create opportunities for advanced antennas and integration.

1.3   Conclusions

Traditional business models are based on a static information architecture, and the IoT is poised to change that. When a customer's buying preferences are sensed in real time at a specific location, dynamic pricing may increase the odds of a purchase. Knowing how often or intensively a product is used can create additional options—usage fees rather than outright sale, for example [12]. To realize the IoT vision, a great number of technical challenges are yet to be defined and solved. This is indeed an exciting time for engineers and researchers putting this ecosystem together.

Bibliography

[1] J. Kurose and K. Ross, *Computer Networking: A Top-Down Approach*, Pearson, Upper Saddle River, NJ, 2013.

[2] D. Yoon, "A New Approach to Low-Power and Low-Latency Wake-Up Receiver System for Wireless Sensor Nodes," *IEEE Journal of Solid State Circuits*, vol. 47, no. 10, pp. 2405–2419, 2012.

[3] Thread Group, "Thread Home Page," [Online]. Available: http://threadgroup.org/.

[4] Thread Group, "Thread Stack Fundamentals," July 2015. [Online]. https://portal.threadgroup.org/DesktopModules/Inventures_Document/ FileDownload.aspx?ContentID=633

[5] "RFC 791- Internet Protocol," 1981. [Online]. Available: http://www.rfc-base.org/rfc-791.html.

[6] IETF, "RFC4944- Transmission of IPv6 Packets over IEEE 802.15.4 Networks," 2007. [Online]. Available: https://www.rfc-editor.org/info/rfc4944.

[7] S. Schmidt, "6LoWPAN: An Open IoT Networking Protocol," in *Samsung Open Source Conference 2015*, Seoul, 2015.

[8] IETF, "RFC6282- Compression Format for IPv6 Datagrams over IEEE 802.15.4-Based Networks," 2011. [Online]. Available: https://www.rfc-editor.org/info/rfc6282.

[9] OASIS, "Message Queuing Telemetry Transport (MQTT) Version 3.1.1," December 2015. [Online]. Available: https://www.oasis-open.org/committees/tc_home.php?wg_abbrev=mqtt.

[10] OASIS, "Advanced Message Queuing Protocol (AMQP) v1.0," October 2012. [Online]. Available: https://www.oasis-open.org/committees/tc_home.php?wg_abbrev=amqp.

[11] IETF, "The Constrained Application Protocol (CoAP) Request for Comments 7252," June 2014. [Online]. Available: https://tools.ietf.org/html/rfc7252.

[12] M. Chui, "The Internet of Things," McKinsey & Company, 2010. [Online]. Available: https://www.mckinsey.com/industries/high-tech/our-insights/the-internet-of-things.

2

Low-Power Wearable and Wireless Sensors for Advanced Healthcare Monitoring

Ifana Mahbub

University of North Texas

Salvatore A. Pullano, Samira Shamsir, and Syed Kamrul Islam

University of Tennessee

Salvatore A. Pullano

University Magna Græcia of Catanzaro

CONTENTS

2.1 Introduction

In recent years, low-power wearable and wireless sensors have become a promising choice for advanced healthcare monitoring. Advancement of sensing technology facilitates continuous health monitoring of patients to detect disorders in the early stage. Early detection with continuous monitoring can prevent potential life-threatening events for the patient without being hospitalized. One of the key points of low-power wearable devices is the successful integration of sensors, which have led to the development of different innovative tools having a wide range of functionalities. Due to limited available power, it is essential to design these sensors for a faster response, preferably without the need of traditional energy supply, while providing greater flexibility in terms of shape and layout. Another important concept is the minimum invasiveness and harmfulness, which are of primary interest in long-term medical monitoring. Different sensor technologies have matured for practical applications, and among them, polymeric sensors represent the fastest growing technologies for practical implementation of biomedical sensors. Consequently, readout circuits for charge-sensitive devices used for biosignal conditioning brought out new design challenges to meet the requirements in terms of sensitivity, noise, and power consumption. Most of the wearable health monitoring systems are battery powered, and the operation of the system should require a minimum amount of power to prevent frequent replacement of the battery. Due to the interruption in health monitoring during battery replacement, batteryless energy harvesting has become an attractive solution for biomedical sensor applications. In this chapter, the reader will be introduced to the emerging low-power circuit design techniques and their applications in wearable healthcare monitoring systems.

Wireless devices for monitoring vital signs and other physiological parameters play a significant role in advancing the modern home-based healthcare applications. In general, biomedical signals have low frequency and thus require a low data rate transmitter for transmitting the data wirelessly. Unlike the traditional radios for cellular application, these radios do not transmit the data with high emission power as they are designed primarily for short-range communication. Most of these radios are either powered by the energy harvested from the ambient sources or tiny Li-polymer batteries. For such an energy-constrained environment, the challenge lies in the design of a low-power radio that can sustain the short-range communication ($\sim$1–2 m) link for a long period of time without compromising the error rate requirement. In this chapter, some of the recently published low-power radio architectures and approaches for wearable low data rate biomedical sensing applications are discussed. The key challenges in establishing a trustworthy communication link for these applications are identified as well.

In summary, the chapter includes various wearable sensors that are currently being used in healthcare monitoring. Methodologies for low-power operation and wireless telemetry of these wearable biomedical sensors are also discussed. Finally, the chapter concludes with a brief discussion of possible future advancements in this area.

2.2   Sensor Materials for Advanced Healthcare

During the last decade, there has been a remarkable development in wearable systems that make the technology less invasive and more user friendly by rapid advances in material sciences, telecommunications, and electronics, particularly integrated circuits and related nanotechnologies. (DeRossi and Lymberis, 2005; Pullano et al., 2016; Mahbub et al., 2017). Figure 2.1 shows the search results for papers on wearable devices indexed in Scopus from 1996 to 2016.

Rapid development of wearable devices is mainly fueled by the need for monitoring of important vital signs, leading to the delocalization of the health system with obvious benefits in terms of its capillary extension across the territory and reduction of costs. In addition, real-time monitoring of the patient for longer periods and during normal activities is an added value in the treatment of chronic and silent diseases. One of the fundamental parts of such devices combines successful integration of sensors and electronic circuits in wearable supports such as smart clothes, and represents a huge advantage (i.e., ~90% of our skin is in contact with the tissues) while playing an important role as interface between the electronic sensory devices and the body (Carpi and De Rossi, 2005). Another key point is the use of flexible support technologies that facilitate adaptability and acquisition of physiological parameters such as pressure and temperature and other biochemical parameters. Applications of these technologies cover the most varied areas from wearable electronics to robotics up to medical implants and generally can be portable, wearable, or implantable (Paradiso et al., 2005; Shepherd et al., 2011; Wong et al., 2012). Another interesting application concerns skin mimicking (e.g., the so-called electronic skin), which is becoming a fundamental aspect of the development of the next generation of robots.

Physical sensors that exploit flexible materials, and in particular flexible substrates, offer an excellent degree of deformability and adaptability on surfaces with various topologies and geometries (Li et al., 2014). Flexible substrates include thermoplastic polymers such as polyurethane and polyethylene

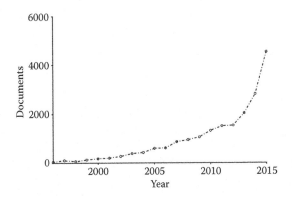

FIGURE 2.1
Publications of papers on wearable devices indexed by Scopus in the last 20 years.

terephthalate (Focke et al., 2010). In addition to these materials, another class of flexible substrates used are soft silicone elastomers, such as polydimethylsiloxane , and are of great interest due to additional benefits such as elasticity and surface adaptability of different textures and geometries (Xu et al., 2014). Moreover, they are chemically inert and biocompatible, which make them particularly suitable, especially in the case of implantable sensors (Lipomi et al., 2011).

Transducers employed can be different depending on the parameters and/or applications that best fit its use, such as piezoelectric (Zhou et al., 2008), capacitive (Matsuzaki et al., 2008), or resistive (Matsuzaki et al., 2008; Zhou et al., 2008). The sensing element can be further classified as solids (e.g., polymers, carbon, semiconductors, carbon nanotubes, nanowires, nanofibers, and metal nanoparticles) or liquids (e.g., ionic fluids and metallic) (Lipomi et al., 2011; Wu et al., 2011; Gao et al., 2012; Pang et al., 2012; Lee et al., 2014; Jung and Yang, 2015; Wang et al., 2015).

Most of the aforementioned transducers are fabricated with flexible substrate materials such as polymers, particularly piezoelectric and/or pyroelectric polymers. These materials belong to a family of polymers whose structures are characterized by the absence of centers of symmetry and may have one or more polar axes. Since they exhibit spontaneous polarization, they are categorized as polar crystals and exhibit both direct piezoelectric effect (a state of electrification caused by a mechanical deformation) and a converse piezoelectric effect (a mechanical deformation caused by the exertion of an external electric field) besides the pyroelectric effect (Jona and Shirane, 1962; Berlincourt, 1981). In nature, materials fall under 32 basic lattice crystal structures. Twenty-one of all crystal structures are noncentrosymmetric and 20 of these are piezoelectric. Among them, 10 also demonstrate pyroelectric properties and are called ferroelectric materials, as shown in Figure 2.2 (Liu and Long, 1978). Ferroelectric crystals belong to the pyroelectric crystal class in which the direction of spontaneous polarization can be reversed by applying an external electric field (Jona et al., 1962). Capacitive transducers are usually fabricated using parylene, polyimide, or polydimethylsiloxane, as monolithic sensors or flexible sensors arrays (Kim et al., 2005; Chang et al., 2009). The conventional way of producing these electronics is to pattern the insulating and piezoelectric polymer substrates. Alternatively, these electronic circuits can also be printed with conductive inks providing a few advantages over the conventional procedures. The printed circuits can provide excellent electrical conductivity, allowing a uniform deposition of conductive ink on the desired polymeric substrates and the subsequent polymerization of the coated substrates at high temperatures. Unfortunately, due to high sintering temperatures, this technique is limited to the use of thermoplastic materials such as polyamide or polyethylene naphthalate (Lechat et al., 2011).

An interesting alternative to these manufacturing techniques is weaving, which requires the use of conductive fibers extruded in thin filaments or used in metallic textile yarns (Büscher et al., 2015). Conductive fabrics are currently

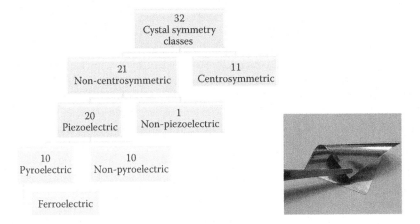

FIGURE 2.2
Classification of crystal symmetry and flexible polymer substrate.

used in various garments such as gloves, T-shirts, and shoes (Tirosh et al., 2013; Li et al., 2015; Sardini et al., 2015). However, there are still technological barriers that prevent the practical implementation of integrated sensors such as high cost and low compatibility for electronics integration.

2.3 State-of-the-Art Wearable Devices for Healthcare Monitoring

Apart from the evaluation of vital signs that has become noticeable due to the increasing need of remote patient monitoring, other interesting advantages of using wearable technologies have been observed in rehabilitation and geriatric care, which employs wearable devices developed to maximize both the effectiveness and efficiency of motor therapy (Mousavi Hondori and Khademi, 2014). Patient-tracking systems capable of detecting movements have facilitated the development of numerous applications in the field of rehabilitation, particularly for the treatment of stroke (Truelsen et al., 2005; Rosamond et al., 2007). These systems also provide feedback to both the patient and the therapist, without the need for constantly monitoring the progress during therapy (Friel and Nudo, 1998; Liepert et al., 2000). Therapies can be carried out at home, i.e., telerehabilitation, as well as in clinical centers, allowing longer therapy sessions for improved effectiveness (Lorussi et al., 2005; Giorgino et al., 2006).

A wearable therapy system can act as a motivational tool for the patient, providing additional help in achieving a functional movement or using repetitive movements. Another important and recent application concerns the control of upper limb exoskeleton devices (Xiao and Menon, 2014, Xiao et al., 2014), where the device activates the exoskeleton even in the presence of minimal movement by the patient. Numerous research projects have been carried out in the development and control of exoskeleton for supporting hand movements, especially for the flexion and extension movements of individual fingers (Englehart and Hudgins, 2003; Castellini and Smagt, 2009; Naik and Kumar, 2012; Naik et al., 2014). The new frontiers of wearable devices are focused on the development of a virtually invisible wearable electronics for the understanding of human activity through the collection of physiological data.

Recognition, tracking, and identification of a gesture are complex and articulated operations, which consist of not only the detection of human movement but also the interpretation of the movement type and the subsequent association with semantic meaningful information. In literature, the commonly used approaches are based on sensorized gloves and vision-based devices (Murthy and Jadon, 2009). The use of sensitive gloves is one of the most common techniques for recognizing hand gestures. Electronic systems integrated into the gloves are typically electromechanical or magnetic systems, such as accelerometers, gyroscopes, and magnetometers (e.g., Cyberglove, CyberGlove Systems LLC, USA) (Sturman and Zeltzer, 1994; Foxlin, 2002; Sánchez-Margallo et al., 2010; Zhou et al., 2010). Sensitive gloves usually include up to tens of integrated sensors (e.g., capacitive, vibrotactile) depending on the application and are used for the detection of flexion–extension movements and adduction–abduction movements of the fingers and the wrist and for monitoring the palm movement. Other approaches include the integration of extensimetric sensors, consisting of conductive mixtures characterized by piezoresistive properties, which are applied directly to the fabric whose mechanical deformations result in variations in their electrical resistance (Lorussi et al., 2005).

Charge or voltage generated by the transducers can be acquired by two circuit topologies: while high impedance voltage preamplifier is preferred for voltage mode, low input impedance is used in the case of current mode (Figure 2.3; Chirtoc et al., 2003). There are several advantages in using current mode

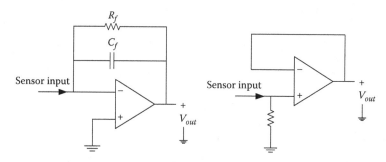

FIGURE 2.3
Readout circuits for current mode (left) and voltage mode (right).

readout circuit, which include low input impedance so that the inherent electrical model of the transducer (e.g., static capacitance and loss resistance C_0, R_0), the connecting cable, and the stray components do not affect the signal even if they are subjected to thermal variations. Moreover, it is less sensitive to electromagnetic interference, and the frequency characteristic is linear up to higher frequency compared with the voltage preamplifier (Chirtoc et al., 2003). Although lots of efforts have been made in the development of these wearable devices, there are still several unresolved issues such as complex calibration, noise, and crosstalk effect when the matrix of sensors is adopted.

Vision-based techniques leave the subjects free to perform gestures in a natural way without the user having to wear any kind of a device. This technique allows for the detection of hand movements and thus recognizes gestures acquired by a camera system with the aid of appropriate image processing software. One of the most widely available and cost-effective technologies called Kinect is developed by Microsoft, which includes a quarter video graphics array infrared camera and a video graphics array (VGA-RGB) camera, both with a frame rate of 30 fps. In addition, the system has an infrared light emitter that projects nonuniformly distributed dots on the scene detected by the depth infrared camera. This allows the user to create a depth map of the observed scene, and an image processing software provides identification of the number, position, and skeletal joints of the subject inside the scene (up to a distance of 4 m). Technologies based on camera, however, have several issues such as camera enclosure, resource management difficulties, costly computing, and high-energy consumption.

Alternative methods to wearable gloves for creating an interactive interface are brain–computer interface and muscle–computer interface. Brain–computer interface allows mapping by means of appropriate electrodes placed on the skin of the user and the electrical activity of the cerebral cortex (EEG) during movement (Wolpaw et al., 2002; Elnady et al., 2015). Muscle–computer interface is an interface where the subject leverages the muscular electrical activity detected by surface electromyographic sensors as a command during the execution of certain movements. In other words, the subject has the ability to control a device or an exoskeleton using their own myoelectric signals (Khokhar et al., 2010; Chowdhury et al., 2013).

2.4   Wireless Data Transmission of Biomedical Signal

As mentioned earlier, biomedical signals, in general, have low frequency and thus require a low data rate transmitter for transmitting the data wirelessly. Unlike the traditional radios for cellular application, these radios can transmit low energy for short-range communication. For an energy-harvested sensor application, the available energy is often less than 100 μW. Assuming the power budget of ∼50 μW for the front-end readout circuitry, the analog-to-digital converter, and the voltage regulator, the transmitter needs to be dissipating less than 50 μW of power. Even though there are many published works regarding the design of traditional narrowband transmitters for low data rate applications, the lowest power-consuming transmitter in that design space still consumes ∼900 μW of power (Cook et al., 2006). For that reason, a duty-cycled radio seems to be appropriate for such an energy-constrained application. Impulse radio ultra-wideband (IR-UWB) architectures are adopted for the implementation of a duty-cycled radio. Much of the work involving IR-UWB transmitters demonstrate excellent energy efficiency for high-data rate applications. However, the energy efficiency is not equally perfect for low data rate applications (Phan et al., 2008; Mercier et al., 2009). Moreover, the power leakage in these works is much higher, which even degrades the energy efficiency at 100 kbps data rate. The aforementioned challenges have inspired the need for the development of a low-power, low data rate IR-UWB transmitter with low leakage power dissipation to achieve longer battery lifetime. Unlike the traditional continuous wave transmitters, the IR-UWB transmitter does not require any mixer or phase locked loop, which minimizes the power consumption of the system. For the wireless transmission of the data, on–off keying (OOK)-based transmitter scheme needs to be analyzed to reduce the complexity of the transmitter architecture.

A system-level block diagram of an IR-UWB transmitter is shown in Figure 2.4. It includes an impulse generator designed with digital gates, a tunable cross-coupled inductor-capacitor based LC oscillator for upconverting the baseband impulse signal to UWB frequency range, and a buffer to drive the 50 Ω antenna. The transmitter is designed using a standard 180 nm complementary metal–oxide–semiconductor (CMOS) process. It follows OOK modulation scheme, where '1' bit is represented by a short pulse and '0' bit is represented by the absence of a pulse signal. The core building block of the impulse generator is a tunable delay block, an inverter, a NAND logic gate, and another inverter to generate the desired impulse signal as shown in Figure 2.5. The input signal, V_{IN}, is the digitized baseband signal acquired by the front-end amplifier. The frequency of this input signal depends on the sampling rate and the resolution of the analog-to-digital converter. The signals in the different stages of the delay block are presented in Figure 2.6. The

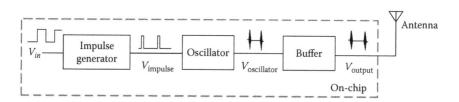

FIGURE 2.4
System-level block diagram of an IR-UWB transmitter.

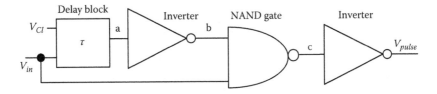

FIGURE 2.5
Schematic of the impulse generator block.

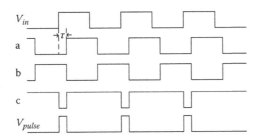

FIGURE 2.6
Signals in different stages of the delay block.

delay is controlled by changing the biasing voltage, V_{C1}. By changing the pulse width of the impulse signal, the spectrum can be modified to make it Federal Communications Commission (FCC) regulation compliant. Noncoherent energy detection receiver with OOK modulation scheme is suitable for a low data rate wireless link because it facilitates simplicity in design, thus ensuring low cost and low power consumption. The bit error rate (BER) performance of the OOK modulation is the error in the number of bits per unit time interval. It can also be computed by dividing the number of error bits by the total number of bits transmitted for a given period. In Figure 2.7, BER for transmitting 10^7 number of bits using OOK modulation scheme is presented. The BER is computed with a sampling rate of 10 kSps, with each sample containing 10 bits, resulting in the data rate of 100 kbps. Considering the duration of the pulse to be 2 ns, the duty-cycling ratio (2 ns/0.01 ms) becomes 0.02%.

For a low data rate application, the target BER can be set as 10^{-4}, which makes the required energy per bit to noise density ratio (E_b/N_o) to be 14. Some of the transmitter design specifications can be acquired based on the power emission constraints, which are presented in Table 2.1.

Clock synchronization between the transmitter and the receiver is the most important factor for any coherent radio communication. Dual-band transmission of the data and the timing of the pulses have been proposed by Dokania et al. (2010), for simplified synchronization and duty-cycling. The transmitter designed using 90 nm CMOS process consumes only 8 μW of power while the

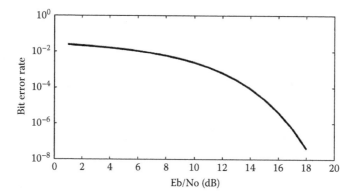

FIGURE 2.7
BER simulation using MATLAB® for OOK modulation.

TABLE 2.1
IR-UWB Transmitter Design Specifications

Parameter	Unit	Equation	Value
Data Rate (R)	kbps	—	100
Distance (D)	m	—	1
Bandwidth (BW)	MHz	—	528
Pulse duration	ns	—	2
Average Tx power (P_{avg})	dBm	$\frac{1}{BW}$	−14.3
Duty-cycling gain (G_{duty})	dB	$10\log_{10}\left(\frac{1}{BW}\right)$	37
Tx peak power	dBm	$P_{\text{avg}} + G_{\text{duty}}$	22.7
Center frequency (f_{c})	GHz	—	3.1–5

receiver consumes 12 μW of power. The receiver dynamically switches between the 3.5 and 4.5 GHz band, and an interband isolation of 30 dB ensures there is no interference between the two data channels. Mahbub et al. (2017) proposed an OOK-based IR-UWB transmitter in the 3.1–5 GHz frequency range in 180 nm CMOS process. With 1.8 V supply voltage, the leakage power is 0.4 μW of power while the transmitter consumes 36 μW of power during the active mode. Kulkarni et al. (2009) proposed an embedded on-chip monopole antenna for an IR-UWB transmitter designed using 0.18 μm CMOS process. Designed in the 6–10 GHz frequency band with a center frequency of 8 GHz, the antenna achieves a simulated maximum directivity of 5.4 dBi and a radiation efficiency of −22 dB. This enabled the possibility of short-distance (cm range) communication using an on-chip antenna. One thing that is common for most of the IR-UWB transmitter and receiver designs is the aggressive duty-cycling and reduction of the supply voltage, which ensure that a competitive energy efficiency is achieved.

2.5 Low-Power Circuit Design and Energy Harvesting

For wearable healthcare monitoring sensors with wireless telemetry, energy and power consumption have become significant design criteria. To address the power constraint, the circuit designers have implemented techniques either to reduce the overall power consumption of the system or to find any suitable replacement for the conventional battery or alternative power source. In general, for an integrated circuit system, total power consumption comprises of two types of power consumption—static and dynamic.

Static power is typically associated with leakage current and DC current sources, while the dynamic power consumption is frequency-dependent and often dominates the total power. It is associated with the charging and discharging of the capacitive loads and the switching action of the transistors connected between the power supplies. In the low-power design approach, the current consumption is reduced by minimizing supply voltage, circuit complexity, clocking frequencies, DC current source values, and capacitance of the switching nodes.

There are several approaches to attain these ultra-low power design objectives, such as supply voltage reduction, subthreshold design, body-driven differential logic, floating-gate transistors, self-cascode structure, and power management.

2.5.1 Supply voltage reduction

The power consumption in analog/RF circuit is proportional to the supply voltage and, therefore, can be reduced by reducing the supply voltage. This technique can also provide long-term device reliability by reducing the electric field. The supply voltage has been reduced with the continuous scaling of MOS devices. However, there is a limitation of continuous reduction of supply voltage as the dynamic range in analog circuit and voltage headroom are decreased with reduced power supply. The main challenge for the reduction of supply voltage with device scaling is that the threshold voltage and the drain-source saturation voltage do not scale down at the same rate, as does the supply voltage.

2.5.2 Subthreshold operation

For extremely low-power operation, the subthreshold operation can be a very suitable option. In this region of operation, the exponential behavior of drain current of a transistor varies with the gate bias voltage and can provide higher transconductance efficiency. Although this operation facilitates very low power consumption, the device is unsuitable for high-frequency applications due to the reduction of the unity current gain cut-off frequency (f_T).

However, this cut-off frequency can be increased with device scaling in deep submicron process.

2.5.3 Body-driven differential logic

In a body-driven differential logic, the body terminal of the transistor is used to design the differential amplifier. By implementing the depletion characteristics of the body-driven metal-oxide-semiconductor field-effect transistor (MOSFET), it is possible to have a low threshold voltage (V_T) requirement in the signal path and increased voltage swing for the lower supply voltage (Rosenfeld et al., 2004). However, the drawback of this technique is that it degrades the transconductance, resulting in lower gain-bandwidth (GBW) and poor frequency response. For this reason, it has larger equivalent input referred noise compared with that of the gate-driven MOSFET (Johns and Martin, 2008).

2.5.4 Floating-gate transistors

Another approach to reducing the supply voltage is to implement floating-gate transistors. A number of low-power analog circuit applications such as CMOS analog trimming circuit, multipliers in neural network, digital-to-analog converters, and amplifiers have been implemented following this technique to achieve low-power operation (Sackinger and Guggenbuhl, 1988; Yu and Geiger, 1993; Mehrvarz and Kwok, 1996; Yin et al., 1997).

2.5.5 Self-cascode structure

In a self-cascode structure, two MOSFETs are connected in series with their gate terminals shorted together. The aspect ratio of each transistor is chosen so that one transistor operates in saturation mode while the other operates in weak-to-moderate inversion. This type of configuration provides high output impedance with larger voltage headroom, which can be further improved by selecting multithreshold voltages for the two transistors (Fujimori and Sugimoto 1998).

2.5.6 Power management

For a wearable sensor with wireless telemetry, a significant portion of the overall power is consumed during signal processing and wireless communication. Simultaneous power management algorithms can play an important role to reduce the power consumption for wireless sensor networks (Karri and Goodman, 2007). The key concept of the power management technique is to reduce the transmission power by turning off the network nodes when they are not in use. If the network node is idle for a specific period of time, it is automatically switched into sleep state, thus consuming a low amount of power. The system should have the capability of waking up from the idle node when it is required to make measurements or transmit and receive data.

2.6 Energy Harvesting

Efficient power management and reduction of power consumption can play an important role in the realization of wearable devices. However, the selection of power supply is also an important factor to determine the overall lifespan of the sensor system, particularly for wearable and wireless applications. Usually standard alkaline and Li-ion/NiMH/Li-polymer are the most common varieties of batteries to power up the sensor circuitries. Due to the limited capacity of these batteries, energy harvesting has generated a significant interest among the researchers. Energy harvesting can be used to recharge the batteries on a regular basis to maximize the lifespan of the system. On the other hand, it can replace the battery altogether to make the overall system self-powered and more convenient with reduced size and weight. However, the current state of energy harvesting is not capable of supplying the required level of power to fully power up the wearable and wireless sensors. Nevertheless, by merging the approaches to reduce the power consumption and increase the harnessed energy, it has the potential to address the tradeoff between performance parameters and lifetime of sensor nodes.

Energy can be harvested from many different sources. However, the selection of energy sources plays a vital role in the availability and controllability of the energy for use. For a controllable energy source, the energy can be harnessed as required by the application. On the other hand, a noncontrollable source allows harvesting energy only when available. Furthermore, the energy sources can also be classified as ambient energy sources, such as wind, solar, etc., and human power, such as the energy derived from human motion. Human power sources can be considered as attractive choices for energy harvesting for wearable sensor and healthcare monitoring, which are most likely to be attached to the human body. Different activities of a human body can be utilized for harnessing the power, and their suitability depends on a particular sensor application. Some of these activities involve human motion, such as arm and leg movement, airflow in breathing etc., and the resulting mechanical energy is transferred to electrical energy by utilizing the piezoelectric property of the transducers used. Other sources, such as heat or friction, incorporate conversion technologies such as thermoelectricity or triboelectricity to generate electrical power (Starner, 1996; Gilbert and Balouchi, 2008).

2.7 Conclusion

This chapter provided a brief overview of different wearable sensor technologies for advanced biomedical applications. Among the different sensor materials reported in literature, a few selected materials such as smart clothes, skin mimicking, flexible substrates, and soft silicone elastomers are discussed in this chapter. However, for any material used for wearable biomedical sensors, it should have the properties of chemical inertness and biocompatibility. In addition, a number of different types of wearable devices and systems for detection of human activity or monitoring of vital sign in advanced healthcare applications have been developed. Many of these systems are based on the evaluation of mechanical, thermal, or vision based data from the human body. Wireless telemetry provides an advanced feature to the wearable devices designed for low-power operation for continuous healthcare monitoring. This chapter particularly emphasized IR-UWB topology for wireless data transmission in the biomedical sensor system. Several circuit-level topologies that can be implemented in the electronic circuitry to further reduce the power dissipation have also been discussed. Finally, it provided some future research direction of harvesting energy from human body activity to make the system self-powered. In summary, this chapter attempted to provide the reader with a brief concept of state-of-the-art energy-efficient sensor topologies with wearable and wireless properties for the application in healthcare monitoring.

Bibliography

Berlincourt, D. "Piezoelectric Ceramics: Characteristics and Applications." *The Journal of the Acoustical Society of America* 70, no. 6 (1981): 1586–95.

Büscher, G. H., R. Kõiva, C. Schürmann, R. Haschke, and H. J. Ritter. "Flexible and Stretchable Fabric-Based Tactile Sensor." *Robotics and Autonomous Systems* 63 (2015): 244–52.

Carpi, F., and D. De Rossi. "Electroactive Polymer-Based Devices for e-Textiles in Biomedicine." *IEEE Transactions on Information Technology in Biomedicine* 9, no. 3 (2005): 295–318.

Castellini, C., and P. van der Smagt. "Surface EMG in Advanced Hand Prosthetics." *Biological Cybernetics* 100, no. 1 (2009): 35–47.

Chang, W.-Y., T.-H. Fang, S.-H. Yeh, and Y.-C. Lin. "Flexible Electronics Sensors for Tactile Multi-touching." *Sensors* 9, no. 2 (2009): 1188–203.

Chirtoc, M., E. H. Bentefour, J. S. Antoniow, C. Glorieux, J. Thoen, S. Delenclos, A. H. Sahraoui, *et al.* "Current Mode versus Voltage Mode Measurement of Signals from Pyroelectric Sensors." *Review of Scientific Instruments* 74, no. 1 (2003): 648–50.

Chowdhury, A., R. Ramadas, and S. Karmakar. "Muscle Computer Interface: A Review." In *Icord'13*, 411–21: Springer, India, 2013.

Cook, B. W., A. Berny, A. Molnar, S. Lanzisera, and K. S. J. Pister. "Low-Power 2.4-GHz Transceiver with Passive Rx Front-End and 400-MV Supply." *IEEE Journal of Solid-State Circuits* 41, no. 12 (2006): 2757–66.

DeRossi, D., and A. Lymberis. "Guest Editorial New Generation of Smart Wearable Health Systems and Applications." *IEEE Transactions on Information Technology in Biomedicine* 9, no. 3 (2005): 293–94.

Dokania, R., X. Wang, S. Tallur, C. Dorta-Quinones, and A. Apsel. "An Ultralow-Power Dual-Band UWB Impulse Radio." *IEEE Transactions on Circuits and Systems II: Express Briefs* 57, no. 7 (2010): 541–45.

Elnady, A. M., X. Zhang, Z. G. Xiao, X. Yong, B. K. Randhawa, L. Boyd, and C. Menon. "A Single-Session Preliminary Evaluation of an Affordable BCI-Controlled Arm Exoskeleton and Motor-Proprioception Platform." *Frontiers in Human Neuroscience* 9 (2015): 168.

Englehart, K., and B. Hudgins. "A Robust, Real-Time Control Scheme for Multifunction Myoelectric Control." *IEEE Transactions on Biomedical Engineering* 50, no. 7 (2003): 848–54.

Focke, M., D. Kosse, C. Müller, H. Reinecke, R. Zengerle, and F. von Stetten. "Lab-on-a-Foil: Microfluidics on Thin and Flexible Films." *Lab on a Chip* 10, no. 11 (2010): 1365–86.

Foxlin, E. "Motion Tracking Requirements and Technologies." *Handbook of Virtual Environment Technology* 8 (2002): 163–210.

Friel, K. M, and R. J. Nudo. "Recovery of Motor Function after Focal Cortical Injury in Primates: Compensatory Movement Patterns Used During Rehabilitative Training." *Somatosensory & Motor Research* 15, no. 3 (1998): 173–89.

Fujimori, I., and T. Sugimoto. "A 1.5 V, 4.1 MW Dual-Channel Audio Delta-Sigma D/A Converter." *IEEE Journal of Solid-State Circuits* 33, no. 12 (1998): 1863–70.

Gao, Q., H. Meguro, S. Okamoto, and M. Kimura. "Flexible Tactile Sensor Using the Reversible Deformation of Poly (3-Hexylthiophene) Nanofiber Assemblies." *Langmuir* 28, no. 51 (2012): 17593–96.

Gilbert, J. M., and F. Balouchi. "Comparison of Energy Harvesting Systems for Wireless Sensor Networks." *International Journal of Automation and Computing* 5, no. 4 (2008): 334–47.

Giorgino, T., F. Lorussi, D. De Rossi, and S. Quaglini. "Posture Classification Via Wearable Strain Sensors for Neurological Rehabilitation." Paper presented at the Engineering in Medicine and Biology Society, 2006. EMBS'06. 28th Annual International Conference of the IEEE, 2006.

Johns, D. A., and K. Martin. *Analog Integrated Circuit Design.* John Wiley & Sons, New York, 2008.

Jona, F., and G. Shirane. *Ferroelectric Crystals.* Vol. 1: Pergamon, New York, 1962.

Jung, T., and S. Yang. "Highly Stable Liquid Metal-Based Pressure Sensor Integrated with a Microfluidic Channel." *Sensors* 15, no. 5 (2015): 11823–35.

Karri, R., and N. Goldman. *System-Level Power Optimization for Wireless Multimedia Communication: Power Aware Computing.* Springer Science & Business Media, US, 2007.

Khokhar, Z. O., Z. G. Xiao, and C. Menon. "Surface Emg Pattern Recognition for Real-Time Control of a Wrist Exoskeleton." *Biomedical Engineering Online* 9, no. 1 (2010): 41.

Kim, S.-H., J. Engel, C. Liu, and D. L. Jones. "Texture Classification using a Polymer-Based MEMS Tactile Sensor." *Journal of Micromechanics and Microengineering* 15, no. 5 (2005): 912.

Kulkarni, V. V., M. Muqsith, K. Niitsu, H. Ishikuro, and T. Kuroda. "A 750 Mb/S, 12 pJ/B, 6-to-10 Ghz Cmos IR-UWB Transmitter with Embedded on-Chip Antenna." *IEEE Journal of Solid-State Circuits* 44, no. 2 (2009): 394–403.

Lechat, C., A. R. Bunsell, and P. Davies. "Tensile and Creep Behaviour of Polyethylene Terephthalate and Polyethylene Naphthalate Fibres." *Journal of Materials Science* 46, no. 2 (2011): 528–33.

Lee, J., S. Kim, J. Lee, D. Yang, B. C. Park, S. Ryu, and I. Park. "A Stretchable Strain Sensor Based on a Metal Nanoparticle Thin Film for Human Motion Detection." *Nanoscale* 6, no. 20 (2014): 11932–39.

Li, R., B. Nie, P. Digiglio, and T. Pan. "Microflotronics: A Flexible, Transparent, Pressure-Sensitive Microfluidic Film." *Advanced Functional Materials* 24, no. 39 (2014): 6195–203.

Li, S., Q. Zhong, J. Zhong, X. Cheng, B. Wang, B. Hu, and J. Zhou. "Cloth-Based Power Shirt for Wearable Energy Harvesting and Clothes Ornamentation." *ACS Applied Materials & Interfaces* 7, no. 27 (2015): 14912–16.

Liepert, J., H. Bauder, H. R. Wolfgang, W. H. Miltner, E. Taub, and C. Weiller. "Treatment-Induced Cortical Reorganization after Stroke in Humans." *Stroke* 31, no. 6 (2000): 1210–16.

Lipomi, D. J., M. Vosgueritchian, B. C. K. Tee, S. L. Hellstrom, J. A. Lee, C. H. Fox, and Z. Bao. "Skin-Like Pressure and Strain Sensors Based on Transparent Elastic Films of Carbon Nanotubes." *Nature Nanotechnology* 6, no. 12 (2011): 788–92.

Liu, S. T., and D. Long. "Pyroelectric Detectors and Materials." *Proceedings of the IEEE* 66, no. 1 (1978): 14–26.

Lorussi, F., E. P. Scilingo, M. Tesconi, A. Tognetti, and D. De Rossi. "Strain Sensing Fabric for Hand Posture and Gesture Monitoring." *IEEE Transactions on Information Technology in Biomedicine* 9, no. 3 (2005): 372–81.

Mahbub, I., S. A. Pullano, H. Wang, S. K. Islam, A. S. Fiorillo, G. To, M. R. Mahfouz. "A Low-Power Wireless Piezoelectric Sensor-Based Respiration Monitoring System Realized in CMOS Process." *IEEE Sensors Journal* 17, no. 6 (2017): 1858–64.

Mahbub, I., S. Shamsir, and S. K. Islam. "A Low-Power Low-Data Rate Impulse Radio Ultra-Wideband (IR-UWB) Transmitter." *International Journal of High Speed Electronics and Systems* 26, no. 3 (2017): 1740013.

Matsuzaki, R., T. Keating, A. Todoroki, and N. Hiraoka. "Rubber-Based Strain Sensor Fabricated Using Photolithography for Intelligent Tires." *Sensors and Actuators A: Physical* 148, no. 1 (2008): 1–9.

Mehrvarz, H. R., and C. Y. Kwok. "A Novel Multi-Input Floating-Gate Mos Four-Quadrant Analog Multiplier." *IEEE Journal of Solid-State Circuits* 31, no. 8 (1996): 1123–31.

Mercier, P. P., D. C. Daly, and A. P. Chandrakasan. "An Energy-Efficient All-Digital UWB Transmitter Employing Dual Capacitively-Coupled Pulse-Shaping Drivers." *IEEE Journal of Solid-State Circuits* 44, no. 6 (2009): 1679–88.

Mousavi Hondori, H., and M. Khademi. "A Review on Technical and Clinical Impact of Microsoft Kinect on Physical Therapy and Rehabilitation." *Journal of Medical Engineering* 2014 (2014): 16

Murthy, G. R. S, and R. S. Jadon. "A Review of Vision Based Hand Gestures Recognition." *International Journal of Information Technology and Knowledge Management* 2, no. 2 (2009): 405–10.

Naik, G. R., A. Acharyya, and H. T. Nguyen. "Classification of Finger Extension and Flexion of EMG and Cyberglove Data with Modified ICA Weight Matrix." Paper presented at the Engineering in Medicine and Biology Society (EMBC), 2014. 36th Annual International Conference of the IEEE, 2014.

Naik, G. R., and D. K. Kumar. "Identification of Hand and Finger Movements Using Multi Run ICA of Surface Electromyogram." *Journal of Medical Systems* 36, no. 2 (2012): 841–51.

Pang, C., G.-Y. Lee, T. Kim, S. M. Kim, H. N. Kim, S.-H. Ahn, and K.-Y. Suh. "A Flexible and Highly Sensitive Strain-Gauge Sensor Using Reversible Interlocking of Nanofibres." *Nature Materials* 11, no. 9 (2012): 795.

Paradiso, R., G. Loriga, and N. Taccini. "A Wearable Health Care System Based on Knitted Integrated Sensors." *IEEE Transactions on Information Technology in Biomedicine* 9, no. 3 (2005): 337–44.

Phan, A. T., J. Lee, V. Krizhanovskii, Q. Le, S.-K. Han, and S.-G. Lee. "Energy-Efficient Low-Complexity Cmos Pulse Generator for Multiband UWB Impulse Radio." *IEEE Transactions on Circuits and Systems I: Regular Papers* 55, no. 11 (2008): 3552–63.

Pullano, S. A., A. S. Fiorillo, I. Mahbub, S. K. Islam, M. S. Gaylord, and V. Lorch. "Non-invasive Integrated Wireless Breathing Monitoring System Based on a Pyroelectric Transducer." *IEEE Sensors Conference*, 2016, Orlando, USA, 30 Oct.–3 Nov.

Rosamond, W., K. Flegal, G. Friday, K. Furie, A. Go, K. Greenlund, Nancy H., *et al.* "Heart Disease and Stroke Statistics A 2007 Update." *Circulation* 115, no. 5 (2007): e69–e171.

Rosenfeld, J., M. Kozak, and E. G. Friedman. "A Bulk-Driven CMOS OTA with 68 dB DC Gain." Paper presented at the Electronics, Circuits and Systems, 2004. ICECS 2004. Proceedings of the 2004 11th IEEE International Conference on, 2004.

Sackinger, E., and W. Guggenbuhl. "An Analog Trimming Circuit Based on a Floating-Gate Device." *IEEE Journal of Solid-State Circuits* 23, no. 6 (1988): 1437–40.

Sánchez-Margallo, F. M., J. A. Sánchez-Margallo, J. B. Pagador, J. L. Moyano, J. Moreno, and J. Usón. "Ergonomic assessment of hand movements in laparoscopic surgery using the CyberGlove®." In Miller K., Nielsen P. (eds) *Computational Biomechanics for Medicine*, 121–128: Springer, New York, NY, 2010.

Sardini, E., M. Serpelloni, and V. Pasqui. "Wireless Wearable T-Shirt for Posture Monitoring During Rehabilitation Exercises." *IEEE Transactions on Instrumentation and Measurement* 64, no. 2 (2015): 439–48.

Shepherd, R. F., F. Ilievski, W. Choi, S. A. Morin, A. A. Stokes, A. D. Mazzeo, X. Chen, M. Wang, and G. M. Whitesides. "Multigait Soft Robot." *Proceedings of the National Academy of Sciences* 108, no. 51 (2011): 20400–403.

Starner, T. "Human-Powered Wearable Computing." *IBM Systems Journal* 35, no. 3.4 (1996): 618–29.

Sturman, D. J., and D. Zeltzer. "A Survey of Glove-Based Input." *IEEE Computer Graphics and Applications* 14, no. 1 (1994): 30–39.

Tirosh, O., R. Begg, E. Passmore, and N. Knopp-Steinberg. "Wearable Textile Sensor Sock for Gait Analysis." Paper presented at the Sensing Technology (ICST), 2013. Seventh International Conference on, 2013.

Truelsen, T., M. Ekman, and G. Boysen. "Cost of Stroke in Europe." *European Journal of Neurology* 12, no. s1 (2005): 78–84.

Wang, J., J. Jiu, M. Nogi, T. Sugahara, S. Nagao, H. Koga, P. He, and K. Suganuma. "A Highly Sensitive and Flexible Pressure Sensor with Electrodes and Elastomeric Interlayer Containing Silver Nanowires." *Nanoscale* 7, no. 7 (2015): 2926–32.

Wolpaw, J. R., N. Birbaumer, D. J. McFarland, G. Pfurtscheller, and T. M. Vaughan. "Brain-Computer Interfaces for Communication and Control." *Clinical Neurophysiology* 113, no. 6 (2002): 767–91.

Wong, R. D. P., J. D. Posner, and V. J. Santos. "Flexible Microfluidic Normal Force Sensor Skin for Tactile Feedback." *Sensors and Actuators A: Physical* 179 (2012): 62–69.

Wu, C.-Y., L. Wei-Hao, and T. Yi-Chung "Integrated Ionic Liquid-Based Electrofluidic Circuits for Pressure Sensing within Polydimethylsiloxane Microfluidic Systems." *Lab on a Chip* 11, no. 10 (2011): 1740–46.

Xiao, Z. G., A. M. Elnady, and C. Menon. "Control an Exoskeleton for Forearm Rotation Using FMG." Paper presented at the Biomedical Robotics and Biomechatronics 2014. 5th IEEE RAS & EMBS International Conference on, 2014.

Xiao, Z. G., and C. Menon. "Towards the Development of a Wearable Feedback System for Monitoring the Activities of the Upper-Extremities." *Journal of Neuroengineering and Rehabilitation* 11, no. 1 (2014): 2.

Xu, S., Y. Zhang, L. Jia, K. E. Mathewson, K.-I. Jang, J. Kim, H. Fu, et al. "Soft Microfluidic Assemblies of Sensors, Circuits, and Radios for the Skin." *Science* 344, no. 6179 (2014): 70–74.

Yin, L., S. H. K Embabi, and E. Sanchez-Sinencio. "A Floating-Gate Mosfet D/A Converter." Paper presented at the Circuits and Systems, 1997. ISCAS'97. Proceedings of 1997 IEEE International Symposium on, 1997.

Yu, C.-G., and R. L. Geiger. "Very Low Voltage Operational Amplifiers Using Floating Gate MOS Transistor." Paper presented at the Circuits and Systems, 1993. ISCAS'93. 1993 IEEE International Symposium on, 1993.

Zhou, J., Y. Gu, P. Fei, W. Mai, Y. Gao, R. Yang, G. Bao, and Z. L. Wang. "Flexible Piezotronic Strain Sensor." *Nano Letters* 8, no. 9 (2008): 3035–40.

Zhou, J., F. Malric, and S. Shirmohammadi. "A New Hand-Measurement Method to Simplify Calibration in Cyberglove-Based Virtual Rehabilitation." *IEEE Transactions on Instrumentation and Measurement* 59, no. 10 (2010): 2496–504.

3

Biomedical Algorithms for Wearable Monitoring

Su-Shin Ang

Inova Design Solutions

Miguel Hernandez-Silveira

Sensium® Healthcare

CONTENTS

CONTENTS

3.1 Introduction

There is no dispute that portable and wearable electronics have gone a long way in changing our lifestyles, whether in communication, travel or finance. It is the opinion of the authors that the widespread use of these devices for health monitoring is only a matter of time. Biomedical algorithms, deployed on these devices, are enabling factors for these solutions, as they are required for the interpretation of the physiological signals acquired from the users.

For many good reasons, the design of biomedical algorithms has often been described as an art. Indeed, the requirements of the problem/s at hand are typically ambiguous. They are not only technical but also physiological in nature. More specifically, designers have to consider the inter- and intra-physiological variations of the device users, which in turn is dependent on factors such as pre-existing medical conditions, concomitant medication and the circadian rhythm of its target users. The myriad factors involved call for the need of clarity and a framework for both the definition of these requirements as well as algorithm design.

To this end, biomedical problems are categorised into three typical types– optimisation of a pre-defined objective function (Type 1), static classification problems (Type 2) and predictive models (Type 3). These types will be delineated in this chapter. In addition, several case studies will be presented to illustrate the different problem types. Here we provide a brief review of state-of-the art approaches, including some created and used by us as part of our current developments. For example, these include the compression of electrocardiogram signals to reduce the bandwidth and energy requirements for cardiac monitoring, calorie expenditure estimation, arrhythmia detection, confidence level for computed physiological vital signs and fall prediction/detection. Although the level of complexity and description of these examples is simple, we hope that these cases provide initial insights for enthusiasts to start developing their own algorithms for wearable devices intended for medical and wellness applications.

In this chapter, we first provide an overview in terms of the problem types and how each can be practically formulated (Section 3.2). Since data is frequently crucial in the design process, we talk about how they can be acquired, and how the data can be used during the training and evaluation process, followed by a series of case studies, where each biomedical problem in question is formulated and solved (Section 3.3). Downstream of the design and prototyping process, the algorithm has to be ported to the target platform. We present such a design flow in Section 3.4. Finally, we conclude in Section 3.5.

3.2   Formulating a Problem Mathematically

From actual problems encountered in the field, we believe that they can be broadly divided into three types. The first type is the generic optimisation problem (Type 1), which is used across multiple application domains. In this case, the objective or cost function and constraints are well-defined. We discuss this at length in Section 3.2.1.

The second type of problem is the static classification problem (Type 2) involving discrete decisions, where the classification function is unknown, but the dataset involving the original measurements and the discrete and expected outcomes are available (Section 3.2.2). The last type is the prediction problem (Type 3) that requires the prediction of random variables based on the statistical distribution of related features (Section 3.2.3).

3.2.1   Generic optimisation problems (Type 1)

The problem that one is trying to solve is usually ill-defined in the initial stages. Examples include the derivation of optimum parameters for a compression algorithm or the derivation of a hyperplane for accurate classification in the feature space. To elucidate the problem, the following steps can be followed.

1. Identify effective metrics used in quantifying how effectively a problem has been solved. This could mean the compression ratio and/or the quality of the reconstructed signal.

2. Identify the parameters with the most substantial impact on these metrics. In the context of signal compression, these parameters could include the quantisation factors for the Discrete Cosine Transform (DCT) coefficients.

3. If possible, define the problem formally as an optimisation problem.

There are two aspects to an optimisation problem – the objective function and the set of constraints. The optimisation problem is defined over the parameter space, where each data point in the parameter space has a certain cost or benefit, which is quantified by means of an objective function. The scope of this parameter space is in turn defined by constraints, which are characterised by a set of equalities and/or inequalities. More formally, the objective function can be specified according to (3.1).

$$\text{Minimise } f \colon \mathbf{R} \to \mathbf{R} \tag{3.1}$$

In this case, the objective function f, defined over a feasible set of parameters, $C \subset R_n$. The domain of the objective function is defined by a tuple with n real numbers - $\{x_1, x_2, \ldots, x_n\}$, and the range consists of single real numbers, which is cost of using the corresponding set of parameters. The set of equalities and inequalities involving the constraint functions, used to define the scope of the parameter space, are shown in (3.2), (3.3) or (3.4).

$$f(x) \leq c_i \tag{3.2}$$
$$f(x) \geq c_j \tag{3.3}$$
$$f(x) = c_k \tag{3.4}$$

In certain cases, it might be useful to convert the inequalities to equalities using slack variables, s_i, to enable certain methods such as integer linear programming [1] to be employed (3.5).

$$g(x) = c_i + s_i \qquad (3.5)$$

A minimisation problem can easily be converted to a maximisation problem by negating the objective function. If the functional surface is convex with a single saddle point, the set of parameters corresponding to this point represent the best solution to this problem. In addition, for many problems, the functional surface has a large number of saddle points (or local minima), and it may not be feasible to carry out an exhaustive search to explore all candidate solutions to arrive at the globally optimum solution.

Ideally, the problem definition should be independent of the algorithm. However, more often than not, the parameters concerned are dependent on the algorithm used. In this case, it is necessary to re-define the problem for each algorithm in the candidate set. The metrics used in determining the outcome should remain uniform so that the performance of different algorithms can be properly compared.

3.2.2   Static classification problems (Type 2)

A specific type of problem, that typically occurs in the biomedical field is classification, where a discrete decision needs to be made – given a measurement, is it possible to *automatically* predict an outcome with a high degree of accuracy? For example, one might like to build a mechanism to identify abnormal patterns within electrocardiograms to determine whether a patient is suffering from arrhythmia. Frequently, the measurements and the expected outcomes (expert opinion by a clinician) are available, but the classification function is unknown and needs to be determined.

More formally, the initial dataset, D, consisting of the measurements, is provided. This dataset contains a set d tuples, $\{t_0, t_2, ..., t_{d-1}\}$. To reduce the dimensionality of the problem, feature extraction is carried out to extract the most pertinent aspects of the data. Therefore, each tuple t_i is associated with feature vector $V_i = \{f_{i1}, f_{i2}, ..., f_{iN}\}$, containing N features. Specifically, $V_i \subset V$, where V is known as a feature matrix. In addition, a corresponding class label C_i should also be available for each t_i and V_i, and $C_i \subset C$ (C is known as the observation vector). The objective is therefore to derive a mapping function, f, from V to a predicted set of outcomes, P, where the difference between P and C is minimised.

A class of solutions, known as supervised machine learning algorithms, has been designed for this type of problem. The steps involved in problem solving are as follows [2].

1. *Feature extraction*: determining and extracting the most relevant aspects from the raw data, to facilitate more effective classification.

2. *Feature selection*: select a subset of the most relevant features from the original set (columns of feature matrix V). This step can either be carried out manually or automatically.

3. *Feature space mapping*: To improve classification accuracy, a transformation is carried out on the original to a new feature space. This might involve increasing or decreasing the dimensionality of the feature space.

4. *Classifiers*: There are many different types of classification techniques, with differing types of classification functions (hyperplanes within the feature space).

5. *Validation/Testing*: This phase involves the methodologies in the training and testing of the classifiers with two objectives – optimising the parameters of the classifier and obtaining representative metrics for the performance of the classifier.

This supervised machine learning framework takes into account human input and experience at the initial stage, specifically at the feature extraction stage. Based on a combination of experience and trial and error, the initial set of features that are likely to enable effective classification are identified. For example, patients suffering from a typical arrhythmia, atrial fibrillation (AF) [3], is known to exhibit substantial variability in the instantaneous heart rate. In consequence, the root mean square of the successive differences of peak-to-peak intervals (RMSSD of the times between each heart beat) is often considered as a candidate feature. Once the feature has been identified, the feature space may be described, as seen in Figure 3.1. This two-dimensional (2D) feature space is populated by two classes (marked differently in the feature space), which are concentrated in distinct clusters, indicating that there is good separability between the classes in this space. The lines separating these clusters

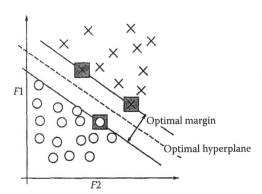

FIGURE 3.1
An example of a feature space and the corresponding hyperplane, derived from the support vector machine [6]. $F1$ and $F2$ are two different features.

are defined by classification functions, or hyperplanes, which partition the feature space into independent components so that any new incoming data may be automatically classified based on its location within this space.

Redundant and irrelevant features may confound the classification problem. Consequently, it is necessary to cull them from the final feature set. The most straightforward manner is to do so manually by visually assessing the separability of the classes in the feature space. There are two limitations to this approach – first, visual assessment is subjective and may therefore be inconsistent between different designers. Second, this approach is limited to at most three dimensions (3D). Potentially, automatic feature selection techniques may be used to overcome these limitations. Two methods that are used for this purpose are the minimum redundancy maximum relevance [4] (mRMR) and the Relief-F [5] algorithms. The mRMR algorithm removes redundant features by locating a subset of features that minimises the metric, mutual information (normalised form of correlation) between different features, and maximising the same metric between selected features and the observation vector. The Relief-F algorithm takes a different approach. In particular, it considers the distance of each feature vector from its two nearest neighbours from the same and the opposite class (assuming a two-class problem) [5], penalising the feature if it is near to the companion in the same class and vice versa for the nearest vector in the opposite class.

Similar to feature selection, the objective of the feature mapping stage is to maximise separability between different classes in the feature space. This can be achieved by increasing the dimensionality of the feature space by means of techniques like kernel substitution [7] or dimensionality reduction techniques like Sammon mapping [8]. In the case of kernel substitution [7], an extra dimension in the feature space is created, and this has been shown to improve classification performance in some applications. An example is $K(x, x') = x^T x$. Another popular kernel is the radial basis function, where the transformed feature space contain euclidean distances between feature vectors, rather than the absolute positions of the vectors themselves.

After the best features and transformation function have been selected for the problem, different classification methods may be applied to the problem. These methods result in different hyperplanes, and the best classifier is dependent on the dataset as well as the problem. The best known classifier is the artificial neural network (ANN), as shown in Figure 3.2. This consists of the basic node or neuron, which is composed of a linear function of adjustable weights and the inputs from other nodes. The output of this function is then fed into a smooth differentiable sigmoidal function. These nodes are organised into input layer, hidden layer and the output layer. The number of nodes in the input and output layers are bounded by the number of features and the classification problem at hand, respectively, whereas the number of hidden layers as well as the number of nodes in each hidden layer are design parameters. In general, a large number of layers and nodes in the hidden portion affords higher degrees of freedom, and subsequently, a more versatile and accurate

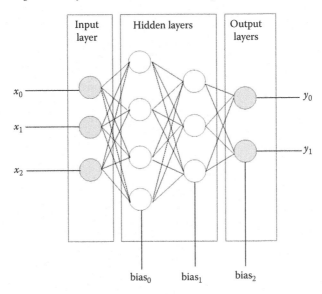

FIGURE 3.2
A back-propagated artificial neural network, with input x and output y. All of these nodes contain adjustable weights, to minimise the errors between the y and the expected outcomes.

hyperplane. On the other hand, it could lead to the problem of over-fitting, or a non-generic hyperplane. Once the topology of the ANN has been fixed, the weights are determined by a technique known as error back propagation [9]. Essentially, training data vectors are fed to the ANN during the forward phase of the algorithm. Subsequently, during the backward phase, the error between the label and the current ANN output is 'back-propagated' and the weights of each node are adjusted using the gradient descent method. With enough representative data, the weights will eventually converge to values where the classification error is minimised.

Two other popular machine learning techniques are Decision trees and support vector machine. An example of a Decision tree is furnished in Figure 3.3, which is designed for arrhythmia detection, based on features extracted from a segment of electrocardiogram (ECG) data – RMSSD, and the average corrected (using an ectopic filter) beat to beat interval. Each node of the tree consists of an inequality, with the feature and a threshold. These parameters are obtained by means of the $C4.5$ algorithm [10], where the feature matrix and observation vector are recursively divided at each node, based on the best feature and the optimal threshold found for that feature and that data subset. The metric used in selecting the feature for each node is the information gain [10]. The support vector machine is another popular machine learning algorithm that has grown in popularity over the recent years. A

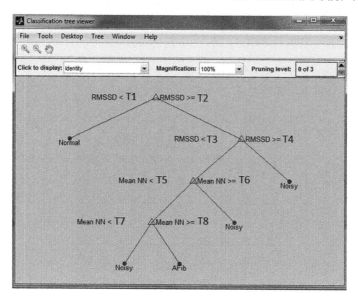

FIGURE 3.3
A binary decision tree for arrhythmia classification, using features RMSSD and mean Normalised R–R interval (NN) (corrected beat-to-beat intervals). T1–T8 represent thresholds derived from the C4.5 algorithm [10].

hyperplane that is derived from this approach can be seen in Figure 3.1. Essentially, this technique derives the optimal hyperplane by maximising the margin between the data points (support vectors) closest to the hyperplane. Kernel substitution is frequently embedded into the objective function, used in the derivation of the hyperplane, to improve classification performance.

As one might expect, the parameters of the classifier are dependent on the data used in the training process. In addition, the initial classifier state and the manner in which data is fed during the training process has an impact as well. These issues will be discussed in Section 3.3. Once the classification parameters are determined, they cannot be changed after training stage. When the expected outcomes or the observation vector are unavailable, un-supervised machine learning algorithms may be used to detect clusters within the underlying data structure to determine the labels. However, we consider that these classes of problems are outside the scope of our discussion.

3.2.3   Prediction problems (Type 3)

In Section 3.2.2, the required outcomes are discrete and the parameters of the classification model are fixed after the training stage, and cannot be altered during run-time. Therefore, it is very important that the true distribution of the feature space is understood and well represented by the data. Frequently, this is not the case. For instance, if one is trying to determine the reliability of heart rate (HR) values produced by an algorithm, the type of artefacts produced by bedridden patients in a general ward would differ from those produced by frequently ambulating patients in the accidents and emergency department, prompting the need for a classifier that is capable of adjusting to the incoming data distribution. Potentially, if the training dataset originates from a single source, it could result in a classifier that is customised to the physical location of the patients as well as their physiological conditions.

Assuming that one is trying to predict the value of a non-deterministic variable random variable, C, which is dependent on a set of observed features, $\{F_j | j = 1, ..., M\}$. By the Bayes theorem, the conditional probability distribution of C is a function of the prior class distribution, $P(C)$, and the likelihood function, $P(F_1, ..., F_{M-1}|C)$, as shown in (3.6).

$$P(C|F_1, ..., F_M) = \frac{P(C)P(F_1, ..., F_M|C)}{P(F_1, ..., F_M)} \tag{3.6}$$

The posterior probability can be factorised and re-expressed using the chain rule, as shown in (3.7). By comparing (3.6) with (3.7), it follows that the likelihood function can be expressed as (3.8). If it can be further assumed that the observed features are independent, so (3.8) reduces to (3.9).

$$P(C|F_1, ..., F_M) = P(C)P(F_1|C)...P(F_M|C, F_1, ..., F_{M-1}) \tag{3.7}$$

$$P(F_1, ..., F_M|C) \propto P(F_1|C_i)P(F_2|C, F_1)...P(F_M|C, F_1, ..., F_{M-1}) \tag{3.8}$$

$$= \prod_{j=1}^{M} P(F_j|C) \tag{3.9}$$

We carry on by making the assumptions that the prior and likelihood distributions are Gaussian. Further, N independent observations of $\mathbf{F}$ were made ($\mathbf{F} = \{F_1, F_2, ..., F_M\}$), so the resulting likelihood function can be formulated, as shown in (3.11). It is assumed that the mean and standard deviation of F are given by μ and *sigma*, respectively. Similarly, the prior distribution is described in (3.12).

$$P(\mathbf{F}|C) \propto \prod_{n=1}^{N} P(\mathbf{F}_n|C) \tag{3.10}$$

$$\propto \frac{1}{(2\pi\sigma^2)^{\frac{N}{2}}} exp \left\{ -\frac{1}{2\sigma^2} \sum_{n=1}^{N} (\mathbf{F}_n - \mu)^2 \right\} \tag{3.11}$$

$$P(C) : N(\mu_0, \sigma_0^2) \tag{3.12}$$

Given that the posterior distribution $P(C|\mathbf{F})$ is a function of both the prior and likelihood function, it is also Gaussian (represented by $N(\mu_N, \sigma_N^2)$). With each incoming datum, both the prior probability and the likelihood functions can be sequentially updated – the likelihood function, $P(\mathbf{F}|C)$, is updated with each new data point. This is known as the inference step. Subsequently, the prediction step is made by taking the product between the prior probability and the likelihood probability (decision step) to obtain the posterior probability. The posterior probability can be re-expressed as shown in (3.14). The term in the square bracket is the posterior probability from the previous step $(N-1)$, and this is re-substituted as the prior probability in the subsequent time step. Clearly, this is an iterative process, with the inference and decision step taking placing cyclically.

$$P(\mathbf{F}|C) \propto P(C) \prod_{n=1}^{N} P(\mathbf{F}_n|C) \tag{3.13}$$

$$= \left[P(C) \prod_{n=1}^{N-1} P(\mathbf{F}_n|C) \right] P(\mathbf{F}_N|C) \tag{3.14}$$

Further, it can be shown that the mean and variance of the resulting distribution are given by (3.15) and (3.16), respectively [7]. The parameter, μ_{ML}, refers to the mean value derived from maximising the likelihood function. From (3.15), it can be seen that for a small number of observations, N, the prior mean has a bigger influence over the posterior mean, compared with the maximum likelihood mean value, and vice versa. In addition, from (3.16), it is clear that the initial variance is high, but falls in magnitude as N grows. This observation implies that the amount of uncertainty in the prediction of C decreases with more incoming data.

$$\mu_N = \frac{\sigma^2}{N\sigma_0^2 + \sigma^2}\mu_0 + \frac{N\sigma_0^2}{N\sigma_0^2 + \sigma^2}\mu_{ML} \tag{3.15}$$

$$\frac{1}{\sigma_N^2} = \frac{1}{\sigma_0^2} + \frac{N}{\sigma^2} \tag{3.16}$$

Thus far, it has been assumed that both the features, and correspondingly the predicted variables, are independently and identically distributed. In many applications, this assumption is unrealistic, and more accurate predictions may be facilitated by capturing the correlation between the predicted variables themselves. For instance, one might be trying to predict if an abnormal rhythm is detected from the ECG of a patient. The most recent prediction is likely to have a bearing on the current prediction, as rhythms tend to be persistent. Potentially, we can capture this temporal correlation using a state space representation of the system. In this paradigm, there are two parameters–observations and latent variables. They correspond to the observed features (F_i) and the parameter that we are trying to predict (C_i, which is assumed to be a discrete random variable in this case), respectively. This particular state space representation is known as a hidden Markov model (HMM), and it is illustrated in Figure 3.4. Note that edges between consecutive values of C_i represent the dependence between them.

   In the case of the HMM, its joint distribution is shown in (3.17) [7]. The conditional parameters for this distribution are captured by θ, which in turn contains $\{\phi, A, \pi\}$, which are explained as follows. ϕ represents the parameters governing $\mathbf{F}$, conditioned upon the state, C. For instance, if $\mathbf{F}$ is a continuous random variable that is normally distributed, ϕ would represent the mean and standard deviation of the distribution. Each element of the transition matrix, A_{ij}, situated in row i and column j, represents the probability that the system will transit to state j, given that it is originally in state i. Assuming that C can take any one of K states, the vector π will be populated by the probabilities of the system being in a particular state (for instance, $\pi = \{P(\text{abnormal rhythm detected}), P(\text{normal sinus rhythm})\}$).

$$P(\mathbf{F}, C|\theta) = P(\pi)\left[\prod_{n=2}^{N} P(C_n|C_{n-1,A})\right]\prod_{m=1}^{N} P(\mathbf{F}_m|C_m, \phi) \tag{3.17}$$

$$\theta = \{\pi, A, \phi\}$$

π : Probability of the system taking a certain state.

A : Transition matrix, $A_{i,j} = P(C_t = j|C_{t-1} = i)$

ϕ : Statistical parameters for the emission distribution

A log-likelihood function can be formed using the 'old' posterior marginal (based on data governed by previous statistical parameters θ^{old}) and the logarithm of the present marginal distribution, as shown in (3.18) [7]. Equation (3.18) can then be combined with (3.17) to obtain (3.19). In this new expression, two new parameters – the marginal posterior probability ($\gamma(C_n)$) and the joint posterior probability of successive latent variables ($\xi(C_{n-1}, C_n)$) are introduced. By maximising the likelihood function with respect to θ, it can be shown that the optimal solutions for π and A are functions of γ and ξ [7].

$$Q(\theta, \theta^{\text{old}}) = \sum_{Z} P(C|\mathbf{F}, \theta^{\text{old}}) \ln P(C, \mathbf{F}|\theta) \tag{3.18}$$

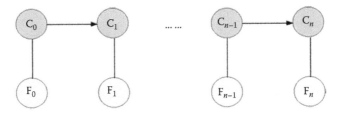

FIGURE 3.4
State space diagram for the hidden Markov model.

$$Q(\theta, \theta^{\text{old}}) = \sum_{k=1}^{K} \gamma(C_{1k}) \ln \pi_k + \sum_{n=2}^{N} \sum_{j=1}^{K} \sum_{k=1}^{K} \xi(C_{n-1,j}, C_{n,k})$$

$$+ \sum_{n=1}^{N} \sum_{k=1}^{K} \gamma(C_{nk}) \ln P(\mathbf{F}_n | \phi_k) \tag{3.19}$$

$$\gamma(C_n) = P(C_n | \mathbf{F}, \theta^{\text{old}}) \tag{3.20}$$

$$\xi(C_{n-1}, C_n) = P(C_n, C_{n-1} | \mathbf{F}, \theta^{\text{old}}) \tag{3.21}$$

The marginal posterior probability, γ, and the joint posterior probability of successive latent variables, ξ, are in turn derived as a function of the forward and backward conditional probabilities – α and β, respectively, based on the Bayes therom, which is seen in (3.22). At time step n, the forward probability is the joint probability of the current prediction and all past observations, while the backward probability is the probability of all future observations, conditioned upon the current prediction. These parameters are, respectively, described in (3.24) and (3.25). Note that they are recursively expressed and are therefore updated at every time step. The additional parameters, $P(F_i | C_j)$ and $P(C_i | C_j)$, are known quantities in each time step and can be derived from θ.

$$\begin{aligned} \gamma(C_n) &= P(C_n | \mathbf{F}) \\ &= \frac{P(\mathbf{F} | C_n) P(C_n)}{P(\mathbf{F})} \\ &= \frac{\alpha(C_n) \beta(C_n)}{P(\mathbf{F})} \end{aligned} \tag{3.22}$$

$$\begin{aligned} \xi(C_{n-1}, C_n) &= P(C_{n-1}, C_n | \mathbf{F}) \\ &= \frac{\alpha(C_{n-1}) P(F_n | C_n) P(C_n | C_{n-1}) \beta(C_n)}{P(\mathbf{F})} \end{aligned} \tag{3.23}$$

$$\begin{aligned} \alpha(C_n) &= P(\mathbf{F}_1, ..., \mathbf{F}_n, C_n) \\ &= P(\mathbf{F}_n | C_n) \sum_{C_{n-1}} \alpha(C_{n-1}) P(C_n | C_{n-1}) \end{aligned} \tag{3.24}$$

$$\begin{aligned} \beta(C_n) &= P(\mathbf{F}_{n+1}, ..., \mathbf{F}_N | C_n) \\ &= \sum_{C_{n+1}} \beta(C_{n+1}) P(F_{n+1} | C_{n+1}) P(C_{n+1} | C_n) \end{aligned} \tag{3.25}$$

Clearly, future data is required for HMMs to function, as seen in (3.25), which may be unavailable for certain applications. The alternative is to make use of predictors like Kalman filters [11], which would require only present and historical data. This predictor and its application will be described in Section 3.3.6. Note that the earlier description of the HMM is by no means comprehensive, and readers are referred to [7] for more details.

3.3   Characteristics of Biomedical Problems

3.3.1   Data types

As mentioned in the preceding sections, data is crucial for both the design and evaluation of the algorithms. In particular, for classification and prediction problems, the results are closely linked to the initial dataset fed to the training algorithms. Since there is no way to constrain the problem directly, the indirect option is to populate the training and testing with carefully specified data. Some suggested guidelines are as follows.

1. *Physiological artefacts*: The dataset should consist of signals reflecting the extreme and typical physiological state of the patients. For example, in the context of arrhythmia detection, signals resulting from different types of arrhythmias (apart from the one that the classifier is trying to detect) should be taken into account. This would include tachycardia (abnormally high heart rate), bradycardia (abnormally low heart rate), various ventricular and atrial arrhythmias.

2. *System artefacts*: Noise is a common occurrence in biomedical signals. Sources originating from the system would include motion hardware noise, mains noise and artefacts introduced by the front-end sensors or other system components such as the radio. These confounding artefacts should be taken into account in the design of the dataset.

3. *Mechanical artefacts*: Mechanical movements on the part of the subjects could introduce motion artefacts in the acquired physiological signals and confound the measurements. Clearly, this is dependent on the type of motion, which is difficult to predict comprehensively during design time. By far, this type of artefacts is the most difficult to characterise.

4. *Prior data distribution*: The distribution of different types of signals should be clearly specified, reflecting that of the target population.

The amount of data available depends on the application being considered. For applications such as ECG signal compression or arrhythmia detection, there are several online repositories [12], which are freely available to algorithm designers. However, these signals are often acquired using front-end sensors, different from those used to collect data for the target application. Consequently, artefacts particular to the target system cannot be accurately captured.

Synthetic data generators, generating particular types of physiological signals, are available. On the other hand, most such tools do not accurately capture the aforementioned artefacts. To resolve this problem, a new tool or framework could be used for synthetic signal generation. As an input, the tool could take in a set of 'seeds' or signals collected from available subjects in a controlled setting (for example, in a laboratory). Motion artefacts could then be independently collected by mounting the target sensor on a rig and subjecting it to representative movements that the subject is likely to engage in, such as walking, eating and talking. In addition, underlying artefacts could be extracted from the seeds using predictors like Kalman filters [11] to obtain the residual noise. These independently collected artefacts can then be superimposed on top of the original seeds to generate further signals. Directly adding physiological artefacts to the signal set could be achieved by means of patient simulators. However, the type of rhythms and the available set of patterns are often limited. Consequently, morphological arrhythmic transformation operators can be applied to the seeds to generate signals with abnormal rhythms. For example, the operator should be able to transform the signal from one shown in Figure 3.5a to Figure 3.5b. These signals should be convincing enough to 'fool' expert clinicians.

Apart from the aggregate classification metrics used in assessing classification accuracy – positive predictivity ($+P$), sensitivity (Se) and specificity (Sp) [2], it is pertinent to accurately specify information about the type of data used in the training and evaluation of data. In addition, tests for statistical significance should be carried out to determine differences between the reference and the algorithmic results. If the distributions of both the reference and

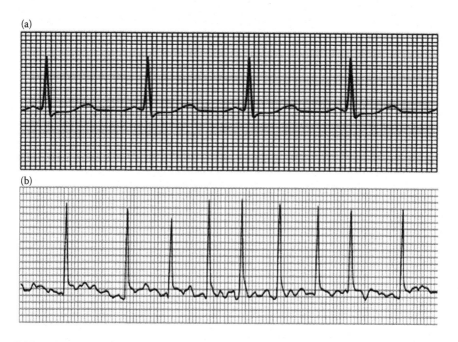

FIGURE 3.5

Examples of ECGs from a healthy patient (a) [14] and a patient suffering from atrial fibrillation (b) [15].

the algorithmic results datasets are Gaussian, the Student t-test [13] should be used to test the statistical significance. Otherwise, non-parametric methods [13] should be used instead. Finally, the differences between the datasets should also be quantified using correlation analysis, as well as the 95% confidence and prediction intervals.

3.3.2 Training frameworks

The classifier parameters are dependent on the data used during the training stage as well as the manner in which the data is fed and used for evaluation. If a high degree of correlation exists between consecutive data vectors, the classifier parameters might get stuck in a local minimum [9]. This suggests that the arrangement of data vectors ought to be randomised before they are being fed for training and evaluation purposes. This is particularly important if the primary objective is to evaluate the relative performance of different classifiers. The collected dataset is partitioned into subsets of two types – training and testing. The metric used to determine the accuracy of classifier is shown in (3.26), for a dataset populated with h data vectors. In most methods, this is a subset of the original dataset of size N.

$$acc_h = \frac{1}{h} \sum_{P_i, C_i} \delta(P_i, C_i) \qquad (3.26)$$

There are many schemes used for training and testing classifiers. Some of these schemes are as follows [16].

1. *Holdout*: This involves partitioning the dataset into mutually exclusive training and testing (or hold-out) sets and subsequently obtaining the classification accuracy using (3.26). This is done iteratively to obtain mean accuracy.

2. *K-Fold cross validation*: Involves partitioning the dataset into K folds or equally sized subsets, using $K - 1$ sets for training, and the remaining set as a test set for accuracy evaluation. This is done iteratively K times, and the overall accuracy is aggregated across the K individual accuracy values which are obtained. A stratified version of this scheme involves adjusting the distribution of each fold such that it has the same distribution as the original dataset.

3. *Leave one out*: This involves leaving one data vector for testing purposes and using the remaining $(N - 1)$ vectors for training purposes, and doing this iteratively N times. In this case, the overall accuracy is aggregated across N values.

4. *Bootstrap*: One data vector is selected for training in each iteration, and the remaining data vectors are used for testing the classifier. This is done b times to obtain a set of b accuracies, $\{\epsilon_i | i = 1, ..., b\}$. In

addition, the resulting classifier is applied to the training set to obtain an accuracy value, acc_s, and the overall bootstrap accuracy is obtained as a function of both ϵ_i and acc_s.

Each of these schemes produces multiple accuracy results [acc_h in (3.26)], which can then be aggregated to obtain the mean error and variance. The ideal scheme would result in a mean error of zero (or low bias) and low variance in errors. According to different reviews, these parameters were dependent on the number of data vectors and the number of features [16]. There was an agreement that stratified 10-fold cross validation was found to produce favourable results [16].

The work mentioned so far makes use of all of the data in the training set. However, for ANNs, it was found that a more representative classifier (lower risk of over-fitting) can be obtained by considering the distribution of the neural network weights, $\mathbf{w}$ [17]. Therefore, given the dataset D, the set of weights chosen at each time instance can be modelled as a random variable, $\mathbf{w}$. Therefore, the objective is to sample from the posterior probability, $p(\mathbf{w}|D)$ using the Metropolis algorithm [18], retaining or discarding each $\mathbf{w}$ heuristically in each iteration, to converge near/at an optimum solution. On the other hand, the Metropolis algorithm is often found to converge much more slowly, so the hybrid Monte Carlo algorithm is used instead [17]. These techniques could be used when direct usage of the data for training causes problems of over-fitting due to the dearth of data. However, training times are often much longer in such cases, compared with the conventional training methods described earlier [17].

3.3.3   Case study 1: Compression of physiological signals (Type 1)

In the context of designing an ambulatory health monitoring device for patients, it is advantageous for the acquired physiological signals to be made available to the attending nurse or doctor, so that abnormalities detected within the signals can trigger early clinical actions such as comprehensive tests or measurements (e.g. a 12-lead ECG). High bandwidth is necessary for the streaming of multi-channel physiological signals, and it is necessary to compress the signal in order to meet the specifications for energy consumption of the product. Clearly, the compression algorithm itself has to be sufficiently energy efficient so as not to negate the benefits from compression.

In this case study, we consider the use of Type II DCT transform for the compression of ECG signals, which are shown in (3.27) and (3.28), respectively, and where block size is N. This transform is used in Joint Photographic Experts Group (JPEG) compression [19]. In addition, it demonstrated competitive compression ratios and distortion errors for physiological signals as well [20]. Specifically, the original ECG signal has a high level of redundancy and spatial correlation. This transform operates on a block-by-block basis and compresses the energy of the signal into the lower frequency bins of the transform. Alternatively, the fast Fourier transform could be used, but it produces a complex output spectrum for a real signal, and is therefore more compute intensive. On the other hand, the DCT produces only real coefficients for a real input signal, as shown in Figure 3.6. While the original signal is observed to vary relatively slowly, signal fluctuations are observed to occur at just the low frequency bins of the spectrum, and extremely low amplitude components (dominated by zeros) towards the higher end of the spectrum.

$$F(k) = \sum_{j=0}^{N-1} f(j)cos\frac{(2j+1)k\pi}{2N} \qquad (3.27)$$

$$f(j) = \sum_{k=0}^{N-1} c(k)F(k)cos\frac{(2j+1)k\pi}{2N} \qquad (3.28)$$

$$j, k = 0, ..., N-1$$

$$c(k) = \begin{cases} \frac{1}{\sqrt{(N)}}, & k = 0 \\ \sqrt{\frac{2}{N}}, & \text{otherwise} \end{cases}$$

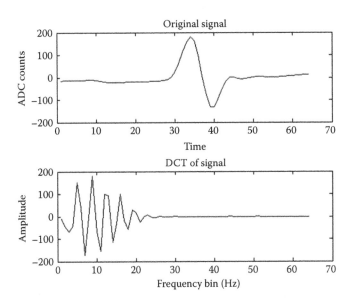

FIGURE 3.6
The chart at the top shows a signal segment containing an ECG QRS complex, while the chart at the bottom shows the compacted spectrum of the DCT.

The flowcharts in Figure 3.7 show the encoder and decoder used for the compression and reconstruction of the ECGs. This algorithm is a lossy compression algorithm – the compression process introduces distortions within the reconstructed signal. Specifically, the lossy aspect of the algorithm is introduced by the quantisation block in Figure 3.7a. Because of the relative lack of sensitivity of the least significant bits of the DCT coefficients to the final quality of the reconstructed signal, they may be truncated or quantised to expose further redundancy within the signal. Downstream of the quantisation module, a lossless compression algorithm – arithmetic compression [21] is used to remove the redundancy within the sequence of quantised DCT coefficients. Using this lossless compression scheme is advantageous because the parameters of the algorithm are dynamically adjusted to the immediate distribution of the signal, based on incoming samples. As can be seen in Figure 3.7b, the decoder is symmetric to the encoder.

To determine the best trade-off between the compression ratio and the amount of distortion in the reconstructed waveform, the parameters within the quantisation module may be adjusted. Each DCT block is modified according to (3.29). The operator '$//$' refers to a division operation followed by rounding.

$$F(n) = \begin{cases} 0, & F(n) < t(n) \\ F(n)//q(n), & \text{Otherwise} \end{cases} \tag{3.29}$$

The set of quantisation factors are defined by $q(n), n = 0, ..., N-1$ for a block size of N. Because frequency components of low amplitudes have little impact on the level of distortion, a separate set of thresholds, $t(n), n = 0, ..., N-1$, is

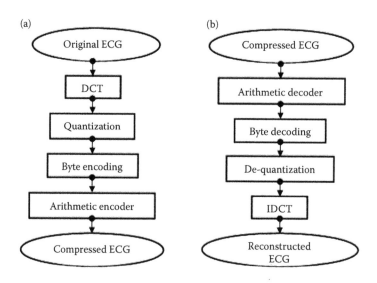

FIGURE 3.7
(a) DCT-based encoder. (b) DCT-based decoder.

used to remove them. The objective is to determine optimum values for these parameters where both compression ratio and reconstruction signal quality are maximised. This problem can be formulated as a Lagrangian [20], as shown in (3.30), where J_n is to be minimised. Entropy is a measure of how effectively data can be compressed, and it is expressed as bits/symbol. To compute entropy, the distribution of the data (in this case, DCT coefficients) has to be available.

$$J_n = H_n(q(n), t(n)) + \lambda D_n(q(n), t(n)) \tag{3.30}$$

H_n : Entropy of DCT coefficient n.

D_n : Mean Squared Error of coefficient n.

λ : Lagrange multiplier.

This is easily obtained by constructing a histogram from existing data. From the histogram, the entropy can be determined using (3.31), where p_i is the approximate probability of the occurrence of value i, $0 \leq i \leq M - 1$.

$$H = - \sum_{i=0}^{M-1} p_i log_2 p_i \tag{3.31}$$

The Lagrangian problem is graphically illustrated in Figure 3.3. Each point in the trade-off curve represents an optimal trade-off, based on the specified level of distortion and entropy. For a specific value of λ, a corresponding trade-off occurs, which is tangential to the trade-off point, as illustrated in Figure 3.8.

Subsequently, the bounds of the algorithmic parameters have to be defined. Intuitively, more compaction can be achieved with a higher DCT block size. However, the complexity of the DCT algorithm varies quadratically with block size, $O(N^2)$, and compaction gains from the transform saturates at a block size of 64 [20]. Therefore, the DCT block size is set to 64. Similar observations were found for the quantisation factors and thresholds. Therefore, the upper bounds for them are set at 64. The lower bound for the threshold is set at $q(n)/2$, since the least significant bits of the corresponding coefficient will be removed during quantisation, and the resolution is set to 0.5. Consequently, the bounds on the algorithmic parameters are shown in (3.32)–(3.34).

$$q(n) = 1, ..., N \tag{3.32}$$

$$t(n) = \frac{q(n)}{2}, \frac{q(n) + 1}{2}, ..., N - \frac{1}{2}, N \tag{3.33}$$

$$N = 64 \tag{3.34}$$

Obtaining the trade-off curve exhaustively will take a long time due to the large number of combinations of quantisation factors, q, and thresholds, t, for each DCT coefficient. In addition, each data point in the trade-off curve has to be computed over a sufficiently large dataset to obtain a representative cost

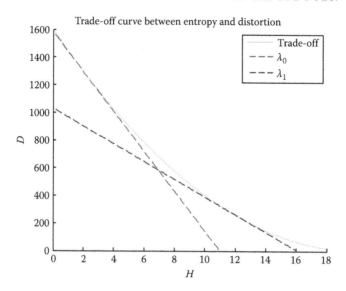

FIGURE 3.8
Illustration of the Lagrangian trade-off curve.

of that particular trade-off. For these reasons, heuristics are used to reduce the time necessary to converge on an optimal solution, for different values of λ, from 0.1 to 1, in steps of 0.1. The simplex algorithm [22] is used to obtain the best trade-off. It is used for the following reasons.

1. The optimisation algorithm avoids the local minimum by periodically expanding the solution space.

2. When moving towards the optimum solution, it takes big steps, and is therefore likely to converge on an optimal solution in a smaller amount of time.

For each value of λ, the solution space, J, can be described by means of a polytope, where each data point is computed for unique values of q and t. The boundaries of the polytope are defined by its vertices. As the algorithm progresses, the polytope gradually 'migrates' to a region where the global minimum occurs, and shrinks in size, as it converges around the optimum solution. At each step of the optimisation process, the vertices are ranked according to its cost, and the vertex with the lowest cost is returned at the end of the optimisation process. There are a few phases in the optimisation algorithm as follows.

1. *Reflection*: The vertex with the highest cost is reflected about the centroid of the polytope, and replaced if the new vertex has a lower cost than the old vertex.

2. *Expansion*: With reference to the centroid, the polytope is expanded in the direction of the vertex with the lowest cost, based on the heuristic that 'good' solutions are likely to occur in close proximity.

3. *Contraction*: The polytope is shrunk during more mature stages of the optimisation process.

4. *Reduction*: The polytope is shrunk around the vertex with the lowest cost, as the optimisation process converges rapidly around the optimum solution.

From this optimisation process, a set of N quantisation factors and thresholds are obtained. These values are stored in the encoder and applied to each coefficient before the application of the next compression stage. The same quantisation factors, $q(n)$, are stored in the decoder, and used in the recovery of the coefficients. Details concerning the byte encoding/decoding and the arithmetic encoder/decoder modules are outside the scope of this chapter. Readers are referred to [20] for further information.

3.3.4 Case study 2: Estimating calorie energy expenditure (Type 1)

Studies have shown that a sedentary lifestyle and other unhealthy lifestyle choices are linked to the occurrence of Type 2 Diabetes [23]. According to the World Health Organisation, 11% of the population is suffering from this disease, and this problem is imposing a substantial burden on national healthcare systems worldwide [24]. Through changes in lifestyle, including regular exercise, it was suggested that these risk factors could be reversed [23]. With the development of low-power physiological and biomechanical sensors, lightweight wearable devices can be used for the estimation of calorie energy expenditure of human subjects. These quantitative measures would allow the consulting doctor to determine whether the patient is exercising at the prescribed level or whether improvements have been made from the previous consultation.

A highly accurate technique of measuring calorie energy expenditure is achieved through the use of indirect calorimetry (an example of an indirect calorimeter is shown in Figure 3.9). This method is premised on the fact that energy is required for muscular contraction in order for the physical activity to occur. The release of energy occurs through the metabolism of fat, carbohydrates and proteins, which requires oxygen and produces carbon dioxide as one of the by-products. Therefore, by allowing the subject to breathe in air of known concentration (20.9% oxygen, 0.03% carbon dioxide, 79.1% nitrogen), and measuring the concentration of gases in exhaled air by means of a face mask, the oxygen uptake and the expelled carbon dioxide can be determined. The calorie energy expenditure can then be worked out as a function of oxygen uptake and carbon dioxide production (Weir formula [25]). Other

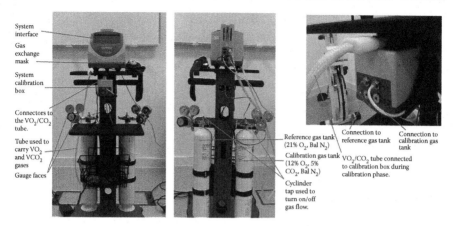

FIGURE 3.9
An indirect calorimeter.

accurate techniques for measuring calorie energy expenditure include direct calorimetry [26] and the doubly labelled water technique [27].

The merits of indirect calorimetry include its accuracy as well as its capability of measuring calorie expenditure on a breath-by-breath basis. However, the need for a face mask and its obtrusiveness precludes it as a device for use in daily living. These disadvantages motivated the development of models for the estimation of calorie expenditure using less obtrusive physiological and biomechanical sensors. Indeed, HR is frequently used as a feature for energy expenditure prediction as it is a good indicator for cardiovascular stress [28]. Despite the linear relationship between HR and oxygen consumption at certain levels of exercise (particularly low and moderate), this parameter should not be used by itself as an indication for energy expenditure. The reason behind this is that HR is also influenced by other factors rather than activity, i.e. emotional stress, anxiety, level of fitness, type of muscular contraction, active muscle group, environment and hydration [29].

Another means of estimating energy expenditure involves the use of motion sensors such as accelerometers. This is based on the fact that energy expenditure increases proportionally with the muscular activity responsible for acceleration and movement of the body and its extremities during physical exercise and locomotion. Recent advances of micro-engineering technologies have enabled the development of small portable and wearable devices intended for measuring physical activity. These lightweight and unobtrusive systems are equipped with accelerometers and data-logging capabilities. Therefore, they can be used in routine clinical practice, and/or at home or elsewhere in the community, i.e. where the patient carries on with his/her usual ordinary activities of daily living. Many of these devices are available in the market, and are described in various studies and comparative reviews [30, 31]. Unfortunately, there are some biomechanical disadvantages impeding the sole use of accelerometers for energy expenditure assessment, as follows.

1. Torso-mounted accelerometers are unable to determine energy expenditure associated with isolated limb motion.

2. Accelerometers may give similar intrinsic outputs (resulting in similar energy expenditure estimates) for activities, which nevertheless have quite differing Physical Activity (PA) levels, e.g. walking level or inclined, ascending stairs, cycling or rowing and walking unloaded or loaded with heavy objects.

3. The relationship between energy expenditure and accelerometer counts is linear and predictable at low and moderate physical activity levels only.

These shortcomings have motivated various researchers to explore alternative solutions, such as fusing accelerometer outputs with other physiological indicators of energy consumption. In 2004, Brage and colleagues [28] developed and evaluated a method [Branched Equation Model (BEM)] for measuring levels of energy expenditure by combining accelerometry with HR monitoring, demonstrating improved estimation of these parameters when tested in 12 normal male subjects. The approach relies on a set of rules, regression equations and thresholds to estimate the energy expenditure. Thus, these parameters are estimated by means of the selected piecewise function (i.e. one out of the four available in the branched model) which best suits the level and intensity of the activity currently performed. This model is illustrated in Figure 3.10. The algorithm does not make a distinction between the type of movement or activity, and assumes that the energy of the signals obtained from the accelerometer will be a 'good-enough' feature for the estimation of energy expenditure. More specifically, the raw tri-axial accelerometers signals (sampling rate of 50 Hz) are segmented into blocks, with a duration of 15 s each. Subsequently, each

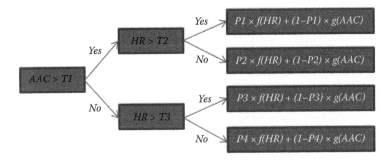

FIGURE 3.10
The branch equation model for calorie energy expenditure estimation.

channel is processed using a band-pass filter in the region of 0.25–6 Hz (the upper limit of the filter bandwidth was chosen to attenuate high-frequency disturbances occurring when the swinging foot impacts the ground during initial contact [32]), rectified, integrated and finally aggregated to obtain a single value – accumulated accelerometer count (AAC). The algorithmic parameters are as follows.

1. *Input features*: AAC and HR.

2. *Conversion functions*: Converts the AAC and HR into calorie expenditure.

3. P_i: Weights for different conversion functions.

4. T_j: Threshold values. The relative efficacy of AAC and HR as features for energy expenditure estimation differs according to the level of physical activity involved. These thresholds facilitate the partitioning of the transfer characteristic (between the input features and the estimated energy expenditure) into different regions where different contributions from AAC and HR are allowed.

More formally, the estimated calorie energy expenditure, $\hat{E}$, is defined in (3.35).

$$
\hat{E} = \begin{cases}
P_1 f(HR) + (1 - P_1)g(AAC), & AAC > T_1, HR > T_2 \\
P_2 f(HR) + (1 - P_2)g(AAC), & AAC > T_1, HR \le T_2 \\
P_3 f(HR) + (1 - P_3)g(AAC), & AAC \le T_2, HR > T_3 \\
P_4 f(HR) + (1 - P_4)g(AAC), & AAC \le T_2, HR \le T_3
\end{cases}
\tag{3.35}
$$

There are several variants of BEM [28, 33, 34], which targets different populations and makes use of different sensors. As such, the algorithmic parameters are different. Therefore, it is imperative that the algorithmic parameters are optimised for our particular accelerometer and HR monitor. Having obtained the best estimates for HR and AAC, using a well-known HR algorithm [35] and the AAC algorithm described earlier, the objective is to estimate $f()$, $g()$, P_i and T_j. To obtain the conversion functions, an experiment is designed to collect sensor data together with reference energy expenditure values, E, from an indirect calorimeter (Figure 3.9). This is carried out for eight healthy subjects, undergoing a variety of activities including walking/running on a treadmill, cycling on an ergometer and stepping exercises. The data are aligned and processed to obtain the HR and AAC, and correspondingly plotted in Figure 3.11. Note the breakpoints for both the HR and AAC transfer characteristics – these points demarcate regions of linearity in the transfer characteristics. Also, in the non-linear regions, the variance is noticeably larger, indicating that better accuracy can be achieved by operating on the linear portion of the transfer functions. Therefore, piecewise regression is carried out to obtain the best approximation for the transfer characteristics. More specifically, second-order

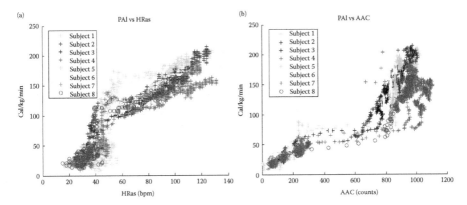

FIGURE 3.11
(a) HR and corresponding E. (b) AAC and corresponding E. The dataset comprises of data collected from an indirect calorimeter, corresponding with HR and AAC values, from eight subjects.

polynomials are used to approximate the non-linear segments of the transfer characteristics.

The objective function, along with the constraints, can then be formulated as shown in (3.36). In this case, the objective is to minimise the sum of absolute error across N data points between the reference value, E, and the corresponding predicted value, $\hat{E}$, which is represented by $\epsilon_{\mathbf{T},\mathbf{P}}$ (Vectors $\mathbf{T}$ and $\mathbf{P}$ are populated by the current thresholds and weights being considered). The parameters AAC_k and HR_k refer to the AAC and HR acquired from the sensors at instance k. The arguments, T_i and P_j, are to be adjusted so to minimise $\epsilon_{\mathbf{T},\ \mathbf{P}}$. The parameter, b, refers to the resolution of the ADC of the accelerometer (which is eight in our case). The parameters, S and I, refer to the sampling rate and the duration of one segment of data (which are 50 Hz and 15 s, respectively).

$$\underset{T_1,\ldots,T_3,P_1,\ldots,P_4}{\text{minimise}} \quad \epsilon_{\mathbf{T},\mathbf{P}} = \sum_{k=0}^{N-1} |E_k - \hat{E}(AAC_k, HR_k, T_1, \ldots, T_3, P_1, \ldots, P_4)|$$

$$\text{subject to} \quad 0 \le T_1 \le 3(2^{b-1} - 1) \times S \times I \qquad (3.36)$$

$$0 \le T_2, T_3 < 300$$

$$0 \le P_i \le 1, \ i = 1, \ldots, 4.$$

It can be observed that the design space is rather large, consisting of seven parameters. In addition, the compute time for $\epsilon_{\mathbf{T},\mathbf{P}}$ varies with the number of data points, N, which is large. Therefore, an exhaustive search for the optimal solution is undesirable. On the other hand, heuristics-based search algorithms are more attractive options. Empirical observations indicate that $\epsilon_{\mathbf{T},\mathbf{P}}$ is a non-convex function. For this reason, a non-greedy optimisation algorithm, simulated annealing (SA), has been selected to search for a pseudo-optimum solution [36]. *SA* has been initially designed to control the annealing process of solids – gradual cooling of a substance from liquid state to solid state at the lowest level of energy. This algorithm has a few critical parameters, including the state of the algorithm, S (consisting of the tunable parameters), the number of points within the design space to search in the current iteration, L, and a control parameter, c. A large control parameter makes it less probable for the algorithm to accept the current state as the best solution so far, and vice versa. The algorithm has a number of phases, as described later [36].

1. *Initialisation*: The algorithm initialises the critical parameters, including S, L and c.

2. *Initial exploration*: Note that each candidate solution (particular values for S) is generated randomly ('perturbation'). Better solutions are immediately accepted. Otherwise, poorer solutions may be accepted depending on c. This non-intuitive approach is carried out to prevent the algorithm from being stuck in a local minima.

3. *Late exploration stages*: L and c are adjusted in each iteration. As *SA* converges, it becomes increasingly greedy and less likely to accept poor solutions.

4. *Stop criterion*: An arbitrarily low value for the optimisation error function, and/or a maximum number of iterations are typically used in stop criterion for *SA*.

To adapt our problem for *SA*, the state S refers to the set of weights and thresholds: $\{P_i, T_j | i = 1, \ldots, 4; j = 1, \ldots, 3\}$, and these are initialised according to parameters found in [28]. In addition, the error function, ϵ, is used to gauge the relative quality of solution candidates, and as the stop criterion for the algorithm. Using this technique, an effective solution tuned to the sensor interface and other algorithmic parameters was located rapidly. The aggregate accuracy of the algorithm (both floating point and fixed point), in comparison with indirect calorimetry can be found in Figure 3.12.

3.3.5   Case study 3: Arrhythmia detection (Type 2)

There are different types of cardiac aperiodic rhythms. AF is perhaps the most commonly seen in general wards, and often leads to severe complications and life-threatening conditions when left untreated. Several experts have pointed out that AF is one the most common types of arrhythmia in clinical practice [3, 37, 38]. The overall prevalence of this condition is between 1% and 1.5% in the general population. The prevalence increases with age, particularly in the elderly – i.e. 10% in those more than 70 years old [15]. A report from the American Heart Association not only revealed an estimate of 2.66 million

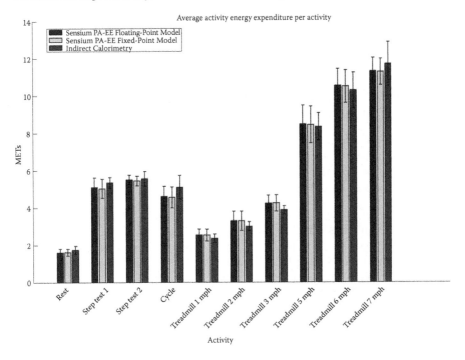

FIGURE 3.12
Bar chart of overall accuracy for a floating-point, fixed-point versions of the calibrated branch equation model in comparison with indirect calorimetry.

people with AF in 2010 but also they predicted that such figure will increase by 12 million in 2050 in the United States [39].

This motivated different scientists and bioengineers to explore and develop effective methods for detecting AF that can be incorporated in clinical monitors. A large body of work has been carried out to distinguish between AF, normal rhythms, as well as other types of abnormal rhythms [37]. However, this method tends to be suitable for patients who are at rest. The objective of this work is to present a three-way classifier that is capable of distinguishing ambulation noise from AF and normal rhythms, thus making it useful in clinical contexts such as the general ward, where patients are ambulatory, but could potentially deteriorate in a short span of time [40].

In the ECG, the electro-ionic activity of the normal heart is represented by the summation of different cells potentials taking place at different parts of the heart. This results in a number of signatures waveforms or complexes, representing the polarisation and repolarisation of different nerve cells across the organ. As shown in Figure 3.5a [14], the normal ECG exhibits a steady and regular rhythm. Deviation from these parameters may indicate the presence of an abnormal rhythm or heart condition. The characteristics of AF, as reflected in the ECG (Figure 3.5b [15]), are as follows:

1. Very irregular R-R intervals (RRIs).

2. Absence of P waves.

3. Abnormal Atrial Activity (F waves – coarse or fine).

4. Although AF is often accompanied by rapid ventricular rates ($>$ 100 bpm), the frequency of ventricular contractions can vary between 50 and 250 bpm depending on the degree of Atrioventricular node (AV) conduction, patient age and medications (such as beta-blockers) [41].

An extensive review of existing literature revealed that determining the extent of irregularity in RRIs will allow the accurate detection of AF. Indeed, extracting/selecting features from RRIs have yielded better performance than those based on hybrid methods or ECG morphological characteristics of AF. One of the most successful methods based on RRI is that proposed by Tatento and Glass [42]. This approach relies on the use of standard density histograms, coefficients of variation of RRIs and their successive differences to detect AF. The authors reported values of Se and Sp of 94.4% and 97.2%, respectively. In addition, Larburu and colleagues tested this algorithm and obtained classification performance values above 90% [37]. Other researchers have also attempted to develop efficient classifiers using RRIs. Moody and Mark developed a method involving the use of Markov's models to detect AF from RRI features. Results from the evaluation of this method using an arrhythmia database created by Massachusetts Institute of Technology and Beth Israel Hospital (MIT–BIH) showed $Se > 90\%$ but $Sp < 90\%$ and $+P < 90\%$. Similar or poorer performances were observed for other RRI based algorithms assessed by Larburu et al. [37].

As mentioned in Section 3.3.1, the types of data used for training the classification algorithm is of particular importance. To capture the impact of different types of artefacts on classification performance, we used a mixture of databases, detailed as follows.

1. *Physiological artefacts*: The MIT/BIH arrhythmia and the MIT–BIH AF [12]. In addition to AF, other atrial and ventricular arrhythmias, including atrial flutter and atrial bigeminy, are included in these databases. In addition, a small database of ECG signals were collected from patients using the target device during a clinical trial in a hospital. The patients suffer from a number of comorbidites including high BMI, diabetes, obesity, pulmonary oedema, AF, congestive heart failure and Wolff parkinson white.

2. *System-generated artefacts*: ECG signals were collected from the laboratory and from the short clinical trial. System artefacts are implicitly included within the signals.

3. *Mechanical artefacts*: Some of the signals within the earlier databases exhibit substantial baseline wander, indicating movement by ambulating patients or the electrodes themselves.

4. *Prior distribution*: A MATLAB graphical user interface (GUI)-based tool has been created for the selection of data, its randomisation, so that the prior distribution of data can be controlled (Figure 3.13).

When the ECG signal is acquired by the front-end sensors, it is divided into segments of 30 s in duration. The RR intervals are then extracted using a R-peak detection algorithm [35]. To enable more effective arrhythmia classification, each block of RR intervals is preprocessed by applying a moving average (median) filter approach [43], as follows. First, this stage calculates the median RRI out of the first nine RRI windows, resulting in the first estimate for x_m. For new windows x_m is updated by the mean of its current value and its previous value. A test statistic is then calculated as shown in (3.37), where med{.} is the median operator applied over the current window and $x(n)$ corresponds to the RRI located in the middle of the window (i.e., fifth element in a window of nine elements, centred around index n). The denominator represents a robust approximation of the standard deviation of an equivalent Gaussian signal. The current RR interval is replaced by the median interpolated estimate based on the criterion specified in (3.38), and this is set to four based on [43].

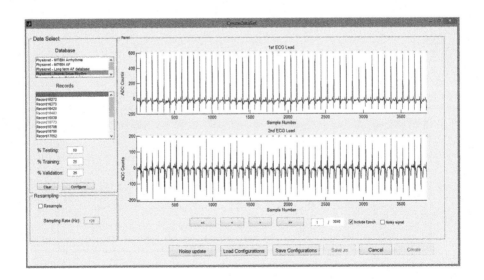

FIGURE 3.13
MATLAB graphical user interface (GUI) used for data selection and the specification of prior distribution.

$$D(n) = \frac{|x(n) - x_m|}{med\{|x(n) - x_m|\}} \qquad (3.37)$$

$$\hat{s}(n) = \begin{cases} x(n), & D(n) < \tau \\ \hat{s}(n), & D(n) \geq \tau \end{cases} \qquad (3.38)$$

It is important to emphasise that the choice of this filter was made bearing in mind its effectiveness removing spurious information from ECG signals or reasonable quality, but without having a significant filter effect in signals fully corrupted by noise. Thus, the filter helps to improve the performance of the classifier when discriminating between different classes.

To understand the data distribution in the context of feature space (defined by features selected by the user) and how certain types of data (particularly noise) can confound the classification process, the MATLAB GUI is extended for the purpose of feature space visualisation (Figure 3.14a). A 'probe' is used

(a)
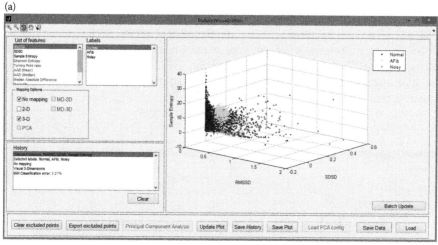

(b)
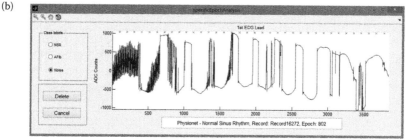

FIGURE 3.14
(a) MATLAB graphical user interface (GUI) used for feature space visualisation. (b) Probe used to investigate the nature of the data point and trace it back to its point of origin.

to investigate particular data points by double-clicking the point of interest – this facility allows the data point to be traced back to the particular database and the patient of origin (Figure 3.14b). This tool has two functions – first, it allows the features that would obtain the best separability between the target classes to be quickly identified. Second, it allows data clusters that could potentially confound the classification results, particularly those on the boundaries of class separation, to be identified in terms of their data type (e.g. noise or the type of arrhythmia) and their location in feature space, and expose potential algorithmic limitations.

The visualisation of the feature space is limited to 3D at a time. Automatic feature selection techniques that can be used beyond 3D, such as Relief-F and mRMR, can be deployed for high dimension feature selection. However, this algorithm is targeted for implementation on an embedded platform, where compute resource is scarce. In addition, the concurrent use of two features– RMSSD and HR (mean NN interval), resulted in an accurate classifier. Consequently, automatic feature selection techniques were not used in this design context. Following the exploration of different classification options (i.e. neural networks, support vector machines and discriminant analysis) using two candidate feature spaces, we found that three-way classifiers based on decision trees met the best compromise between classification performance and the low computational cost.

Among the data chosen, the start and end times of each AF episode are recorded. However, some of them are too short to be meaningfully detected. For that reason, an episode is considered to contain an AFib rhythm only if it contains 15 or more seconds of the rhythm. Other characteristics of the database are as follows. Table 3.1 shows the relative sizes of the randomly created partitions used in training and testing the AF classification algorithm.

1. A vast majority of signals corresponded to normal rhythms.

2. A large proportion of AF signals encompasses rapid rhythms between 120 and 190 bpm.

3. The proportion of noise signals is considerably smaller in comparison with the other two classes.

The tree classifier used in the determination of AF rhythms is derived, as shown in Figure 3.3. The hold-out method described in Section 3.3.2, with

TABLE 3.1

Dataset Partitions for Training and Evaluating the AF Classification Algorithm

Partition	Size	Proportion(%)
All	2246	100
Training	1568	69.8
Testing	678	30.2

TABLE 3.2
Results for the AF Classifier Using the Hold-Out Method

+Class	Se (%)	Sp (%)	+P (%)
AF	97.04	96.08	94.24
NSR	98.11	99.76	99.62
Noise	90.21	98.13	92.81

the test set being used for the accuracy evaluation of the AF classifier. Since this problem consists of three classes – AF, Normal Sinus Rhythm (NSR) and noise, the classification results for each class (defined as the +Class) is shown in Table 3.2.

3.3.6   Case study 4: Probabilistic scores for signal quality assessment (Type 3)

In the early days, algorithms for detection and quantification of physiological events were designed mainly for bedside monitoring. These devices are usually found in hospital areas such as surgical theatres, post-surgical recovery rooms and critical care units. In these dependencies, the levels of noise are often mild, since patients are passive mostly lying in their beds. In addition, there are less issues related to poor electrode-skin contact as nurses check patients more frequently than in other less critical areas. Therefore, most algorithms included relatively simple preprocessing stages mainly developed to remove periodic noise – such as mains interference, baseline wandering and cardiac artefacts; and some stochastic noise – such as mild motion artefacts and myoelectric noise. In addition, the incidence of false alerts due to noisy signals is low as patients in these areas are resting quietly in their beds.

In contrast, environments where patients are more dynamic introduce more difficult challenges. Patients are often moving, and thus the strength and incidence of motion artefacts and other types of noise are higher.

This motivated different researchers to investigate solutions for cancelling out motion artefacts (adaptive filtering, data fusion with Kalman filters, independent component analysis and frequency or time-frequency domain techniques). However, the situation becomes even more complex when patients are monitored using ultra-low-power wearable technologies. Despite their lightweight, unobtrusive and wireless nature; these devices are limited in processing power when compared with their bedside counterpart. Note the solutions mentioned earlier usually require simultaneous acquisition of multiple physiological signals and significant processing power; thus, increasing the hardware complexity, power consumption and cost of the wearable devices.

More recent approaches rely on probabilistic machine learning techniques to generate scores relative to the quality of the signals processed at a given point of time. Overall, Bishop [7] divided the probabilistic classification problem into two stages – inference and decision. Bearing in mind the Bayes' rule, the inference stage uses a training dataset to learn a model for estimation of the posterior probability $P(C_i|x)$ of an observation belonging to a class C_i after observing the data (x input vector from feature space) [7]. These classification models first require initial knowledge of the prior probability $P(C_i)$, which could be estimated from the available data. Later, it is necessary to estimate the class-conditional probability $P(x|C_i)$. The latter expresses how likely the observed data x corresponds to class C_i, and it is estimated by maximising the likelihood (or minimising the error) of a cost function. Subsequently, the posterior probability is used in the decision stage, as it represents the likelihood of a pattern belonging to a class after observing the input vector of features x. Hence, the probability of misclassification is reduced by selecting the class with the highest posterior probability, as shown in (3.39) [44].

$$P(C_i|x) > P(C_j|x), \forall i \neq j \tag{3.39}$$

This two-phased approach is related to Bayesian inference and decision. In addition, approaches that model the distributions of the input and output data are known as generative models, since they allow the generation of synthetic feature space data by sampling from the distribution [7].

Researchers at Oxford University [45] developed a generative approach to extract respiration rates (RRs)and probabilistic scores from multiple respiration signals. The respiration envelops were derived from photoplethysmography. Resultant RRs from the multiple derived signals were accompanied by a probabilistic score, enabling the algorithm to report only valid RRs above a certainty threshold. Based on their scores, each of the RR estimated can be fused. To achieve this, the researchers applied a Gaussian process (GP) framework. The GP method infers the distribution over the functions that represent the possible waveforms given the training dataset of signals, and then use this prior to make predictions on new/unseen data (testing dataset). Thus, a GP defines a prior over functions that is used to generate posterior over functions once new data is observed [46]. Using this approach is advantageous, since rather than adopting a fixed parametric function (e.g. linear or quadratic), GP generates models that learn general functions from a Gaussian distribution. The explanation and formulation of GPs are outside the scope of this chapter, but we recommend the reader to look at references [7, 30, 45] for further information. Finally, the authors developed a signal quality index based on a set of hard-coded rules that involve the use of logical expression and simple statistics for combining all the probabilistic scores obtained from each one of the multiple respiration waveforms derived by different techniques. The mean absolute error of the predicted against the reference values was obtained for different ranges of age, and the results showed low errors (or high accuracy) between both methods in general. Another alternative consists of a more direct approach, whereby an unknown function $y = f(x)$ is approximated using a discriminative function trained with a partition D_{train} of the whole dataset (i.e. training set), to map the feature vector x into decisions using (3.40) [46].

$$\hat{y} = \hat{f}(x) = \arg\max_c P(y = c|x, D_{train}) \tag{3.40}$$

Such decisions can be converted directly into posterior probabilities, specifically using a squashing function that ensures that the results always lie between 0 and 1 [7, 46]. In the case of classifiers of some discriminant classifiers [e.g. feedforward neural networks, support vector machines and logistic regression (LR)], it is common to adopt the sigmoidal function to achieve this effect. Finally, the same rule portrayed in (3.39) is used to make classifications on new incoming data.

Based on our experience at Sensium®, discriminative models are easier to integrate with the wearable sensors, require less processing and demand less battery power. At Sensium®, we developed and evaluated discriminative models for probabilistic scores for SensiumVitals®. Two illustrative examples are outlined later in this chapter; first, an overview about the method the patch uses to measure respiration activity is given.

The SensiumVitals® patch is a lightweight device that is attached to patients at the chest using two conventional ECG pre-gelled electrodes (www.sensium.co.uk). During each patch cycle, 30 s of ECG is obtained first, followed by the acquisition of one minute of respiration signal obtained via impedance pneumography (IP). Subsequently, axillary temperature is measured, and a battery check performed at the end of the cycle.

IP involves the application of a tiny but high-frequency current (e.g. 10 μA at 32 kHz) to the thoracic area through the electrodes placed at the thorax. This constant current is adequate for measuring continuous fluctuations of the impedance in the lungs that stem from cyclical changes in their volume during inspiration and exhalation. Changes in impedance are measured with the same electrodes used to apply the AC current to the chest. With appropriate circuitry, the signal is filtered and amplitude-modulated to obtain the envelope corresponding to the actual respiration activity. It should be borne in mind that the fluctuating impedance of the lungs measured at the electrodes is typically not larger than 3 Ω, which represents a challenge when mixed with motion artefacts or cardiogenic noise (known as 'heart bumps' that stem from dynamic changes of the heart volume and inflow/outflow of blood during each contraction). Although the patch incorporates hardware and software for attenuating these sources of noise, the technique is still sensitive to motion artefacts (Figure 3.15) due to the ambulatory nature of the target patient population [47]. Therefore, a probabilistic indication of the quality of the signal is compulsory in this scenario, to report RRs only when there is enough confidence (certainty) they are clinically valid.

We have developed and evaluated signal quality scores (SQS) models for the assessment of the respiration signals, following the process shown in Figure 3.16. The models are discriminative in nature and can be easily

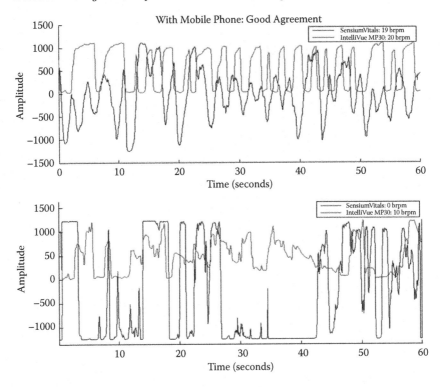

FIGURE 3.15

IP signals simultaneously recorded with SensiumVitals® and a reference bed-side monitor. The top figure corresponds to a good quality respiration signal. Respiration events (inspiration and exhalation) can be seen in the waveforms due to their quasi-periodic cyclical nature so that valid and accurate RRs can be obtained. In contrast, the bottom figure shows poor quality IP signals for both devices, which were severely corrupted by motion artefacts. It is evident that the periodicity of the signals is lost, and RRs are inaccurate and invalid.

integrated as part of any of our RR algorithms, leading to rejection of invalid RRs resulting from signals corrupted by noise.

First, we developed a suite of applications in MATLAB for preparation and partitioning of data; generation and selection of features and automatic creation-spanning of probabilistic classification models. The data comprised a balanced set of about 2500 respiration signals (sampled at 25.6 Hz), together with their Sensium® and reference RR values (range 5–60 brpm); both recorded simultaneously from patient simulators, acute patients with comorbidities and healthy volunteers using the patch and a bedside monitor [47, 48]. These signals were conditioned using band-pass filters to attenuate components outside the respiration bandwidth (0.08–1.1 Hz). A self-tunable filter was applied to remove heart contaminants from all signals. Then, a rule-based

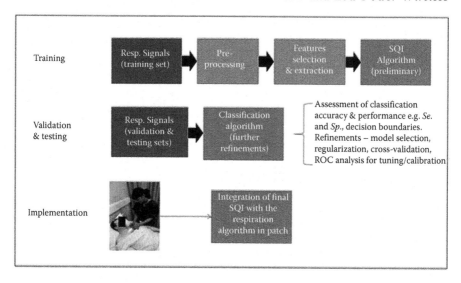

FIGURE 3.16
Process for the development and evaluation of probabilistic machine learning
models for inspection of respiration signals acquired with the SensiumVitals®
patch.

logic extracted respiration events – i.e. inhalation peaks, exhalation troughs,
zero-crossings – from each signal for subsequent RR estimations and features
generation. All these stages are part of the actual Sensium respiration algorithm (outside the scope of this chapter).

Next, all the signals were visually inspected and labelled as valid (1) or
invalid (0) depending on whether the signal was corrupted by noise, or reasonably or perfectly clean. Criteria such as ability to extract an RR from the
signal, timefrequency spectrogram and energy dispersion in the spectrogram
were considered when assigning a class to each signal. Thus, the subjective
opinion of the examiner was compensated by using these objective measures.
In addition, a semi-supervised technique (combination of clustering and an
instance-based non-parametric classifier) was used for automatic labelling of
an important subset of unlabelled signals (about a third of the total dataset).
Finally, the data was split into training (65%), validation (15%) and testing
(25%) sets.

Following data preparation, the feature space was generated. For this
purpose, we calculated a number of statistical, quantitative and complexity-
measures features. Then, the features were normalised (i.e. Gaussian stan-
dardisation – zero mean and unit variance) to avoid bias due to disparities in
different feature scales. Thus, a 44-dimensional feature space was formed.
When developing machine-learning models for ultra-low-power wearable
devices, it is important to bear in mind the trade-off between computational
complexity and classification accuracy. Thereby, we selected the less redundant

and most relevant features using a specialised feature selection algorithm. The mRMR [4], eliminates the redundant features by examining the mutual information between each pair of features i and j. Then, mRMR selects those features that are most relevant by examining the mutual information between the features i and the target labels for each feature. In simple terms, this framework processes feature pairs and features against target classes by means of operators such as the mutual information difference and the mutual information quotient; so that it performs an exhaustive search over all the features, leading to the creation of the best feature subset (eight features in our case) that yields good generalisation performance during the classification task. In summary, the whole dataset, D_{all}, encompasses all the observations of the 8D feature space with their corresponding labels. This can be described as: $D_{all} = \{D_{training}, D_{validation}, D_{testing}\}$. The contents of D_{all} are described in (3.41), where x_i and y_i are the ith 8D input vector and target label, respectively, among a total of N observations.

$$D_{all} = \{(x_i, y_i) | i = 1, ..., N\} \qquad (3.41)$$

With the resultant eight-dimensional feature space, we generated 28 two-dimensional and 56 three-dimensional discriminative models from all possible combinations of pairs and triplets of features, respectively. We selected binary LR classifiers because they are simple models containing a single processing unit, and its output can be used directly as probabilistic scores. Therefore, this type of classifiers is suitable for integration into ultra-low-power wearable devices.

LR is a discriminative technique, which aims to model posterior probabilities of different classes using a linear function of w. In its simplest form, the function is shown in (3.42), where x is the input vector of features corresponding to the observed respiration signal, and w is the vector of parameters (coefficients), used in the model. It is important to note that for 2D and 3D LR models are $\mathbf{x} = (1, x_1, x_2)^T$ and $\mathbf{x} = (1, x_1, x_2, x_3)^T$, respectively.

$$p(y = 1|\mathbf{x}) \approx f(\mathbf{x}, w) = w^T \mathbf{x} \qquad (3.42)$$

For two-class classification problems, a simple linear regression model can be generalised by [46], as follows.

1. Replace the Gaussian distribution of the output by a Bernoulli distribution with random variable y, where $y = 1$ and $y = 0$ represent two realisations –'Valid' and 'Invalid'. It follows that the probabilities for each realisation are p and $(1 - p)$, respectively. The conditional probability of y, given an observation vector $\mathbf{x}$

$$p(y|\mathbf{x}, w) = Ber(y|\mu(\mathbf{x})), \forall y \in 0, 1 \qquad (3.43)$$

$$\mu(\mathbf{x}) = E(y|\mathbf{x}) = p(y = 1|\mathbf{x}) = w^T \mathbf{x} \qquad (3.44)$$

2. Squashing can then be applied to (3.42) using the sigmoidal function, as shown in (3.45) and (3.43).

$$sigm(\alpha) = \frac{1}{1 + e^{-\alpha}}, \alpha = w^T \mathbf{x} \qquad (3.45)$$

$$p(y|\mathbf{x}, w) = Ber(y|sigm(w^T \mathbf{x})) \qquad (3.46)$$

Bearing in mind that after this step $\mu(x) = sigm(\alpha)$, the model can be fitted (or trained) via maximum likelihood estimation. This is simplified by minimising its negative-log likelihood or cross-entropy error function [46], as shown in (3.47).

$$NL(w) = -\sum_{i=1}^{N} [y_i log\mu(\mathbf{x}) + (1 - y_i)log(1 - \mu(\mathbf{x}))] \qquad (3.47)$$

Error minimisation is done by differentiating NL with respect to its parameters, and setting the partial derivatives equal to 0. However, the expression is not in closed form, and specialised numerical methods are therefore required to solve its parameters. There are different methods for which computer programs are available, such as the gradient descent, Newton's method and iteratively re-weighted least squares.

Using our MATLAB tool features, we generated various 2D and 3D LR models after applying the steps described earlier to all possible combinations of features. Once the models were fitted, the tool allowed us to predict the classes and assess their accuracy using the training data first.

Figure 3.17 shows an example of a 2D LR linear model, which is useful when classes are linearly separable. Following Receiver Operator Characteristic (ROC) analysis with training data, model number 1 outperformed the rest of the models, resulting from combinations of 2D features. The model yielded reasonable values of sensitivity and specificity ($Se = 89.9\%$ and $Sp = 86.6\%$), with a re-substitution (training) error of 11.6%. It can also be seen that the resultant 3D sigmoidal directly maps each entry of the 2D input vector x into probabilistic output scores.

Likewise, we also generated and evaluated 3D LR models. This time the decision boundaries were 2D hyperplanes (surfaces), as shown in Figure 3.18. It is also possible to develop LR models with non-linear fitting functions. This is possible by applying a vector of basis functions on the input vector of features – e.g. $\theta(x) = \{1, \mathbf{x}, \mathbf{x}^2\}$. For this case, the separation hyperplane is a parabolic surface (Figure 3.18).

So far we have seen the creation of 2D, and 3D LR models fitted with linear and non-linear functions using the training data only. But, how well these models perform when tested with unseen data? In machine learning, it is well known that increasing the complexity of the model in terms of the number of parameters and non-linearities may lead to over-fitting. This is the model is unable to generalise – i.e. poor classification performance with new data regardless of its good performance with training data. Under-fitting occurs

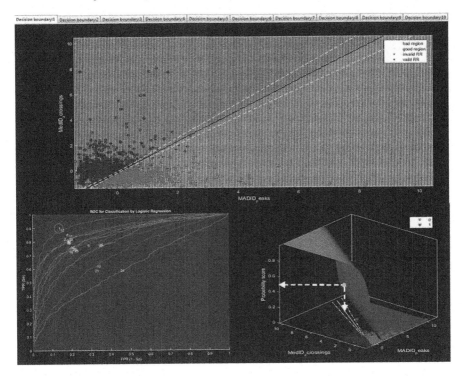

FIGURE 3.17

Two-dimensional logistic regression model. The top graph shows a rectilinear decision boundary with its 95% confidence intervals that separates vectors corresponding to 'good quality' from 'bad quality' signals. Note that both classes are not 100% linearly separable, as some A vectors overlap the 'good region' and some B vectors overlap the 'bad region'. The bottom-left plot corresponds to the ROC analysis for all the models created from all possible combinations of the eight features contained in the training dataset.

when the model is too simple, so it is not flexible enough to account for all the features and it fails to fit the data well. These models show poor classification performance with both training and new data. To minimise these problems, appropriate model selection is important. The aim is to select one or a small group of candidates that perform equally well with both, training and new data (i.e. low training and testing errors). Sometimes this process can be tedious and long-lasting as different models could have different features, dimensionalities, fitting functions and regularisation terms. Fortunately, machine learning programming packages offer different methods to automate this process, such as grid search.

Since in our case the total number of LR models was relatively small ($n = 84$), we automated model selection using k-fold cross-validation only.

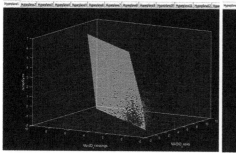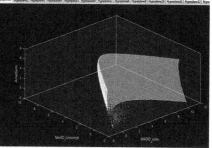

FIGURE 3.18

(Right) separation hyperplane for 3D model fitted with a linear function. Note that the hyperplane separates very well valid inputs (B) from invalids. (Left) A 3D model fitted with a quadratic function performing almost as good as its linear counterpart. Note that the separation hyperplane is now a parabolic surface.

This technique splits the validation set (15% of the total data) into k folds of the same size. These k partitions are obtained by randomising the data whilst keeping the class ratios equal in all partitions. One of these partitions is used to train an LR model with its associated features, whereas the remainder $k-1$ partitions are used to test the resultant trained model. This is done k times, such that each partition is used once for training a new model. The cross-validation performance measures are the average of the measures obtained for each one of the k folds.

The best selected models (not disclosed for confidentiality reasons) were tested with the remainder testing set (25% of the total data), achieving sensitivity and specificity values of 89% and 86%, respectively. It should be borne in mind that these results were obtained when using the default classification threshold for categorising a class (3.48).

$$C_i = \begin{cases} Valid, & SQS \geq 0.5 \\ Invalid, & SQS < 0.5 \end{cases}$$

Finally, the best selected models were integrated as part of our latest respiration algorithm, showing statistically significantly accurate results (not shown in this chapter).

This chapter revisited the use of probabilistic machine learning approaches to create SQSs to assess the quality of respiration signals obtained from wearable devices used by ambulatory patients. Based on our own experience and awareness of the state of the art, we conclude that these probabilistic models can be created with high classification accuracy.

Regarding our own developments, we believe that the classification performance can be further improved by either tuning the classification threshold or adding a threshold band with upper and lower probabilistic bounds – e.g. signal is valid if $SQS > th1$, and invalid when $SQS < th2$; such that signals with SQS falling within the interval $[th1, th2]$ can be classified as unknown. Thus, the respiration algorithm can interpret this information and either take further processing actions or just show appropriate messages to inform clinical decisions.

3.3.7 Case study 5: Fall detection/prediction (Type 2/3)

The remote monitoring of health and wellness extends beyond the physiological status of the subjects concerned, to their biomechanical activity. Indeed, in certain cases such as serious falls, clinically significant events can often be detected more quickly through biomechanical sensors, as a time lag typically exists between falls and actual physiological deterioration. Given the World Health Organisation findings that the second leading cause of accidental injuries worldwide is attributed to falls [49], it follows that a rapid and reliable technique for fall prediction and detection is likely to have a large and positive impact on general health and safety, whether for remote workers, first responders (e.g. firefighters and paramedics) or for the elderly living alone in their homes.

There are three broad approaches for fall detection – camera vision, ambient sensor and wearable sensor based detection methods. More specifically, the camera vision based method identifies fall events by processing captured image frames, whereas the ambient method makes use of vibrational sensors on the floor to detect impact events [50]. The last approach makes use of wearable sensors that are mounted on individual subjects – these typically involve motion sensors like accelerometers and gyroscopes. A problem with the first two approaches is that they are severely limited in scope, where the subjects can actually be monitored. On the other hand, wearable sensors are mounted on the subject, potentially affording more comprehensive monitoring. For this reason, we will be focusing on the wearable sensors in this section, particularly for a tri-axial accelerometer that is mounted on the subject's hip.

It is useful to divide individual falls into different phases, as shown in Figure 3.19. The chart in the middle of the figure shows the root sum of squares (RSS) variation from a tri-axial accelerometer that is mounted at the right hip of the subject (Figure 3.20), as a frontal fall is performed onto a crash mat. At position (A) of Figure 3.19 or the *pre-fall phase*, the subject is standing in a stationary position, and the RSS is 1 g, corresponding to the magnitude of the gravitational vector. The subject commences the fall just before (B), and is 'in flight' at (B). Hence, RSS falls towards zero. The *impact phase* occurs at the trough between (B) and (C). Subsequently, the subject sinks further into the mat and decelerates rapidly, bringing *RSS* back to 1 g at (D). The subject further decelerates to a maximum values at (E). The *post-fall phase* of the fall then occurs, where RSS eventually returns to 1 g. The final phase – *Recovery* may or may not exist as the subject returns

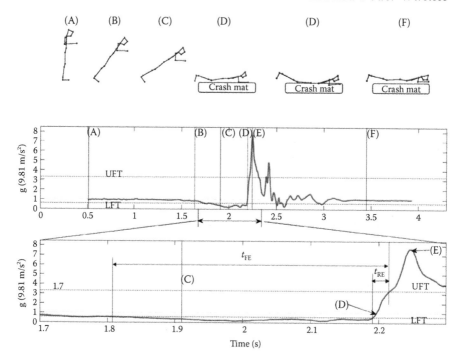

FIGURE 3.19

The chart in the middle shows the RSSs derived from a tri-axial accelerometer, and illustrates the signal variation corresponding to the biomechanical movements of the stick figure at the top; the figure at the bottom shows a sub-segment of the signal showing the pre-fall and fall phases, extracted from [51].

back to the initial state/posture after a fall. Upper and lower RSS amplitude thresholds and the timing thresholds where amplitude thresholds are exceeded, t_{FE} and t_{RE}, could be used to help identify transition points between different phases of the fall.

The fall detection algorithm should be capable of identifying critical phases of a fall, and subsequently fusing the information in a manner that will minimise errors. Based on a fall detection system with a tri-axial accelerometer feed, a top level system diagram of the fall detection algorithm may be found in Figure 3.21. The control flow diagram in Figure 3.21a illustrates the three principal stages of the algorithm, including pre-fall, impact and posture detection, and the dependencies between them. The structure of the algorithm and the data flow between individual modules are depicted in Figure 3.21b. In [51], simple threshold based implementations of the three sub-modules – pre-fall, impact and posture detection – are used and found to produce good results [51].

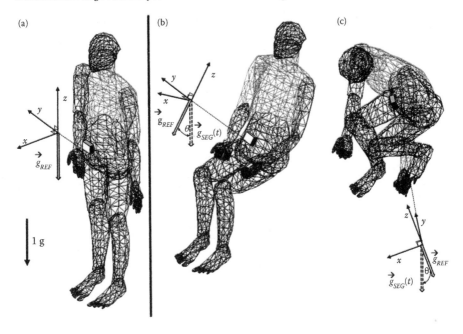

FIGURE 3.20
Position of a hip mounted accelerometer in the initial reference standing position (a), sitting (b) and kneeling position (c), extracted from [51].

Whilst the raw accelerometer signal values may be converted to universal gravitational units (or 'g's), different motion sensors typically have different signal characteristics, necessitating the use of custom algorithms. For this Type 2 classification problem, it follows that the application of machine learning techniques is likely to help in the derivation of hyperplanes for all three modules, as opposed to simple threshold-based algorithms [51]. The only module that will be discussed in this section is impact detection, as there are marked similarities between them.

Most prior works involving fall detection algorithms [51, 52] involve two stages – scripted falls and the activities of daily living. It is important that the algorithm is able to accurately detect falls (Type 2 errors or false negatives). At the same time, the algorithm should not falsely detect falls that could be triggered from activities of daily living as well (Type 1 errors or false positives), as the resources deployed in response to each detected fall could be considerable, and monitoring personnel could be de-sensitised to actual fall events. During the design of the fall detection data collection protocol, the difficulty faced by researchers is the simulation of falls in a realistic, safe and ethical manner. Important considerations include the type of surface that subjects will be falling on, and re-producing natural behaviour during slips, trips and falls. A number of creative attempts have been made to get around

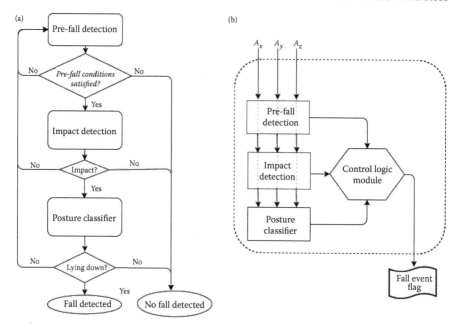

FIGURE 3.21

Top-level diagram of the fall detection algorithm: (a) Control flow diagram (b) Structural data flow diagram, involving data from a tri-axial accelerometer - A_x, A_y and A_z [51].

these issues, including the involvement of martial artists [54]. The position/s of the accelerometers vary as well, and they include the head, chest and torso regions, as well as different combinations of these sensors. From the tri-axial accelerometer data that is collected, the three axes, A_x, A_y and A_z may be aggregated into a single dimension, A_T, by using the Euclidean norm operator (3.48).

$$A_T = \sqrt{A_x^2 + A_y^2 + A_z^2} \qquad (3.48)$$

From the aggregated accelerometer signal, pertinent features may be found in Table 3.3 [53]. Note that the first 11 coefficients of the autocorrelation function and the first five peaks of the Discrete Fourier Transform (DFT) are considered in Ref. [53].

Five different classification techniques are considered, including the K nearest neighbour classifier, least means square method, support vector machine, Bayesian decision making, dynamic time warping and ANNs [53]. The ROC analysis of each of these techniques is found in Figure 3.22. The ROC for the K nearest neighbour classifier was found to have the best performance, with $Se = 100\%$, $Sp = 99.8\%$ and $Acc = 99.9\%$, as shown in Figure 3.22.

TABLE 3.3

Useful features for fall impact detection. Features are extracted from a window of N samples, where s_i is the ith sample, by [53] is licensed under CC by 2.0.

Feature Type	Definition
Mean	$\sum_{n=1}^{N} s_n$
Variance	$\frac{1}{N}\sum_{n=1}^{N}(s_n - \mu)^2$
Skewness	$\frac{1}{N\mu^3}(s_n - \mu)^3$
Kurtosis	$\frac{1}{N\mu^4}(s_n - \mu)^4$
Autocorrelation	$\frac{1}{N-\Delta}(s_n - \mu)(s_n - \Delta - \mu)$
DFT	$\sum_{n=0}^{N-1} s_n e^{-\frac{2\pi qn}{N}}, 0 \leq q \leq N$

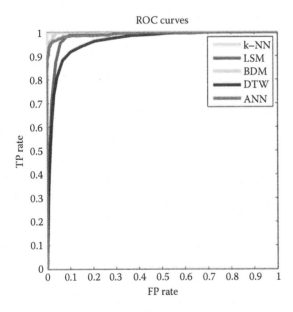

FIGURE 3.22

ROC analysis of the five different impact classification techniques, by [53] is licensed under CC by 2.0.

While high level of classification efficacy is achieved using the algorithm described earlier, there are two major limitations. First, fall notifications are retrospective. Second, this algorithm does not adapt to the fall patterns of individual users and attempt to make improvements to fall detection accuracy for the particular individual. An ideal algorithm should be able to predict falls accurately well in advance of the actual event (Type 3 problem), so as to facilitate early intervention. Indeed, fall prediction could allow airbags on the subject to be deployed to buffer the impact of the fall [55]. In the proposed system, the fall prediction algorithm estimates the probability of a fall occurring, up to 400 ms, in advance based on a combination of factors, including the extent of fluctuations within the accelerometer signal, as well as the sequence in which they take place [55]. The algorithm then makes use of two thresholds, T_1 and T_2, to determine whether a fall has been predicted and/or detected. The intention of this algorithm is to minimise injury caused by actual falls by deploying safety mechanisms such as an airbag to buffer falls. However, robust fall prediction is a far more difficult problem than fall detection as it is more tightly coupled to behaviour prior to a fall, which is highly variable among individuals. Frequent errors are undesirable as erroneous airbag deployment are inconvenient at best and could lead to injury, which would then be counter-productive.

An alternative method that could be potentially more robust and adaptive is shown in Figure 3.23. More specifically, features are extracted and fed to both a supervisor fall detection and fall predictive algorithm. The supervisor could be the fall detection algorithm described in Figure 3.21 – this algorithm produces generalised but reliable fall detection results. These results, alongside other intermediate data generated by the fall detector sub-modules $(S_0, ..., S_{N-1})$, such as the pre-fall probability and posture, can then be fed into fall predictor. The fall predictor then attempts to make an early assessment of whether a fall is about to take place, based on these observations as well other features retrieved from the feature bank.

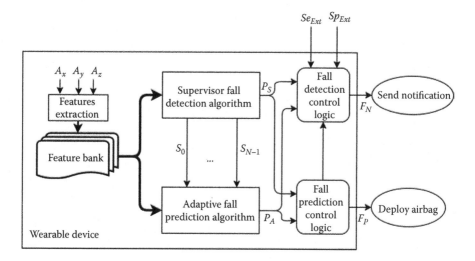

FIGURE 3.23
Data-flow diagram of the adaptive fall prediction algorithm.

Initial attempts by the fall predictor to predict falls may produce sub-optimal results due to aforementioned reasons. However, Type 1 and 2 errors may be reliably found by waiting for actual results from the supervisor. As such, during the initial deployment of the algorithm, when the predictive algorithm is undergoing 'training' by its supervisor, less weight should be assigned to its decision output by the fall prediction control logic module. Specifically, the weight should be computed as a function of the sensitivity (Se) and specificity (Sp) of the fall prediction classification module, as shown in (3.49). Subsequently, assertion ('1') and dis-assertion ('0') of F_p of the control input to the airbag will then be dependent on the weighted probability of the predictor output (wP_A). Note that the parameters, λ and T_1, are tunable parameters in this case.

$$w = \lambda Se + (1 - \lambda)Sp, 0 \leq \lambda \leq 1 \tag{3.49}$$

$$F_p = \begin{cases} 1, & wP_A \geq T_1 \\ 0, & wP_A < T_1 \end{cases}$$

It is anticipated that beyond a certain point, the adaptive fall prediction algorithm outperforms its supervisor. Consequently, the final fall detection decision can be computed as a weighted sum of the fall predictor and supervisor algorithms. The accuracy of the notifications may not be locally known within the device. However, these notifications are typically sent to a server, where intervention may be triggered by means of a web application interface. These decisions and the subsequent follow-up actions for each individual subject are then logged within the database server. The accuracy of these notifications can then be distilled into corresponding sensitivity (Se_{Ext}) and specificity (Sp_{Ext}) values, and fed back to the device. Based on these values, the final decision regarding the requirement for notification can therefore be computed as a weighted function of P_A and P_S, as shown in (3.50). Similar to the prediction control logic module, γ and T_2 are tunable parameters.

$$v = \gamma Se_{Ext} + (1 - \gamma)Sp_{Ext}, 0 \leq \gamma \leq 1 \tag{3.50}$$

$$F_N = \begin{cases} 1, & vP_S + (1 - v)P_A \geq T_2 \\ 0, & vP_S + (1 - v)P_A < T_2 \end{cases}$$

There are a variety of methods to implement the adaptive fall prediction module. The HMM is suggested as a possible method [55]. In this case, the observations are pertinent features derived from the feature bank as well as the outputs from the supervisor algorithm ($S_0, ..., S_{N-1}$), and the latent states in this case correspond to whether a fall is going to occur or not. Subsequently, the total probability, P_A, may be determined as a function of the state transition and emission matrices; these matrices may be updated in each time step using Baum–Welch maximum likelihood estimation method – the reader is referred to Section 3.2.3 for further information.

3.4   Discussion

The main objective of this chapter is to create a framework, allowing problems and their constraints to be formulated clearly. Once the problem has been properly distilled, the metrics for evaluation defined, and the corresponding set of candidate solutions or algorithms identified, how should one go about implementing the solution? A design and build flow, which we found useful, is shown in Figure 3.24.

The exploration of new or existing approaches involves a comprehensive review to identify existing algorithms, which can be used to solve the formulated problem in the light of existing constraints. Some of the approaches may be unsuitable due to these constraints. For example, a real-time solution might be required to solve the problem at hand, implying that future data is not available to produce a result at each time instance. This might rule out approaches such as the HMM as a means of solving the problem. Once the potential algorithm candidate set has been determined, they are prototyped in flexible programming languages such as MATLAB [56], which allows rapid implementation and comprehensive analysis of their relative performance. Further, a representative dataset needs to be collected. The guidelines for its collection and evaluation are detailed in Section 3.3.1. In the case of classification problems, training methods for determining the classifier parameters are discussed in Section 3.3.2. Candidates who do not meet performance requirements are culled from the solution set at this point.

The most promising candidates are then ported over to a programming language compatible with the target device. An example is C, which is the language of choice for many embedded processors. The programming constructs used in this case are kept as generic as possible so that the algorithm candidates are portable across multiple platforms. Subsequently, an initial algorithm profile is carried out to determine the resource requirements, including the code, data memory space requirements, as well as its energy consumption. Algorithm candidates that do not meet these platform constraints are removed. At this stage, an additional data collection stage might be carried out to further test the robustness of the algorithm. For instance, in the case of a HR computation algorithm, ECG signals might be collected from ambulating patients who are known to have a variety of arrhythmias and subsequently applied to the candidate algorithms. If the algorithmic performance or its computational requirements are deemed to be too expensive, further optimisation may be necessary to ensure that the platform or performance constraints are met. This is often an iterative process that would proceed until these constraints are met.

Downstream of the necessary checks for regulatory compliance, the selected algorithm has to be integrated into the target embedded platform. At this stage, the application programmer interface functions for the algorithm and test vectors, and the corresponding results should be made available. These

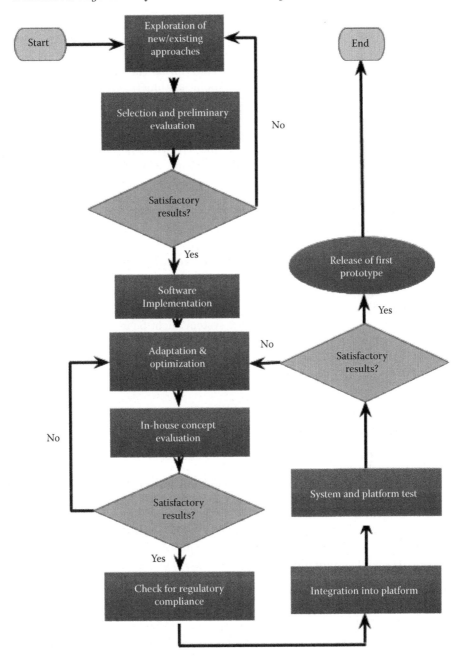

FIGURE 3.24
Design and build flowchart for biomedical algorithms.

application programmer interface functions allow the algorithms to be easily deployed without the need for changing the underlying source code. The test vectors and their expected results provide the means of checking whether the functionality of algorithm has been preserved during the integration process. As an example, double precision floating-point arithmetic operations that are common in MATLAB are implicitly reduced to single precision in the Keil environment [57] (Programming environment for several embedded processors including the Intel 8051 eWarp processor [58]). In applications such as high-order Infinite Impulse Response Filter (IIR) filters, this might create unacceptable differences in algorithmic behaviour. Therefore, the integration dataset should be large and representative enough to expose these differences and any algorithmic limitations.

An overall system test is then carried out to determine whether the system works well with the newly integrated algorithm. System-level bugs might be revealed at this stage. For instance, intermediate buffers for dynamic data storage might not be properly allocated and removed within the algorithm during run-time, potentially causing memory corruption in other modules, such as the software radio module responsible for the wireless transmission of the payload. Therefore, a comprehensive set of test cases should be defined to represent realistic scenarios in which the system might be used, so as to expose system-level bugs, which can then be corrected. Finally, after the system has passed the battery of tests, the first prototype can be released for deployment in the field.

3.5   Conclusion

On the basis of our observations, we found that problems encountered in the healthcare monitoring industry can each be categorised into one of the three types – generic optimisation problem, static classification problem and prediction problem. There are substantial overlaps between these problems, and algorithms are often suited for multiple problems. At times, the problem type is made based on certain application constraints. For instance, by formulating a problem as one of a static classification task, rather than a dynamic predictor, one is able to create a more compact and efficient classifier that can be realised on the target platform.

In addition, the dataset used in the training and evaluation of algorithm is crucial, as it impacts the algorithmic parameters, the relative results between algorithm candidates and potentially the choice of algorithm. A representative and comprehensive dataset is often difficult to come by, and relevant data is often dependent on the sensor frontend used for data collection as well as the physiological condition of the patient. To solve this problem, guidelines for initial data collection in controlled laboratory conditions, together with methods to augment the dataset were presented in this chapter. This is particularly important for static classification problems. For predictive problems, the initial dataset and algorithmic parameters are important because the predictive outcomes of such algorithms are dependent on prior data distribution during early runs of the algorithm and they may take a longer time to converge on the correct results should the initial dataset be of bad quality.

To demonstrate the formulation framework and the thought processes involved during algorithm design, four case studies involving different types of problems were provided. Note that, in each case, only a cursory discussion on the statistical evaluation of the results was given because they are outside the scope of this chapter. Apart from the algorithm design, its implementation aspects were also discussed in Section 3.4, with emphasis on algorithmic portability across different platforms and a robust testing framework – from the initial unit testing, integration testing, to the final system testing.

Consequently, it can be concluded that in conjunction with a representative dataset, problem formulation is an especially critical part of the design process. They provide a means by which problems can be described concisely as well as a framework within which to automate problem solving. The authors acknowledge that there may be problems that cannot be categorised into any of the three types described, but this work is the result of the actual problems encountered in the field, and our insights and attempts at solving them.

Bibliography

[1] R.S. Garfinkel and Nemhauser G.L. *Integer Programming*. John Wiley & Sons, London, 1970.

[2] V. Kumar, P. N. Tan and M. Steinbach. *Introduction to Data Mining*. Addison-Wesley, Boston, MA, 2005.

[3] L. Sornmo, M. Stridh, D. Husser, A. Bollmann and S. B. Olsson. Analysis of atrial fribillation: from electrocardiogram signal processing to clinical management. *Philosophical Transactions of the Royal Society of London A*, 28:235–253, 2009.

[4] F. Long, H. C. Peng and C. Ding. Feature selection based on mutual information: criteria of max-dependency, max-relevance, and min-redundancy. *IEEE Transactions on Pattern Analysis and Machine Intelligence*, 8:1226–1238, 2005.

[5] M. Robnik-Šikonja and I. Kononenko. Theoretical and empirical analysis of relieff and rrelieff. *Machine Learning*, 1–2:23–69, 2003.

[6] C. Cortes and V. Vapnik. Support-vector networks. *Machine Learning*, 3:273–297, 1995.

[7] C. M. Bishop. *Pattern Recognition and Machine Learning*. Springer, Cambridge, 2006.

[8] J. W. Sammon. A nonlinear mapping for data structure analysis. *IEEE Transactions on Computers*, 18:401–402, 1969.

[9] J. Principe. Artificial neural networks, in Richard C. Dorf (ed.) *The Electrical Engineering Handbook*. CRC Press, Reading, MA, 1997.

[10] J. R. Quinlan. *C4.5: Programs for Machine Learning*. Morgan Kaufmann Publishers, Boston, MA, 1993.

[11] L. Mason. *Signal Processing Techniques for Non-Invasive Respiration Monitoring*. PhD thesis, Oxford University, UK, October 2002.

[12] A. L. Goldberger, L. A. N. Amaral, L. Glass et. al. PhysioBank, PhysioToolkit, and PhysioNet: components of a new research resource for complex physiologic signals. *Circulation*, 23:e215–e220, 2000.

[13] C. Redmond and T. Colton. *Clinical Significance Versus Statistical Significance*. John Wiley & Sons, London, 2001.

[14] S. Navas. Atrial fribillation: part 1. *Nursing Standard*, 17:45–54, 2003.

[15] S. Goodacre and R. Irons. Abc of clinical electrocardiography: atrial arrhythmias. *British Medical Journal*, 324:594–597, 2002.

[16] R. Kohavi. A study of cross-validation and bootstrap for accuracy estimation and model selection. In *International Joint Conference on Artificial Intelligence*, pp. 1137–1143, Canada, August 1995.

[17] W. D. Penny, D. Husmeier and S. J. Roberts. An empirical evaluation of bayesian sampling with hybrid monte carlo for training neural netowork classifiers. *Neural Networks*, 12:677–705, 1999.

[18] M. N. Rosenbluth, N. Metropolis and A. W. Rosenbluth. Equation of state calculations by fast computing machines. *Journal of Chemical Physics*, 21:1087–1092, 1953.

[19] JPEG. Jpeg homepage. http://www.jpeg.org/jpeg/index.html. Accessed July 2015.

[20] E. U. K. Melcher, L. V. Batista and L. C. Carvalho. Compression of ecg signals by optimized quantization of discrete cosine transform coefficients. *Journal of Medical Engineering and Physics*, 23:127–134, 2001.

[21] R. M. Neal, I. H. Witten and J. G. Cleary. Arithmetic coding for data compression. *Communication ACM*, 6:520–540, 1987.

[22] J. A. Nelder and R. Mead. A simplex method for function minimization. *The Computer Journal*, 4:308–313, 1965.

[23] F. B. Hu. Sedentary lifestyle and risk of obesity and type 2 diabetes. *Lipids*, 2:103–108, 2003.

[24] WHO. Global status report on ncds, 2010.

[25] J. Weir. New methods for calculating metabolic rate with special reference to protein metabolism. *Journal of Physiology*, 109:1–9, 1949.

[26] D. C. Simonson and R. A. Defronzo. Indirect calorimetry: methodological and interpretative problems. *American Journal of Physiology-Edocrinology and Metabolism*, 258:E399–E412, 1990.

[27] K. R. Westrup and G. Plasgui. Physical activity assessment with accelerometers: an assessment against doubly-labelled water. *Journal of Obesity*, 15:2371–2379, 2007.

[28] N. Brage, S. Brage and P. W. Franks. Branched equation model of simultaneous accelerometry and heart rate monitoring improves estimate of directly measured physical activity energy expenditure. *Journal of Applied Physiology*, 96:343–351, 2004.

[29] F. Slinde, H. P. Johansson and L. Rossander-Hulthen. Accelerometry combined with heart rate telemetry in the assessment of total energy expenditure. *Journal of Nutrition*, 95:631–639, 2006.

[30] S. Murphy. Review of physical activity measurement using accelerometers in older adults: considerations for research design and conduct. *Journal of Preventive Medicine*, 48:108–114, 2009.

[31] D. Andre and D. Wolf. Recent advances in free-living physical activity monitoring: a review. *Journal of Diabetes Science and Technology*, 1:760–767, 2007.

[32] E. Antonsson and R. Mann. The frequency content of gait. *Journal of Biomechanics*, 18:39–47, 1985.

[33] J. Churilla, S. Crouter and D. Bassett. Accuracy of the actiheart for the assessment of energy expenditure in adults. *European Journal of Clinical Nutrition*, 62.6:1–8, 2007.

[34] J. McClain and C. T.-Locke. Objective monitoring of physical activity in children: considerations for instrument selection. *Journal of Science and Medicine in Sports*, 12:526–533, 2009.

[35] J. Pan and W. J. Tompkins. A real-time qrs detection algorithm. *IEEE Transactions of Biomedical Engineering*, 3:230–236, 1985.

[36] J. Korst, E. Aarts and W. Michiels. *Search Methodologies*. Springer, New York, 2005.

[37] N. Larburu, T. Lopetegi, and I. Romero. Comparative study of algorithms for atrial fribillation detection. In *Proceedings of IEEE Conference on Computing in Cardiology*, pp. 265–268, Hangzhou, China, September 2011.

[38] F. Yaghouby, A. Ayatollahi, R. Bahramali, M. Yaghouby and A. H. Alavi. Towards automatic detection of atrial fribillation: a hybrid computational approach. *Computers in Biology and Medicine*, 40:919–930, 2010.

[39] D. L.-Jones, R. J. Adams, et. al. Heart disease and stroke statistics - 2010 update. *Circulation*, 121:e46–e215, 2009.

[40] S. Hugueny, D. A. Clifton and L. Tarassenko. Probabilistic patient monitoring with multivariate multimodal extreme value theory. *Biomedical Engineering Systems and Technologies*, 127:199– 211, 2011.

[41] S. Kara and M. Okandan. Atrial fribillation with artificial neural networks. *Pattern Recognition*, 40:2967–2973, 2007.

[42] K. Tateno and L. Glass. Automatic detection of atrial fibrillation using the coefficient of variation and density histograms of rr and deltarr intervals. *Medical and Biological Engineering and Computing*, 39:664– 671, 2001.

[43] J. McNames, T. Thong and M. Aboy. Impulse rejection filter for artifact removal in spectral analysis of biomedical signals. In *Proceedings of the IEEE Medical Biology Society*, pp. 145–148, Haifa, Israel, September 2004.

[44] C. M. Bishop. *Neural Networks for Pattern Recognition*. Clarendon Press, Oxford, 1995.

[45] M. Pimentel. *Probabilistic Estimation of Respiratory Rate from Wearable Sensors*. Springer International Publishing, Switzerland, 2015.

[46] K. Murphy. *Machine Learning: A Probabilistic Perspective*. MIT Press, Cambridge, 2012.

[47] M. H.-Silveira. Assessment of the feasibility of an ultra-low power, wireless digital patch for the continuous ambulatory monitoring of vital signs. *British Medical Journal*, 5:1–9, 2015.

[48] M. H.-Silveira. Preliminary assessment of the sensiumvitals®: a low-cost wireless solution for patient surveillance in the general wards. In *Engineering in Medicine and Biology Society (EMBC)*, pp. 4931–4937, 2015.

[49] WHO. Who global report on falls prevention in older age. http://www.who.int/ageing/publications/Falls_prevention7Marc.pdf.

[50] L. Seed, M. Mubashir and L. Shao. A survey on fall detection: principles and approaches. *Neurocomputing*, 100:144–152, 2013.

[51] G. M. Lyons, A. K. Bourke and J. V. O'brien. Evaluation of a threshold-based tri-axial accelerometer fall detection algorithm. *Gait & Posture*, 26:194–199, 2007.

[52] D. Chang, J. Luk, R. Bajcs, J. Chen and K. Kwong. Wearable sensors for reliable fall detection. In *Engineering in Medicine and Biology Society, 2005. IEEE-EMBS 2005. 27th Annual International Conference of the*, pp. 3551–3554, 2006.

[53] B. Barshan and A. T. Özdemir. Detecting falls with wearable sensors using machine learning techniques. *Sensors*, 14:10691–10708, 2014.

[54] A. Cappello, L. Chiari, K. Aminian, J. M. Hausdorff, W. Zijlstra, J. Klenk E. Bagalà, C. Becker. Evaluation of accelerometer-based fall detection algorithms on real-world falls. *PloS one*, 7:e37062, 2012.

[55] Y. Ge, M. Liu, L. Tong and Q. Song. Hmm-based human fall detection and prediction method using tri-axial accelerometer. *IEEE Sensors Journal*, 13:1849–1856, 2013.

[56] Matlab. The mathworks, inc.

[57] ARM. Keil compiler, tools by arm. http://www.keil.com/. Accessed July 2015.

[58] A. C. W. Wong, D. McDonagh, O. Omeni, C. Nunn, M. Harris, M. Silveira and A. J. Burdett. Sensium: an ultra-low-power wireless body sensor netowrk platform: Design and application challenges. In *Engineering in Medicine and Biology*, pp. 6576–6579, Minneapolis, MN, September 2009.

4

Approaches and Techniques for Maintenance and Operation of Multisink Wireless Sensor Networks

Miriam Carlos-Mancilla

Universidad del Valle de México Unidad Guadalajara Sur

Ernesto López-Mellado and Mario Siller

CINVESTAV Unidad Guadalajara

CONTENTS

4.1   Introduction

A Wireless Sensor Network (WSN) is one of the most challenging areas of Internet of Things (IoT), the reason why this chapter is focused on them. A WSN is composed of a set of hundreds of devices deployed in a common area, with communication between each other and a common purpose. Each device is able to transmit/receive data to a set of devices or to all devices. These devices are connected according to a topology, either be fixed or not, at the beginning of the process.

There are a wide variety of current research problems in this area, such as tracking, surveillance, military applications, building automation, disaster management, agriculture, among others, in which each device has specific tasks, or the tasks are defined according to the area where the device is deployed. The applications are designed to obtain precise information from the environment regardless of the human presence. The applications have to be efficient to take note of the limited energy and memory that the sensor devices have. Radio communication, processing, and transmission/reception are generally the main causes of power consumption.

A single sink is enough when the density of the network does not cause a bottleneck during the data collection period. However, a single sink scenario becomes inefficient when the number of nodes increases and there is a lot of information waiting to be processed or collected. In this chapter, distributed and centralized techniques using multisink scenarios are presented and analysed.

A multisink approach is becoming an efficient scheme to reduce the energy consumption in the whole network, either in a centralized or distributed application. The analysis focusses on the used strategies in a multiple sink environment, whether the nodes are static or mobile, and how applications can be improved by mixing more than one strategy; the objective of the application such as event detection, following a target, finding the best route, collection and data aggregation, among others.

The chapter covers recent and traditional works and presents a discussion of their advantages and drawbacks. This chapter is organized as follows: Section 4.2 presents the generalities and features of centralized and distributed WSNs and a proposed classification of them. Section 4.3 presents the main strategies in multisink environments; Section 4.4 describes a theoretical comparison between techniques. Finally, Section 4.5 presents the concluding remarks that drive further research in the area.

4.2   Generalities of Wireless Sensor Networks

WSNs with a single sink node are sensible to suffer attacks and losses as the density increases; one of the main problems are the named bottleneck problem (Hochbaum and Shmoys, 1986) and hotspot problems (Sungjin and Kim, 2006) presented during a data collection period. Another limiting factor emerges when the purpose of the application is finding a target and the sink is far away from it; in this case, the routing can require great expenses of energy or even additional resources to send the information. Another drawback is presented when the sink requires responses in a short period of time and the routing is not efficient, the sink node is not available, or there is a high demand for transmissions and one sink is not enough.

Multisink platforms arise in response to this kind of problem; a multisink platform provides redundancy, avoids problems in the data collection, and increases the efficiency of the application. A set of sinks can be strategically or randomly distributed in the coverage area to collect and process the information from the environment. The sinks can be static or mobile according to the available resources of the application. The density of the network and the number of sink nodes are specified at the beginning of the application.

Multisink scenarios that improve the performance of a network based on different techniques are analysed. In multisink techniques, every device is able to collect information from the area surrounding through one or more sensors, process this information in a local manner and send it through other devices.

Data collection is usually made by a sink node or base station. A WSN can be used to control multiple tasks in one environment; the sensors require data sense from multiple sources by delivering information to multiple sinks at the same time.

Once the topology strategy is specified, the information can be collected following a *distributed* or *centralized* policy via multihop mode. In this chapter, we present a classification of distributed and centralized applications of multisink environments. The analysis is focused on whether the nodes are static or mobile, the formation is event detection based or not, and network backbone is formed or not. The survey is devoted to recent works and presents a discussion of their advantages and drawbacks.

4.3 Classification of Wireless Sensor Network Techniques

The connections and communication of the devices in WSNs are defined according to the needs of every application. The protocols must be able to provide an improvement whether from energy consumption, nodes mobility, environment noise, limited battery charge and loss of messages, among others.

Figure 4.1 shows the taxonomy of our proposed classification. It can be seen that all WSN organization techniques can be classified into one of the discussed groups: *centralized* or *distributed.*

Centralized networks take directions from a unique device. This central node is responsible to provide network operation services, such as node

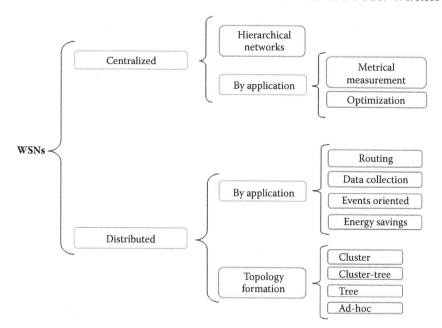

FIGURE 4.1
WSN classification.

localization, event detection, traffic routing, etc. A suitable logical topology for this approach is a *start* in which the nodes are connected to a specific device with additional duties and the total control of the network. The centralized networks can be classified according to how the information is processed.

- **Hierarchical networks**: The aim is to gather the sensor information in one or several base stations to reduce the time of data collection or data processing. This classification is based on a specific area in the network.

- **By application**: Many works suggest using the network for a specific purpose, such as finding the best route based on event detection, data collection, among others.

A further classification can also be made according to the environment of the network. The most important topologies used in centralized approaches are cluster, grids, and trees; while the application division differentiates **metrical measurement** and **optimization** classification. A brief discussion of this type of network is given in Section 4.4.

Distributed techniques allow the nodes to take their own decisions in a local way and dispense the information to a limited set of neighbor nodes within a scope area, which provides energy savings and spreads the information in a uniform manner to avoid the overhearing or hotspot problems. The flexibility of distributed systems allows targeting random environments without complications during information collection. Devices can be connected using either ad hoc or strategy based (self-organization, clustering, pheromone tracking, and so on; Carlos-Mancilla et al., 2016). The distributed classification can be derived into the following categories:

- **By application**: oriented to solve one problem at a time or solving a specific problem defined by available resources in every application. The applications are divided into the following categories: *routing, data collection, event-oriented,* and *energy saving.*

- **Strategies**: it is based on the communication between the nodes to achieve a common goal.

A third-level classification is described in Figure 4.1. *By application* class, considers routing, data collection, event-oriented and energy savings as subdivisions, while the topology formation only considers the most common topologies for WSNs such as *cluster, cluster tree, tree,* and *ad hoc* strategies. This sub classification is described in Section 4.5.

4.4   Centralized Techniques Using Multisink Environments

Centralized formation techniques are suitable for networks in which the processing power capacity relies mostly on a unique device. In such cases, this device is responsible for the processing, coordination, and management of the sensed information activity. It also forwards this data to multiple sink nodes (Figure 4.2).

The main advantages of this approach are:

- Centralized schemes allow more efficient energy management.

- Roaming is allowed inside the network.

- Network coverage analysis is simplified.

- Context information availability allows a better application design (placement of nodes, application awareness, etc.).

A brief discussion for every type of classification is given as follows.

4.4.1   Hierarchical networks

A sensor defines priorities according to its role in the network. Traffic forwarding nodes have a lower precedence than fully functional nodes (sense,

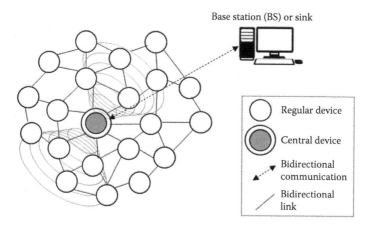

FIGURE 4.2
Centralized strategy.

coordinate, process, and forward information). The control of the network is performed in a hierarchical way and is defined based on the roles. This kind of networks is usually implemented using the 802.15.4 (IEEE, 2005) protocol.

For instance, Silva (2009) presents a multisink environment architecture (ICatchYou) (Silva et al., 2009) based on the 802.15.4 protocol. It employs a multihop forwarding strategy and addresses the sensor localization problem. They proposed a centralized technique to guarantee high mobility between the sink nodes.

Self-configuration is used to find the appropriate sink for the registration process, and there are some metrics used for choosing the appropriate sink, how the information is gathering, and so on. Each sensor node receives all messages directly through the sink node. They consider two scenarios; the first one is a closed one with obstacles and interference and the second one without obstacles or interference. The second scenario presents better results, because nodes achieved a better performance with a higher distance.

In this proposal, the authors do not present final results and they do not guarantee the full functionality of the sensor in the environment. Although multihop offers some advantages, and it is also easier to guarantee an efficient fast handover between sink nodes. Their algorithms are inefficient because all nodes send broadcast messages and may cause a flooding of the network. The technique can be applied to mobile scenarios, but an implementation is not reported in this chapter. Besides, energy consumption or scalability are not taken into account. In this work, only the link quality is considered, which makes the decision making for choosing a sink node inefficient.

In other proposal, Singh et al. (2010) presents a tree-based routing protocol where every node has the capability of environmental sensing, computation tasks, or keeps communication with other nodes in the network. Nodes are mobile; the node movements are defined following a target. The routing algorithm is composed of different stages: the first one is the formation of the tree by broadcast messages, the second one is collection and transmission data; it is handled by Time Division Multiple Access (TDMA) schedule. The last one considers failures, energy level, or movement of the parent node.

A tree-based routing protocol improves the lifetime of nodes and the network by moving the nodes to the next higher level when a threshold of energy has been reached. This algorithm works in a centralized way, and it does not consider energy consumption when a node sends messages. This algorithm is compared with Low Energy Adaptive Clustering Hierarchy (LEACH) (Heinzelman et al., 2002), and even this algorithm does not work with tree formation and Threshold sensitive Energy Efficient sensor Network protocol (TEEN) (Manjeshwar and Agrawal, 2001) protocol.

4.4.2   By application networks

4.4.2.1   Metrical measurement

Usually, nodes are placed in strategic positions before or during the procedure. The aim is to provide better data collection, energy savings, or processing performance. Formation techniques are discussed later.

Tzong (2012) proposes an adaptive learning scheme for load balancing schemes with zone partition in multisink WSNs Q-learning based Adaptive Zone Partition scheme (QAZP). A centralized Mobile Anchor (MA) agent, which is equipped with a directional antenna and Global Positioning System (GPS) device, is introduced, and the MA is orientated at the intersection by drawing a Voronoi diagram (Fortune, 1992). After the location, the information of the MA is determined; the MA sends beacon signals to different sink nodes through directional antenna, and the sensor nodes that receive the beacon signal can transmit the follow-up collecting sensing data to the nearest sink through hotspot devices; the hotspots are devices that notify other sensors that the sensor is able to transmit its information. A machine learning process is applied to the MA to make it adaptable to any traffic pattern.

This proposal consists of a large number of sensor nodes, several sink nodes, and one MA. The characteristic of the MA affiliation makes the network capable of being partitioned into several regions according to the number of sinks, and its location information is attained through GPS devices. There are defined movements for the MA: *upper left-hand corner* (highest residual energy) and *lower right-hand corner*; after the movement is done, the sensor node chooses another route to balance the load based on parameters like residual energy and hop distance.

The assumptions used in this proposal are difficult to fulfill in a real implementation; the environment is always observable; good decisions are made and good behavior is expected, ideal environments without traffic or loss of messages are designed, and there are sensors with infinite energy. The MA is controllable and always predefined. Hotspots concentrate a large amount of information from the whole environment, and data collection is taken from specific zones.

In Gandhman et al. (2003), an energy scheme for WSNs with multiple mobile base stations is proposed. The authors use an Integer Linear Program (ILP) to determine new locations of the base station and propose a flow-based routing protocol. The purpose of this approach is to prolong the lifetime of the sensor network. ILP divides the lifetime into equal periods of time (rounds). The solution of ILP indicates the base station's next location. Sensor nodes use a multihop routing protocol, e.g., minimum cost forwarding. ILP is formulated to minimize the maximum energy spent by a sensor node in one round. One of the disadvantage is that at the beginning of every round the location of every base station is calculated, and the node needs to access the complete information for calculating the new position; the energy consumption for this process is excessive, and the nodes require location information, distances, and neighbors. If one base station is isolated, it will be costly to send the information.

Kim (2009) proposed a multiple sink WSN and a topology configuration scheme that automatically reconfigures the network in case of node failures. The number of retransmissions caused by random losses of messages in wireless communication is calculated. The authors consider a static network without having any previous information. The data are collected by the sink nodes, and they do not consider the transmission or reception between the regular nodes. Furthermore, fault node events originated by an environment incident or an enemy attack occur at a random; consequently, sensors have to sense continuously by causing the depletion of energy.

For a better performance, the path cost is calculated using the Signal-to-Noise-Ratio model (SNR is a measure used in science and engineering that compares the level of the desired signal to the level of background noise). The wireless spanning tree protocol (IEEE, 1998) is used for routing and is divided into different stages; the reconfiguration is proposed when a node detects a failure when some parent node dies, and then the affected node searches a new parent and connects to it. A main drawback of this proposal is that the wireless spanning tree protocol does not consider the total communication cost, provoking a loss of communication with sink nodes; this implies a high consumption of energy for this activity.

4.4.2.2 Optimization

The concept of *Dynamic Convoy Tree-Based Collaboration* (DCTC) is introduced by Zhang (2004); the strategy works in a centralized way, and it is stated as a multiple objective optimization problem. The aim is to find a convoy tree sequence with high tree coverage and low energy consumption. It finds a minimal sequence with a maximum coverage, in which they assume ideal communication. The authors proposed a conservative and prediction-based scheme for tree expansion and pruning, and a sequential and localized reconfiguration scheme for tree reconfiguration is introduced.

Convoy tree is a moving tree that tracks a target; this target can move along the environment, and the tree is dynamically configured to add or prune some nodes as the target moves. The overall function of the proposal resumes as follows: first, the target goes into the detection region; then the sensor nodes detect the target and collaborate with each other nodes to select a root and construct an initial convoy tree.

This proposal presents two algorithm versions: DCTC and O-DCTC (*Optimal DCTC* Solution with Dynamic Programming), the first one is used when convoy tree is reconfigured and the target moves and the second one consists in the formulation of the optimization problem, which finds a min-cost convoy tree sequence with high tree coverage.

In Deepak and Deshmukh (2013), an energy balancing multisink optimal solution is presented. The total number of displayed nodes is denoted as n. The location of every node is known, and the whole network is partitioned into k disjoint clusters. The centroids of the clusters are considered in place of sink positions. The node position is chosen according to some metrics. The clustering formation is based on particle swarm optimization (Venter and Sobieszczanski-Sobieski, 2003). The sensor network is represented as a connected graph $G = (V, E)$, where V is the set of n vertices (sensors nodes) and E is the set of edges (transmission links); sinks are predefined and non-mobile.

The objective is to find the optimal locations for every sink node by minimizing the average sensor distance from the sink, and maximizing one-hop connectivity of each sink placed in the network to reduce the consumed energy in the network and extend the lifetime of the network. The authors use the K-mean algorithm for the clustering process, which is iteratively applied from a k initial cluster center. Furthermore, an iterative parallel search algorithm based on a particle swarm optimization strategy is used in which agents are defined as particles.

The set of experiments involves different quantities of sink nodes for computing the average sink nodes degree and the average hop count. The disadvantages of this algorithm are: it spends a lot of energy for every position known in the environment, and the increasing number of clusters is proportional to the number of sinks.

These are some of the most representative proposals in WSNs. In the next section, distributed techniques are presented, including some relevant results.

4.5   Distributed Techniques Using Multisink Environments

In *Distributed* formation techniques, the information is managed by each node, and decisions are locally taken and limited to its neighborhood (single-hop neighbors) (Figure 4.3). The main characteristics of distributed networks include

- Autonomous devices.

- Each node shares information to its neighborhood.

- Suitable for distributed applications (multiagent systems, self-organized systems, etc.)

- The information is mainly forwarded to a single node.

- Interconnection devices (routers, bridges, etc.) are not required.

- Their flexibility allows targeting harsh environments.

A WSN aims to gather environmental data, and the placement of node devices may be known or unknown a priori. Network nodes can have *a driven* Internet Protocol (IP) communication or *logical* communication (topological communication defined by users or by the application) with all devices; such a communication defines the topology according to the application. The logical topology is mainly defined based on the communication between the nodes such as a cluster, tree formation, self-organization, among others.

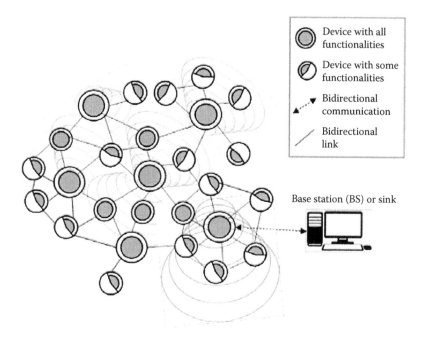

FIGURE 4.3
Distributed strategy.

Distributed protocols in WSNs must provide a reduction in energy consumption, robustness, and reliability despite the limited energy, noise, loss of messages, and obstacles. Some proposals are analysed later; the classification considered is the presented in Figure 4.1.

4.5.1 By application networks

An application is defined by considering the available resources and the environment where the devices will be deployed. According to the application, the nodes have an assigned task; every task of a node has a main purpose, such as find the best available route, discover a specific target, collect data, etc., to reduce the energy consumption, prolong the lifetime, generate a robust network, or reduce the use of resources.

Recently, many works focus on reducing energy consumption, minimizing the number of hops, providing robustness to the network, increasing the throughput, and so on. However, a route with a minimum number of hops may not always be the optimal and reliable route. A network with a large number of devices is not always the best option, or even minimizing the number of transmissions can have adverse effects. For the aforementioned reasons, according to the nature of the application, it is important to define the set of metrics, constraints, and features to evaluate the performance of a network (Muhammad et al., 2015). We consider that the available applications can be classified into four categories: *routing, data collection, event oriented,* and *energy savings.*

4.5.1.1 Routing strategy

In order to find the best route inside the network for transmitting the information to the sinks, consider the specified metrics to achieve it (Figure 4.4). Examples of these are presented as follows.

In Foster et al. (2008), the results and experiences further from implementing the reinforcement learning based multicast routing protocol Feedback Routing for Optimizing Multiple Sinks (FROMS) in a test bench of Scatter-Web nodes are evaluated and discussed. This work can be classified for using *routing* and *energy-saving* techniques.

FROMS is a multisource multisink routing approach that uses Q-learning to identify network routes to optimize the shortest path or the best energy efficiency. A good exploration/exploitation ratio ensures that routing costs are kept low by often using the currently best available route, which corresponds to the local minimum.

Also, they use different techniques with methodologies like Ant Colony Optimization (ACO) (Dorigo et al., 1992). The node device only uses the local information available to the number of hops to the sink node and finds the global optimum route. Even they have a good application, i.e., they use one predefined node device for transmitting information to the whole network and they do not use node reconfiguration.

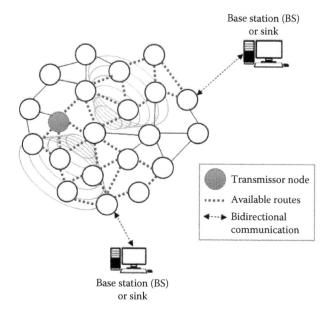

FIGURE 4.4
Routing protocol generalization.

Another work presented by Meng et al. (2007) proposes a new system architecture for a multisink environment and two new routing algorithms: Energy Level Based Routing (ELBR) and Primary-Based Routing (PBR). A new definition for energy level is introduced; it is related to the number of times a node can transmit data, considering the energy consumption for every message sent. The first routing algorithm (ELBR) calculates the path energy level and chooses the maximum energy level path to transmit data; PBR algorithm takes into account the energy level and the energy cost of the routing path. Some drawbacks are: only one sink node can receive information at one time, reconfiguration is not considered, and a node is preselected to send information.

A proposal about a multisink and load balance routing algorithm is presented by Wang and Wu (2009) and it is compared with ELBR and PBR (previous work). The authors have introduced a new concept for nodes in one hop communication with the sink nodes called deputies. They consider that deputies nodes are one kind of sink nodes and that every deputy node sends a broadcast message in a round robin way. The algorithm is based on hardware implementation. A bottleneck problem can exist in deputy nodes caused by flooding of messages, and their energy can be quickly consumed.

Reconfiguration is not considered and nodes can die soon. The same data are transmitted to all sink nodes; this may cause an excessive energy loss and data redundancy; the energy consumption is not considered.

A path bottleneck oriented and energy cost-based routing scheme is proposed by Lin and Wu (2010); in this work, (a) a centralized version for sensor networks with global topology information, (b) a semi-distributed version to further improve energy efficiency, and (c) a fully distributed version to support large-scale sensor networks are designed. Sensor nodes dissipate energy only in data transmission; a multisink environment is considered. The architecture is responsible for sending data via the internet to the base station and assigns activities to sink nodes. Consecutively, sink nodes send instructions to nodes in the environment via multihop paths.

The base station has the complete knowledge of the topology and the remaining energy of all the sensors; sink nodes calculate the appropriate routing to a specific node. A modified version of Bellman–Ford Algorithm (Sedgewick and Wayne, 2015) is used for routing, and the k-means algorithm for deployed sink nodes. It is not explained how exactly the distributed algorithm works and the authors do not take into account reconfiguration; data seem to be forwarded always following the same route.

A routing protocol to balance the loads in the network with multiple sinks to prolong the lifetime of the system is presented by Zhang (2015). Nodes choose a parent node according to a probabilistic list that presents the most preferable node to be the next-hop destination to send a packet. The sink nodes select a sink to be the head of the network named as an *arbitrator*.

The arbitrator sinks decide the assignment of nodes to the sinks based on two criteria: the hop number and the number of connected nodes to the sinks. In some situations, the tree structure cannot guarantee the balancing of the loads, some collisions can appear, and messages can be lost. Data aggregation is not considered.

An intelligent agent-based routing protocol that provides data delivery to mobile sinks is presented by Kim et al. (2010). The communication with the agents is performed through a broadcasting method. In this protocol, the messages can be lost during the data collection process and does not support channel variability. The energy consumption is also related to the source data rate, which is independent of the protocol.

4.5.1.2 Data collection strategy

In this kind of application, the formation of the network is not important; the main goal is the way the data is collected. Some applications look for spreading the data collection through defined nodes to get to a sink node, while other applications focus on the energy conservation, improving the performance, increasing the throughput, and so on (Figure 4.5).

An example of data collection through defined nodes is presented by Winston and Seah (2006). In this work, a virtual sink architecture for WSN that mitigates the near-sink contention by defining a group of spatially physical sink is proposed. A multipath routing is adopted to provide alternative paths to increase the probability of a successful delivery.

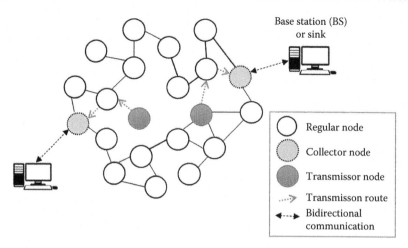

FIGURE 4.5
Data aggregation and collection through the network.

The authors propose a multitier topology by introducing local aggregation points and distributing them amongst sensing nodes and multiple local sinks. A sensor sends a packet simultaneously using diverse routes to deliver the packet until the sink node. The probability for delivering a message increases, but the energy consumption also increases. The data transmission cost is high and the channel conditions are assumed to be ideal. It is assumed that the local links are connected via high-speed links to a network where the resources are enough to support communications needs.

Data gathering for large-scale sensor networks with multiple sinks (M-collectors), which traverse several shorter subtours concurrently to satisfy the distance/time constraints is presented by Ma et al. (2013). The objective is to find a set of data-gathering subtours in the network, such that the number of M-collectors can be minimized.

An M-collector is responsible for gathering data from local sensors in the subarea, and the information is forwarded to only one M-collector that has a connection with a final data sink. An available subtour is not guaranteed always, the energy depletion during the subtour searching is not considered, and the transmission would not be possible if the M-collector, which has direct communication with the data sink, dies.

Cardei and Marta (2009) proposed a data gathering algorithm employing sink mobility with the pre-established path. The heterogeneous WSN consists of a large number of sensor nodes with limited capabilities and multiple mobile sinks with unlimited capabilities. The authors suppose the interconnection of sink nodes all the and the design of a network that moves the sink nodes along the environment to maximize the lifetime of the network. Nodes are uniformly and randomly distributed.

The movement to the sink nodes is predefined into places with major quantity of energy inside a hexagon perimeter; nodes with low energy can be left uncovered. Every sink remains as the center of every cluster with a big depletion of energy. Data are forwarded to the closest sink using multihop communication based on the collection trees formed by the clustering algorithm. In simulations, each sink is connected with each of its six neighbor sinks.

4.5.1.3 Event-oriented strategy

In this kind of work the events define the behavior of the complete network, and the sensor that detects the event respond to this event with defined activities and sends the information to the corresponding sensors (Figure 4.6).

A tolerance control for enhancing a WSN using a multisink environment is presented by Huang et al. (2010). In this work, a tolerance area is defined; the nodes out of this area select the shortest path to route their data from source nodes to the nearest sinks. The nodes are deployed uniformly in one area, and they might switch to different paths to balance the load of sensors in the area.

Drawbacks of this approach appear when the loading of sensors in the network is not balanced; thus, sensors should sometimes select a candidate route to bypass the sensors with too much load or low battery level. The energy depletion is high when the nodes look for different routes based on a tolerance area τ, which is defined according to a maximum threshold. If the threshold is not well calculated, the tolerance area can cause a big energy depletion.

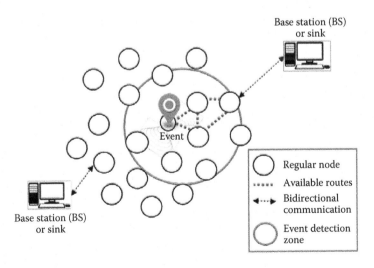

FIGURE 4.6
Topology formed from an event detection.

A virtual infrastructure based on honeycomb tessellation for data dissemination is presented by Erman et al. (2012). This work can be classified as a *routing* or *event-oriented proposal.*

The virtual infrastructures support mobile sinks in a WSNs and acts as a rendezvous region for storing and retrieving collected event data. When the mobile sink crosses the network, the sensors in the rendezvous region are requested to notify the event data. The virtual infrastructure is based on honeycomb tessellation, and the data dissemination is hexagonal cell-based. A drawback in this proposal is when an event is detected and there is no sink near the sensed area; data aggregation and failures are not considered.

A bioinspired self-organized algorithm (Saleem et al., 2009) would also meet the enhanced sensor network requirements, including energy consumption, success rate, and time. This article also used an ACO algorithm, which is a probabilistic technique for solving computational problems that can be reduced, to find good paths through graphs for an optimum route discovery in a multihop WSN. This proposal works in a distributed way to collect data and/or detect an event; a routing strategy is used to control the sent messages.

Results are based on an NS2 simulator (NS2, 1995); nodes have bidirectional communication; the weight of a link to transmit information is proportional to the power consumption of a node; thus, a large amount of energy can be depleted. The convergence time is not considered. Every path has a unique ID to avoid cycles. The route is selected according to the following strategy: when one packet passes through a node with a certain speed, the node meets the ant agents into a buffer and stores the routes using a table, and the best route is chosen from this table to transfer packets.

From the table the best node to start the transmission to the neigbor nodes to determine the best succesor node in the route. It can be noticed that this strategy may need a lot of memory, which is one of the main constraints in sensor networks. The article does not include graphical results about the performance of the algorithm or comparative results.

4.5.1.4   Energy-savings strategy

In these proposals, the objective is saving energy to prolong the lifetime of the whole network; usually, these proposals combine other strategies to complete the objective.

For instance, a multiple data sink node election in a multisink environment is presented by Pietrabissa et al. (2016). This algorithm works in a distributed and iterative manner, in which the sink role is periodically reassigned to avoid the hotspot problem and increase the lifetime of the network. The network partition is done according to a centroidal Voronoi tessellation method, which leads to a spatially well-balanced distribution. This proposal neither takes into account the energy depletion in the topology formation nor the overhearing problem.

Algorithms that build a backbone connecting mobile devices in a distributed fashion using an adaptive learning scheme are presented by Tzong

(2012). The authors consider a load balancing strategy with a zone partition in a multisink WSN environment (QAZP). The movements of a node are predefined, and reconfiguration to failures is not addressed.

A reliable energy-efficient multilevel routing algorithm for Wireless Networks is presented by Yu et al. (2011). The authors proposed an approach to reduce the energy consumption by introducing fuzzy Petri nets. The algorithm includes a clustering strategy and a multihop routing procedure; the cluster formation is based on residual energy, number of neighbors; centrality parameters such as the sum of distances of a node from all their neighborhood; nodes calculate a degree and choose the best for transmitting the data to the defined sink node. A node is susceptible to be isolated if it is far from the sink nodes.

4.5.2 Topology formation

The topology formation is essential when the sensor network application looks for having a better resource control. In this chapter, the main topologies used in this kind of networks are considered. Furthermore, some metrics are presented as the emergency property of this behavior, such as reliability, energy consumption, and latency.

4.5.2.1 Cluster formation

Cluster-based control structures allow a more efficient use of resources. A hierarchical view of the created network through clustering decreases the computational complexity in the formation of an underlying network. This is especially true in sensor networks that are expected to involve of a large number of individual nodes.

On a topological level, clustering is achieved by grouping nodes inside a certain transmission area. A designed leader node controls this group of nodes, usually known as Cluster Head (CH) or a leader node (Figure 4.7). A leader node is selected according to the weight that may correspond to a node

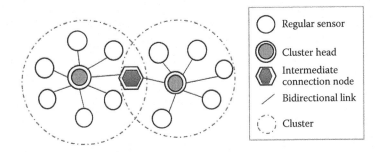

FIGURE 4.7
Cluster-based formation.

capability to perform additional duties. It can be determined by taking into consideration, issues such as node residual energy, memory amount, processing capabilities, the number of neighbors, etc. Usually, the weights are computed locally in each node, and they may depend on the application where the structure is used.

For instance, an Energy-efficient Multisink Clustering Algorithm (EMCA) is proposed by Xu et al. (2012) to solve the *energy hole* problem. Its intercluster and intracluster routing algorithm focus on energy consumption and ensure the optimal path from nodes to its CH or sink. The network is divided into n equal clusters; each cluster has a CH chosen by considering the residual energy and sends aggregated data to the relevant sink.

The routing is used to calculate the nearest distance between the CH nodes and the sink nodes. The data are sent in a multihop manner; the sink nodes increases while the energy is consumed fast. No single reconfiguration method is considered.

Two algorithms for sink mobility based on clustering strategy and reduction of the energy consumption are proposed by Wang (et al., 2013). The objective of this work is to study the influence of fixed and mobile sink strategies on home network performance in terms of energy consumption, network lifetime, and mitigation of hotspot problems. The authors consider the study of diverse parameters, including number of mobile sinks, velocity, location, and moving trajectory. Simulations are presented and compared with LEACH (Heinzelman et al., 2000).

In Wang et al. (2015), a routing protocol algorithm that combines ACO (Dorigo et al., 2006), clustering, and sinks mobility techniques for home automation networks is presented. The network is divided into several clusters; each CH has communication with a mobile sink. ACO is applicable to the guidance of sink mobility; if the ACO fails, the mobility of the sink is affected. Obstacles and reconfiguration are not considered, and the cluster formation is energy expensive.

4.5.2.2 Cluster-tree formation

Cluster-tree formation is one of the most recent approaches, which allows combining two strategies and grabbing the best of both strategies. The cluster nodes inside a transmission range or nodes that have similar characteristics choose a leader node called cluster head (CH); once the formation of the cluster finishes, the tree formation procedure deletes redundant links and controls how the information is collected.

The tree strategy is defined once the cluster has been formed usually, the tree formation is launched over the formed cluster, while the activities and purpose of the nodes are established by the application (Figure 4.8).

An example for this formation is presented by Cuomo et al. (2008). The main contribution is the simulation and analysis of a formation under IEEE 802.15.4 protocol using different network settings with single-sink and

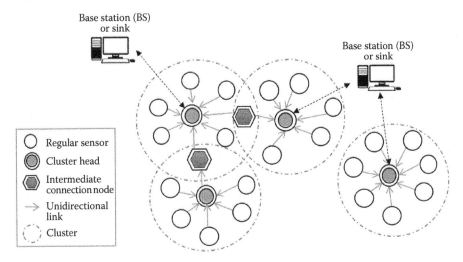

FIGURE 4.8
Cluster-tree based formation.

multisink scenarios. Hops define the levels of the tree from a node to the relevant sink. Figure 4.9 presents a cluster-tree topology with the IEEE 802.15.4 protocol. This protocol uses three different kinds of roles: Personal Area Network (PAN) coordinator, Full Function Device (FFD), and Reduced Function Device (RFD). The PAN coordinator acts as a router and manages the

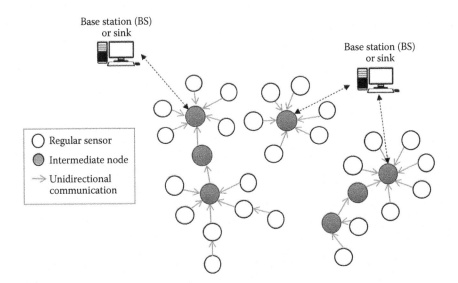

FIGURE 4.9
Tree-based formation.

network load; it is the root and the sink of the network; the FFD device is able to perform any task and can have communication with all nodes in the network, and only RFD devices can have communication with FFD devices while their activities are limited.

Sensors are static and deployed in a simulated square area using the NS2 platform. The authors suppose an indoor scenario. Sink nodes are located in the center of the area. The number of generated hops relies on factors such as collisions, link quality, and sensory data. In this strategy, the number of children per node by adding some coordinator rules and an appropriate value of the maximum tree depth for better performance are controlled.

There are some drawbacks in the reported simulations: the complete connectivity of the network is not assured, results or implementations are not shown, and reconfiguration or energy constraints are not taken into account.

In Buratti et al. (2007), a mathematical formulation used to optimize the average number of children per parent and the number of levels in one tree (*tree height*) through maximization of the network association probability is proposed. The topology formation is based on the IEEE 802.15.4 protocol. The authors run the algorithm M times; each time an independent topology is created, having fixed N nodes and S sinks on the network (multisink environment). A completely random distribution in space, i.e., Complete Spatial Randomness (CSR), is assumed to derive an average number of children per node and maximize the number of levels. The energy consumption and reconfiguration are not considered.

In Aslam et al. (2011), a novel energy-efficient cluster formation algorithm based on a multicriterion optimization technique is presented, which uses multiple individual metrics in the CH selection process as input while simultaneously optimizing the energy efficiency of the individual sensor nodes as well as the overall system. However, the multicriterion optimization technique involves some complex matrix operations. As sensor nodes are significantly constrained in computational capacity, it is difficult for them to execute complex matrix operations. The nodes are susceptible to remain without memory.

4.5.2.3   Tree formation

The objective in tree formation is to find a setting of a configuration for each group of nodes to maximize a defined metric in the topology. An example of this formation is presented in Figure 4.9. A tree formation strategy is a known technique to reduce the overhead and the energy consumption; it also helps to create routes to send data through the network.

One technique for tree formation is presented by Mottola (2011), where distributed heuristics are embodied to minimize the number of nodes involved in routing; furthermore, there is load balancing strategy using information available in a one-hop neighborhood. A tree formation strategy is used to build the best available route to the sink nodes; for this purpose, multiple routes are mixed into a single one to send information to different sink nodes and to assure loaded balanced routes. The technique neither considers the density of the network nor the transmission of excessive messages, nor depletion of energy in the routing formation nor reconfiguration.

Guan (2006) proposes an Information Selection Branch Grow Algorithm, which achieves a higher network lifetime and reduces the end-to-end network delay. The base station is defined as a sensor node that is connected to a gateway (data sink) with a wired cable; the nodes cannot be selected as the base station; it is assumed that every base station has an unlimited supply of power and possesses high computational capabilities. Every base station is treated as the center of a grid and the connection is made by cables; in a real scenario, this can be done only when there is an ideal environment.

The proposed algorithm uses a tree topology and develops branches where leaf nodes are closer to the base station according to the minimum number of hops; the base station is chosen according to metrics considering lightest weight, the smaller number of child nodes, the minimum degree of freedom, and so on.

The aim is to build a balanced tree to achieve energy balancing. For this purpose, a balanced criterion is considered when all branches of the base station have an equal number of child nodes. The algorithm selects the potential branch to grow, and all nodes send a broadcast with its neighboring information eventually. In this proposal, loss of messages is not considered.

In Bandara and Jayasumana (2007), a concept of Virtual Sensor Networks (VSNs) to provide a protocol to support the formation, usage, adaptation, and maintenance of sensors is presented. These sensors collaborate on specific tasks. A VSN is a subset of sensor nodes dedicated to a certain task or application defined at a given time. Thus, the remaining nodes which do not belong to the formation provide support functionality to create, maintain, and operate the VSN. As the nodes in a VSN may be distributed over the virtual network, they may not be able to communicate directly with each other.

The formation of a VSN is a tree, where the root is a defined sink and the major functions of VSN can be divided into two categories: VSN maintenance and membership maintenance. The membership in a VSN is dynamic, and the communications among VSN nodes frequently rely on whether or not it is currently a member of a VSN. The VSN maintenance functions include the following: the management of nodes entering and leaving VSN, broadcast joining two VSNs, splitting VSNs, and originating contours of boundaries.

Many nodes can remain without performing any activity in any stage of the network; also, the energy consumption for VSN formation is not taken into account. The broadcasting method is expensive and the authors do not consider reconfiguration.

4.5.2.4 Ad hoc formation

Certainly, one of the most important advances in the current proposal has been the *self-organization* approach in which the nodes are considered autonomous and they decide the actions to perform according to its local information; the nodes also interact with the neighbor nodes. The aim is to achieve collective tasks that exceed its individual capabilities (Figure 4.10).

The examples of these techniques are found in nature (insect colonies, biological cells, the flock of birds, the foraging behavior of ants, etc.) (Mamei et al., 2006; Schmeck et al., 2010). Finally, distributed multisink approaches support large density scenarios without compromising the network performance.

A multihop architecture *ICatchYou* is proposed by Silva et al. (2009); it provides self-configuration abilities and allows easy and quick integration of nodes in the network. Self-configuration procedures are required to support the deployment of a random multisink.

ICatchYou performs two processes: *registration* and *updation* of information. The sink nodes should be accessible from conventional IP networks, providing the measured data in real-time for remote destinations, and allowing a simpler management and interoperability with the WSN. Depletion of the energy is not considered and the efficiency of the nodes depends on the link quality.

In Kiri et al. (2007), a system for gathering data from sensor networks with multisink configurations is presented; the strategy is inspired by the swarm intelligence of ants, in which each sensor node determines its next action through repeated interaction with its neighbors. Clustering and routing strategies emerge in a self-organized manner. The authors focused on the problem of the communication from the sensor to the sink nodes; the number of sensors is equal to the number of clusters in the environment. Data processing and energy consumption are not considered. If a sink fails, the cluster is unable to collect data. Data collection over an area may be impossible due to cluster-level power depletion, where a number of events occur in a particular area.

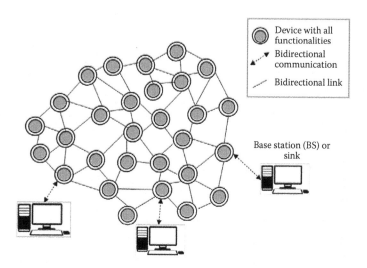

FIGURE 4.10
Ad hoc formation.

In Carlos-Mancilla et al. (2015), a technique for formation and networking operation in multisink environment algorithm is proposed. The nodes only use local information in the whole process. The algorithm is divided into three main stages: First, the backbone formation, which is based on self-organization and clustering strategies. This stage allows low energy consumption, information redundancy, and robustness. In the second stage, the sink nodes launch tree formation using a load-balanced strategy. Finally, in the third stage, the nodes sense and collect the data from the environment according to the role of the node, and the data are sent to the assigned sink node. It is assumed that a finite number of static nodes are randomly deployed in a multisink environment without any a priori information.

4.6 Theoretical Comparison of Multisink Techniques

4.6.1 Advantages and drawbacks in centralized multisink approaches

The advantages of centralized techniques are: the ability to handle an entire network from a unique device. This device has access to the information and usually to the location of all nodes. Normally, whether an event emerges, this can be controlled by easily interacting the surrounding nodes and sending the data to a target or sink node; the conflicts that may arise during transmissions and reception are avoided.

The routing is chosen considering the properties and behavior to the whole network and environment; it is easily calculated. The optimal sink positions can be known, taking into account metrics such as distance between nodes, network density, amount of energy, hops, and location between the nodes, among others. The energy consumption can also be measured using an optimization method and controlling the number of transmitted messages. Reconfiguration in centralized techniques is easy to implement, and the resources are synchronized to be available for other nodes.

The drawbacks of these techniques include excessive energy consumption, given that every time the node has to transmit something because nodes need to know the destination node of the message. Usually, GPS or triangulation techniques are used on every node to localize them, which implies a lot of processing and energy depletion. Memory constraints are not taken into account, and usually, these techniques suppose ideal behaviors of nodes in the network, which implies a loss of messages; besides, obstacles and interferences are not considered.

In these techniques, the reconfiguration requires more network resources with a high energy cost. The network does not support high density of nodes, because of the large amount of information generated in the network. The connectivity of the network is not always assured, because in this kind of application, it is usually decided to connect defined nodes instead of the complete network; robustness and reliability depend on the application. Robustness is easy in regular nodes, but it becomes difficult when the central device suffers a failure; if this happens, the network collapses.

The central device is responsible for repairing a failure. Failure recovery is difficult for some nodes since full recovery is required; besides, the central device requires all information in the environment.

4.6.2 Advantages and drawbacks in distributed multisink approaches

Nowadays, distributed approaches are the most used techniques; they have been characterized to be one of the robust and reliable approaches in WSNs. The main difference with the previous technique is that every node in this approach is responsible for its actions.

The main advantages are: the information is local, which means that a node only keeps information of its neighborhood (one or two hop neighbors). Distributed algorithms are considered scalable. Reconfiguration is made locally on the affected part, since nodes are autonomous; the decisions are made by every node according to its position, its activity, or the roles defined in the network; some of these roles include event detection, routing, data collection, topology formation, sensing and information processing, among others. All this process is based on the available information in the coverable area, be either the distance between nodes or the number of hops.

When a node dies, the network will remain in operation and the performance is not affected considerably. The distributed approach allows dealing with noisy environments including obstacles. The energy consumption is reduced by every node; usually, the routing starts if an event is detected or there is a target to follow, which implies there is no unnecessary depletion of energy. One of the best results has been obtained for the combination of the best features of distributed and multisink approaches.

The drawbacks in distributed techniques are: the nodes can deplete a lot of energy if the organization is not well established. Also some routing mechanisms are susceptible to suffer loss of information if the route does not consider recovering from failures. Reconfiguration is complex because the nodes only have local information and no single global optimization can be measured. Also, the connectivity of the entire network cannot be assured, and node mobility requires more energy.

4.7 Concluding Remarks

In this chapter, relevant works on distributed and centralized multisink WSNs have been reviewed. It presents the evolution, design, and implementation of some important WSN techniques in the last years and the most used protocols and standards to improve the sensor applications.

Some metrics can affect the performance of centralized and distributed networks, such as the number of hops to get to a target or specific device, the number of retransmissions, the flow rate, the link quality, and the number of devices. For this reason, it is possible to claim that the application and a good technique rely on the available resources. Thus, the description of a good strategy that increases the performance of the network reduces the number of available resources to prolong the lifetime of the network according to the environment where the network will be implemented.

One has remarked that distributed solutions are preferred over centralized ones, since distributed techniques support scalability, autonomous nodes, deployment, and elimination of nodes; also, it is possible to use self-organization strategies inspired from nature, in which the interaction is made only with neighbor nodes.

Bibliography

Aslam N., W. Philips, W. Robertson and S. Sivakumar (2011). A multi-criterion optimization technique for energy efficient cluster formation in wireless sensor networks. *Journal Information Fusion*, Volume 12, no. 3 pp. 202–212.

Bandara H.M.N.D. and A.P. Jayasumana (2007). An Enhanced Top-Down Cluster and Cluster Tree Formation Algorithm for Wireless Sensor Networks. In Proceedings 2nd International Conference on Industrial and Information Systems (ICIIS 2007), Sri Peradeniya, Sri Lanka, August 2007, pp. 565–570.

Buratti C., F. Cuomo, S. Della luna, U. Monaco, J. Orriss and R. Verdone (2007). Optimum Tree-Based Topologies for Multi-Sink Wireless Sensor Network Using IEEE 802.15.4. Vehicular Technology Conference, 2007. VTC2007-Spring. IEEE 65th. Italia, April 2007, pp. 130–134.

Cardei M. and M. Marta (2009). Improved sensor network lifetime with multiple mobile sinks. *Pervasive and Mobile Computing*, Elsevier, ScienceDirect, Volume 5, pp. 542–555.

Carlos-Mancilla M., E. López-Mellado and M. Siller-Gonzalez (2015). A localized multi-sink multi-hop algorithm for Wireless Sensor Networking. In *IEEE Global Information Infrastructure and Networking Symposium (GIIS)*, Guadalajara, Jalisco, México. IEEE.

Carlos-Mancilla M., E. López-Mellado and M. Siller (2016). Wireless sensor networks formation: approaches and techniques. *Journal of Sensors,*

Volume 2016, Article ID 2081902, Hindawi Publishing Corporation, pp. 1–18.

Cuomo F., S. Della Luna, E. Cipollone, P. Todorova and T. Suihko (2008). Topology Formation in IEEE 802.15.4: Cluster-Tree Characterization. Sixth Annual IEEE International Conference on Pervasive Computing and Communications, 2008. PerCom 2008. University of Roma "Sapienza", Rome, 17–21 March 2008, pp. 276–281.

Deepak R.D. and P.R. Deshmukh (2013). Energy Balancing Multiple Sink Optimal Deployment in Multi-hop Wireless Sensor Networks. Advance Computing Conference (IACC), 2013 IEEE 3rd International, Ghaziabad, India, 22–23 February 2013, pp. 408–412.

Dorigo M., M. Birattari and T. Stützle, (2006) Ant colony optimization – artificial ants as a computational intelligence technique. *IEEE Computational Intelligence Magazine.*

Dorigo M., A. Colorni and V. Maniezzo (1992). Distributed Optimization by Ant Colonies. Appeared in Proceedings of Ecal91 – European Conference on Artificial Life, Paris, France, Elsevier Publishing, pp. 134–142.

Erman A.T., A. Dilo and P. Havinga (2012). A virtual infrastructure based on honeycomb tessellation for data dissemination in multi-sink mobile wireless sensor networks. *EURASIP Journal on Wireless Communications and Networking* 2012:17.

Fortune S. (1992). Voronoi diagrams and Delaunay triangulations. In D.-Z. Du & F. Hwang (Eds.), *Computing in Euclidean Geometry.* Lecture notes series on computing (Vol. 1, pp. 193–233). Singapore: World Scientific.

Foster A., Murphy A.L., Schiller J. and Terfloth K. (2008). An Efficient Implementation of Reinforcement Learning Based Routing on Real WSN Hardware. Networking and Communications, 2008. WIMOB '08. IEEE International Conference on Wireless and Mobile Computing. University of Lugano, Lugano, 12–14 October 2008, pp. 247–252.

Gandhman S.R., M. Dawande, R. Prakash and S. Venkatesan (2003). Energy Efficient Schemes for Wireless Sensor Networks with Multiple Mobile Base Stations. GLOBECOM 2003, IEEE.

Guan G.H., M.L. Sim and H.T. Ewe (2006). Energy Efficient Routing for Wireless Sensor Networks with Grid Topology. IFIP International Federation for Information Processing LNCS 4096, pp. 834–843.

Heinzelman W.R., A. Chandrakasan and H. Balakrishnan (2000). Energy Efficient Communication Protocol for Wireless Microsensor Networks. Proceedings of the 33rd Annual Hawaii International Conference on System Sciences (HICSS), vol. 2, Maui, HI, pp. 1–10.

Heinzelman W., A. Chandrakasan and H. Balakrishnan (2002). An application specific protocol Architecture for wireless microsensor networks (LEACH). *IEEE Transactions on Wireless Communications*, Volume 1, no. 4, pp. 660–670.

Hochbaum D.S. and B.D. Shmoys (1986). A unified approach to approximation algorithms for bottleneck problems. *Journal of the ACM (JACM)*, Volume 33, no. 3, pp. 533–550, New York.

Huang C., R.-H. Cheng and S.-R. Chen. (2010). Enhancing network availability by tolerance control in multi-sink wireless sensor network. In *IEEE Conference on Information Technology Convergence and Services (ITCS)* (pp. 1–7). Cebu: IEEE.

IEEE Standard Association (1998). IEEE Standard for Local Area Network MAC (Media Access Control) Bridges 802.1D, MAC Bridges section 1998.

IEEE Standard Association (2005). IEEE STD. 802.15.4: Wireless Personal Area Networks (PANs), 2005.

Kim J. (2009). Spanning Tree Based Topology Configuration for Multiple-Sink Wireless Sensor Networks. First International Conference on Ubiquitous and Future Networks, 2009. ICUFN 2009. South Korea Sungchang Lee, 7–9 June 2009, pp. 122–125.

Kim J.W., J.S. In, K. Hur, J.-W, Kim and D,-S, Eom (2010). An intelligent agent-based routing structure for mobile sinks in WSNs. *IEEE Transactions on Consumer Electronics*, Volume 53, no. 4, pp. 2310–2316.

Kiri Y., M. Sugano and M. Murata (2007). Self-Organized Data-Gathering Scheme for Multi-sink Sensor Networks Inspired by Swarm Intelligence. Published in First International Conference on Self-Adaptive and Self-Organizing Systems (SASO 2007), Cambridge, pp. 161–172.

Lin Y. and Q. Wu (2010). Energy-conserving dynamic routing in multi-sink heterogeneous sensor networks. In *IEEE International Conference on Communications and Mobile Computing (CMC)* (Vol. 3, pp. 269–273). Shenzhen: IEEE.

Ma M., Y. Yang and M. Zhao (2013). Tour planning for mobile data-gathering mechanisms in wireless sensor networks. *IEEE Transactions on Vehicular Technology*, Volume 62, no. 4, pp. 1472–1483.

Mamei M., R. Menezes, R. Tolksdorf and F. Zambonelli (2006). Case studies for self-organization in computer science. *Journal of Systems Architecture*, Volume 52, no. 8, pp. 443–460.

Manjeshwar A. and D.P. Agrawal (2001). TEEN: A Routing Protocol for Enhanced Efficiency in Wireless Sensor Networks. IPDPS'01 Proceedings of the 15th International Parallel & Distributed Processing Symposium, IEEE Computer Society Washington, DC, USA, 2001. p. 189.

Meng M., X. Wu, H. Xu, B.-S. Jeong, S. Lee and Y.-K. Lee, (2007). Energy Efficient Routing in Multiple Sink Sensor Networks. Fifth International Conference on Computational Science and Applications, Kuala Lumpur 2007, pp. 561–566.

Mottola L. and G.P. Picco (2011). MUSTER: adaptive energy-aware multisink routing in wireless sensor networks. *Mobile Computing on IEEE Transactions*, Volume 10, no. 12, pp. 1694–1709.

Muhammad M.A., W.K.G. Seah and I. Welch (2015). Reliability in wireless sensor networks: a survey and challenges ahead. *Computer Networks*, Volume 29, pp. 166–187.

NS2 (1995). "Network Simulator" Development was Supported by DARPA through the VINT Project at LBL, Xerox PARC, UCB, and USC/ISI in 1995. [Available online]: http://www.isi.edu/nsnam/ns/.

Pietrabissa A., F. Liberati and G. Oddi (2016). A distributed algorithm for Ad-hoc network partitioning based on Voronoi Tessellation, *Ad-Hoc Networks Journal*, Volume 46, pp. 37–47.

Saleem K., N. Fisal, S. Hafizah, S. Kamilah and R.A. Rashid (2009). Ant based self-organized routing protocol for wireless sensor networks. *International Journal of Communication Networks and Information Security (IJCNIS)*, Volume 1, no. 2, pp. 42–46.

Schmeck H., C. Muller-Schloer, E. Cakar, M. Mnif and U. Richter, (2010). Adaptivity and self-organization in organic computing systems. *ACM Transactions on Autonomous and Adaptive Systems*, Volume 5, no. 3, pp. 10: 1–10:32.

Sedgewick R. and K. Wayne (2015). Chapter 4.4 "Shortest Paths". In R. Sedgewick & K. Wayne (eds). *Algorithms*, 4th Edition, , Editorial Pearson.

Silva, R. (2009). A New Approach for Multi-sink Environments in WSNs. IFIP/IEEE International Symposium on Integrated Network Management, 2009. IM '09. Simek, M. Boavida, F. June 2009, pp. 109–112.

Silva R., J.S. Silva and F. Boavida (2009). Youcatchme, icatchyou – two approaches to node and service discovery in 6lowpan. Unpublished European Wireless Sensor Networks (EWSN), 2009.

Singh M., N. Lal, M. Sethi and S. Poonia (2010). A tree based routing protocol for Mobile Sensor Network (MSNs). *International Journal on Computer Science & Engineering*, Volume 1, Supplement 1, p. 55.

Sungjin A. and D. Kim (2006). Proactive Context-Aware Sensor Networks. Proceedings of European Workshop on Wireless Sensor Networks, Zurich, Switzerland.

Venter G. and J. Sobieszczanski-Sobieski (2003). Particle swarm optimization. *AIAA Journal*, Volume 41, no. 8, pp. 1583–1589.

C. Wang and W. Wu, (2009). A load-balance routing algorithm for multi-sink wireless sensor networks, in Proceedings of the International Conference on Communication Software and Networks (ICCSN '09), pp. 380–384, Macau, China, February 2009.

Wang J., J. Cao, L. Bin and S. Lee (2015). Bio-inspired ant colony optimization based clustering algorithm with mobile sinks for applications in consumer home automation networks. *Consumer Electronics on IEEE Transactions*, Volume 61, no. 4, pp. 438–444.

Wang J., Y. Yin, J. Zhang, S. Lee and R.S. Sherratt (2013). Mobility based energy efficient and multi-sink algorithm for consumer home networks. *IEEE Transactions on Consumer Electronics*, Volume 59, no. 1, pp.77–84.

Winston K.G. and Seah, H.P. (2006). Multipath Virtual Sink Architecture for Wireless Sensor Networks in Harsh Environments. Proceedings of the First International Conference on Integrated Internet Ad Hoc and Sensor Networks. New York, USA, 2006. Article 19.

Xu Z., Y. Yin, J. Wang and J.-U. Kim (2012). An energy-efficient multi-sink clustering algorithm for wireless sensor networks. *International Journal of Control & Automation*, Volume 5, no. 4, pp. 131–142.

Yu A.H., X.A. Fu, Y.L. Cai and M.C. Vuran (2011). A reliable energy-efficient multi-nivel routing algorithm for wireless sensor networks using fuzzy Petri nets. *Sensors*, Volume 11, no. 3, pp. 3381–3400.

Zhang W. (2004). DCTC: dynamic convoy tree-based collaboration for target tracking in sensor networks. *IEEE Transactions on Wireless Communications*, Volume 3, no. 5, pp. 1689–1701.

Zhang C., Y. Wang and Y. Liang (2015). Load-balancing routing for wireless sensor networks with multiple sinks. 12th International Conference on Fuzzy Systems and Knowledge Discovery (FSKD), Zhangjiajie, China. IEEE.

5

Energy-Efficient Communication Solutions Based on Wake-Up Receivers

Heikki Karvonen and Juha Petäjäjärvi

University of Oulu, Centre for Wireless Communications

CONTENTS

5.1   Introduction

Wireless sensor networks (WSN) and wireless body area networks (WBAN) have enormous amount of possible use cases in various fields. For example, environmental monitoring, healthcare, smart buildings, and smart cities are taking advantage of low-power sensor nodes that provide data to internet of things (IoT) as a part of fifth generation systems. Typically, wireless transceiver consumes most of the sensor nodes' energy resources [1–6]; therefore, a careful design must be executed for communication techniques and protocols, to enable that the WSNs and WBANs can be deployed for long periods of time without battery replacement or recharging. Indeed, a huge amount of research work has been carried out to improve the energy efficiency by optimizing protocols using a layered and cross-layer approach [7].

In this chapter, the focus will be on energy-efficient communication solutions that can be achieved by using an intelligent hierarchical network architecture that can be used to effectively utilize heterogeneous devices, collecting different types of sensor data from the patient's body or environment, performing autonomous networking, and providing data for the databases of the IoT. In the hierarchical network case, energy consumption can be decreased by utilizing a wake-up concept that enables to keep the devices at a sleep mode as long as possible. Hierarchical architecture will be introduced in Section 5.2. There are two different types of concepts that can be used to enable wake up between different hierarchical layers of the architecture: duty-cycling based radios and wake-up receiver (WUR) usage. The wake-up concept needs a joint design of physical and medium access control (MAC) layers. The wake-up concept design issues will be discussed in Section 5.2.2, and a generic wake-up radio-based MAC (GWR-MAC) protocol will be introduced in detail in Section 5.2.3.

Different types of state-of-the-art WUR solutions that can be used to enable wake-up concept will be introduced, and their future research directions are outlined in Section 5.3. Energy efficiency comparison results for a GWR-MAC-based hierarchical WSN architecture and conventional duty-cycle MAC-based WSN are introduced in Section 5.4 to show that WUR-based networking has a remarkable potential to improve energy efficiency particularly in the target scenarios where events occur rarely. The long lifetime of WSN and WBAN devices will make the applications more user-friendly, and therefore it will foster widespread deployments. Once the number of WSNs providing sensor data for the databases of the IoT is increased, the possibilities of horizontal deployment of different types of applications will be enormous. It has been estimated that 50 billion devices and objects will be connected to the IoT by 2020 [8]. For example, wireless medical networks have enormous possibilities to improve quality and effectiveness of healthcare [9]. Examples of envisaged applications that can be built using the described intelligent hierarchical architecture will be outlined in Section 5.4.

5.2 Hierarchical Architecture

The WSN architectures can be roughly divided into two categories: flat and hierarchical. In the flat architecture case, which is the traditional approach, all the network nodes are at the same level, and they have similar roles from the communication point of view and, typically, also similar characteristics. In the hierarchical network case, the nodes have different characteristics and roles at different hierarchical layers. The flat network structure is simpler, but it cannot provide efficient communication, especially when the network is composed of a large amount of nodes. The hierarchical network structure has been found to provide more efficient communication in the case of heterogeneous networks, since the operation of the nodes is designed so that the overall performance will be improved in comparison with a flat architecture.

At the early stage of WSN research, the target applications included a homogeneous set of sensor devices, which were performing a simple sensing task and reporting sensor observations to a central (sink) node. That star-topology is still valid for many WSN applications. However, the development has led to an emergence of heterogeneous networks that include different types of devices with varying capabilities, enabling the implementation of more versatile application scenarios. Support for heterogeneous devices is needed in WSNs for energy efficiency, scalability, and quality of service purposes. The network of a heterogeneous set of devices must be designed carefully to enable efficient and reliable operation. The previously proposed hierarchical architectures can be divided into intranetwork and in-network approaches. In the intranetwork approach, the main design goal is efficient communication between the WSN and backbone [10, 11]. The in-network approaches of WSNs typically limit to two-tier topologies, where the higher tier (gateway) collects data and forms connection to a backbone network. The lower tier nodes can be simpler, and they save energy by communicating directly, in a star-topology fashion, only with the gateway [12]. Several WSN protocols, e.g., ZigBee [13] and Z-Wave [14], make a distinction between a routing (full function device) and a nonrouting (reduced function device) device. However, that approach also allows clustering of nodes and designating only one node at a time as an energy-consuming higher tier node (cluster head) [15, 16]. The multitier architectures are typically designed for a specific application, e.g., hospital environment [17], traffic monitoring system [18], surveillance [19], environmental monitoring [20], smart home [21], or underwater acoustic sensor networks [22].

Hierarchical network's total energy consumption can be decreased by carefully taking into account the characteristics of the heterogeneous devices at different layers of the architecture. Different types of functionalities of devices must be designed so that communication and sensing requirements can be met while maintaining low energy consumption. From the energy consumption point of view of communication, it is important to maximize the length of the sleep mode of nodes and minimize the number of retransmissions. That is particularly important in a heterogeneous network for the higher tier nodes, which are the most power consuming. In addition, long sleep modes are also important for a simple low-power node to improve their lifetime.

Intelligent hierarchical architecture for heterogeneous WSNs will be introduced in Section 5.2.1. To enable energy efficiency by putting the nodes into sleep mode, the heterogeneous hierarchical network requires a method for awakening the nodes when required from the application point of view. For that purpose, duty cycle-based radio is the traditional approach while WURs have started to gain more and more research attention in recent years. Both approaches will be discussed in Section 5.2.2. A generic wake-up radio-based MAC protocol will be introduced in Section 5.2.3.

5.2.1 Architecture for a heterogeneous network

Network architecture based on hierarchical levels of intelligence and usage of wake-up concept has been introduced [23]. It can be used in various WSN and WBAN application scenarios that include different types of devices. The wake-up functionality is seen as a long lifetime enabler, particularly for the wireless long-lasting (e.g., medical, surveillance, structural, environmental, and industrial) monitoring systems, which have many possible application scenarios in both private and public sectors. Due to a very wide application space, the hierarchical architecture is designed to be flexible and scalable to varying configurations [23]. Therefore, a high-level architecture which is independent of the specific implementation techniques (e.g., radio interfaces) is defined to enable that different application developers have the flexibility to choose the most suitable implementation technique. The proposed high-level architecture can be used as a starting point for the network design. In the discussed hierarchical network case, the nodes are categorized into different architectural layers based on their capabilities and functionalities. The high-level architecture and the functionalities offered at the different layers are illustrated in Figure 5.1.

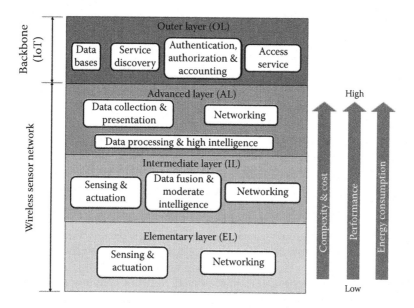

FIGURE 5.1
High-level architecture for a hierarchical network with heterogeneous devices.

In this design, the hierarchical layers of the architecture are named to be meaningful as follows: elementary layer (EL), intermediate layer (IL), advanced layer (AL), and outer layer (OL). The complexity, cost, capabilities, and performance of the devices at different layers increase from the bottom to the top. Consequently, the devices at the lowest layer consume less energy than the devices at the higher layers. The objective of the intelligent hierarchical design is to decrease the overall energy consumption by using the low-complexity devices for continuous event monitoring and data collection while keeping the more power-consuming higher layer devices in the sleep mode as long as possible. The EL nodes will wake up the higher-complexity devices only when required. The EL devices are usually simple sensor nodes providing basic sensing and possibly also actuation services. However, the devices at the lowest layer are elementary from the application service point of view, since they provide the essential data about the monitored event or object. The networking services offered by the EL nodes are communication with the devices at the IL, and in the mesh network case, the EL sensor nodes can communicate with each other. The simplest service which the EL node can offer is just to send a simple message containing, e.g., a body temperature or environment humidity value sensed by the node. The IL nodes have more capabilities, and they can offer higher performance functionalities and sensing, such as performing electrocardiography or recording a video. The IL nodes may also perform data aggregation for information collected from EL nodes and make decisions based on the data by using, e.g., pattern recognition algorithms. Therefore, the IL nodes can decide whether a sensed event is so critical that the AL also needs to be awakened. The IL nodes offer important networking services by communicating both with upper and lower layer devices. The AL nodes are the most intelligent devices in the architecture, and they will eventually collect all the relevant data from the lower layers and make intelligent decisions, process data, and act as gateways between the WSN and the backbone network. Therefore, AL devices must provide adequate service interfaces so that the application data can be offered to the end-user through backbone. The OL is the backbone (public or private) infrastructure providing wireless and/or wired communication (e.g., broadband or cellular network) back-end systems and applications servers. Since this layer is not part of the actual WSN architecture, it is called OL in this design.

Figure 5.2 illustrates an example of application scenarios that was implemented, by University of Oulu and Technical University of Tampere, to verify the introduced hierarchical architecture principle. Subnetwork 1 illustrates a WSN where the sensor nodes (TelosB [24]) collect data from the office environment and send it to the embedded computer (FriendlyARM [25]), which forwards the data to the desktop computer with an Ethernet connection to the backbone network (Internet). Subnetwork 2 illustrates a surveillance WSN where the sensor nodes (TUTWSN [26]) detect movements in the monitored area and wakes up a wireless local area network camera node to record a video once the moving object has been sensed. The wireless local area network

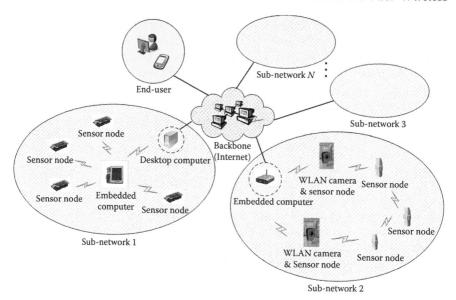

FIGURE 5.2
Distributed heterogeneous network example.

camera node sends the video to an embedded computer, which forwards the
data to the backbone network (Internet). An authenticated end user can
access the data from different subnetworks through the Internet and receive,
e.g., automatic alarms about intruders in the monitored area. In medical
applications, this type of network with tailored sensors could, for example,
monitor elderly or Parkinson disease patient at home environment and provide alarms for the patient itself and for the nursing staff at hospital when
movements are detected in unusual or dangerous areas. Patient's WBAN can
be also implemented using hierarchical architecture principle. The information
collected by WBAN would provide useful information about the patient state,
for example, when the monitoring network has triggered the alarm and caught
medical persons to pay attention to the patient due to a detected event (DE)

5.2.2   Wake-up concept

As was described earlier, the idea of the energy-efficient hierarchical architecture is to keep the most power consuming nodes in a sleep mode as long
as possible. For that purpose, there is a need for a low-power wake-up concept, which can be implemented using a specific WURs or using a duty-cycled
MAC protocol.

The design of wake-up concept must take into account physical and MAC
layer characteristics. The wake-up signal (WUS) transmitted through the

physical medium must be designed to enable that it can be detected by using low-power consuming components at the WUR, whose design options are discussed in detail in Section 5.3. There are different types of MAC protocols that try to improve energy efficiency. Their efficiency depends on the application characteristics, since different types of networks require different types of solutions. Indeed, the scalability and adaptability to network changes are important MAC protocol design objectives, because in that way the protocol performance can be ensured in many types of scenarios. MAC protocols must be designed to take care of the packet collision avoidance, idle listening, and overhearing with minimum control overhead. Packet collisions must be avoided to keep the number of retransmissions low. Idle listening occurs when radios listen to the channel redundantly, and when there are no incoming transmissions. Retransmissions and control overhead decrease data throughput and increase energy consumption because redundant bits need to be transmitted. Consequently, the available sleep time of the sensor nodes also decreases. Overhearing should be avoided so that only the target nodes will receive and decode the packets.

Most of the proposed sensor network MAC protocols are duty-cycle based, i.e., the radios have a definite sleep/awake schedule that is followed. The MAC protocols can be divided into synchronous and asynchronous categories. Duty-cycling principle is illustrated in Figure 5.3a for synchronous case, and in Figure 5.3b for asynchronous case. Synchronous protocols schedule the sleep/awake periods so that the nodes which are expected to communicate with each other are awake at the same time. Synchronous protocols typically require a centralized control and use clustering of nodes, and inside each cluster, there is a common sleep/awake schedule which is controlled by the cluster head. Asynchronous protocols include methods for communications between the nodes that have different sleep/awake schedules.

Asynchronous communication is enabled using a sender- or receiver-initiated communication. In the sender-initiated case, the data source will send a preamble before data transmission. Once the receiver detects the preamble, it will continue the listening to receive the data packet that will follow the preamble, as illustrated in Figure 5.3b. If the receiver does not detect a preamble, it will go back to the sleep mode. In the receiver-initiated case, the receiver will use probing to query for potential transmissions. In each case, the idle listening will occur if there is no incoming transmission when the nodes wake up to listen to the channel according to their schedule. Idle listening is expensive from the power consumption point of view because the transceivers should be in the sleep mode as much as possible to save power.

Recently, wake-up radio-based MAC solutions have gained attention due to their energy efficiency superiority, particularly in applications with rare events and transmissions [27–30]. The wake-up radio principle has been illustrated in Figure 5.3c. In this case, the data transceiver can be in the sleep mode until there is incoming data packet to be received from some other node. The WUR of the source node will notify the target node(s) by sending a WUS.

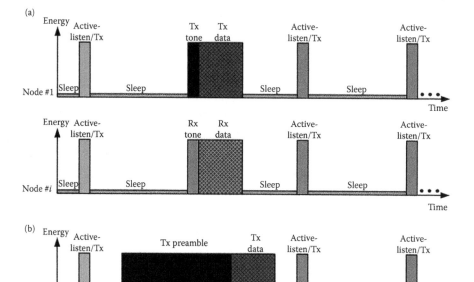

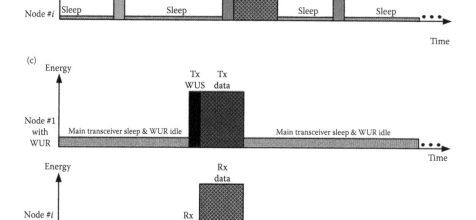

FIGURE 5.3

Principle for (a) synchronous duty-cycling, (b) asynchronous duty-cycling, and (c) wake-up radio-based MAC.

After receiving the WUS, target node's WUR will trigger wake up of the main transceiver for data packet reception.

Duty cycle-based MAC protocols work well in many WSN applications but not in applications where the monitored events, and communication, occur rarely. A drawback of the duty-cycle approach is idle listening, which increases energy consumption and should thus be avoided if possible, i.e., the duty-cycle radios (DCRs) will listen to the channel unnecessarily if the event-reporting frequency is lower than the duty cycle. Idle listening can be decreased by setting the duty cycle low when the traffic load is low. Adaptive duty-cycle protocols have been proposed for that purpose. However, low duty cycle will increase the communication delay and may not be able to satisfy the application requirements.

WURs [31–62] can be used to avoid the idle listening problem while allowing energy-efficient and low latency operation. The WUR can continuously detect the WUS when it is in the ultra-low-power standby mode. The communication delay can be avoided when the event occurs and there is data to transmit. Therefore, WURs have the potential to decrease energy consumption in comparison with DCR-based networks. WUR design approaches will be discussed in more detail in Section 5.3.

5.2.3   A Generic-level WUR-based MAC for hierarchical architecture

A generic WUR-based MAC (GWR-MAC) protocol, which is based on dual-radio approach, is introduced to enable idle listening avoidance in sensor network applications [27]. In the dual-radio node architecture case, nodes include WUR and main data radio, as illustrated in Figure 5.4. The GWR-MAC protocol is not restricted to any specific WUR technology or data radio technology. Two different options for the wake-up procedure are defined for the GWR-MAC protocol: source-initiated and sink-initiated. The data transmission period of GWR-MAC can be implemented by using different types of channel access methods, as will be explained later.

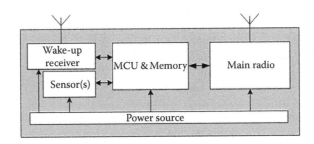

FIGURE 5.4
Sensor node architecture for dual-radio approach.

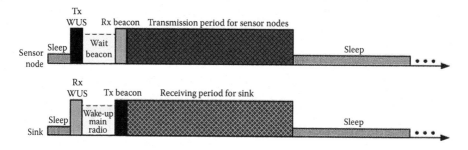

FIGURE 5.5
Source-initiated mode of the GWR-MAC protocol.

In the source-initiated mode, the sensor node(s) will wake up the sink node from the sleep mode by transmitting WUS, as illustrated in Figure 5.5. To decrease the probability of WUS collisions, a random (or predefined) delay for WUS transmissions can be used. The WUR of the sink node will receive the WUS and generate a wake up via microcontroller (MCU) to the main radio. The sink node's main radio will then broadcast a beacon (BC) message to initiate transmission period for the sensor node(s) according to the channel access procedure, which can be based on different methods suitable for different scenarios. The beacon message is, at the same time, an acknowledgement (ACK) to the sensor node(s) that the WUS has been received by the sink. If the sensor node does not get the beacon message, it will retransmit the WUS after a random back-off period. The WUS transmission procedure is therefore similar to the Aloha channel access with a random (or predefined) delay for the first transmission. Once the beacon message is received, the sensor node(s) will send the data packet(s) to the sink during the transmission period using the channel access method informed in the beacon message. This mode of the GWR-MAC protocol is therefore a combination of the source-initiated wake-up procedure and the following channel access control method for the transmission period.

In the sink-initiated mode, illustrated in Figure 5.6, the sink node will wake up the sensor nodes from the sleep mode by sending the WUS using broadcast, unicast or multicast. It depends on the used WUR technology, whether it is possible to use addressing to wake up only certain sensor nodes or whether broadcast should be used. When the sensor node receives the WUS, it will send an ACK message to the sink. The sink node knows that the WUS has been detected correctly and sends the beacon-containing information about the following transmission period. Data transmissions are then performed during the transmission period, and once they are finished, all the nodes have entered the sleep mode.

Typically, the data flow is from the sensor node to the sink node. However, the transmission period can also be dedicated to the sink node to transmit

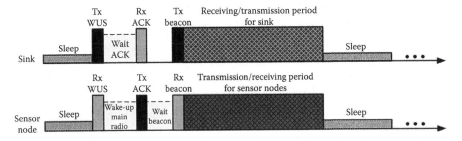

FIGURE 5.6
Sink-initiated mode of the GWR-MAC protocol.

data to the sensor node(s), if, for example, a wireless software update or reconfiguration of sensor node(s) requires it. Therefore, in Figure 5.6, it is illustrated for the sink-initiated case that the transmission period can be dedicated to the sensor node(s) or to the sink node to transmit data packets. The sink node can determine in the beacon the upcoming transmission period channel access mechanism, timing, and scheduling information. Different channel access methods can be used for the transmission period management. When the transmission period is finished, all the nodes will enter the sleep mode and the next transmission period will take place after the next wake-up procedure.

For the transmission period channel access, one option is to use a contention-based MAC. In that case, the nodes compete for channel access and transmit packets according to the contention-based MAC principle. For example, in the Aloha case, nodes transmit when they have a packet to transmit. If the packet is not successfully received, then the retransmission policy defines either that the packet is discharged or retransmitted. For example, in the Aloha case, the unsuccessful packet will be retransmitted again after a random back-off period if the ACK is not received during a certain time. In addition, other contention-based methods, e.g., carrier sensing multiple access with collision avoidance, etc., can be used during the transmission period. Another option is to use contention-free scheduled methods, e.g., time division multiple access (TDMA)-based protocols, guaranteed time slots defined in the IEEE Std. 802.15.4 [63, 64] and scheduled access mode of the IEEE Std. 802.15.6 [65] or ETSI SmartBAN MAC [66]. The requirement for the usage of contention-free methods is that the sink node assigns dedicated time slots for each sensor node. The sink node does not have information about which nodes have a packet to transmit, and therefore, the channel resources may be wasted. In an ideal contention-free case, collisions will not occur if the nodes are perfectly synchronized and follow the schedule.

The described GWR-MAC protocol principle is suitable for different types of application, since it defines a bidirectional wake-up procedure between the sensor nodes and the sink. In addition, it enables the usage of different channel access methods for the transmission period. The described GWR-MAC is a general-level framework for short-range networks that take advantage of the wake-up radios. It depends on the application scenario which mode of the GWR-MAC protocol should be used. For example, in some applications, only the source-initiated case may be used, and then only the sink node should be equipped with a WUR. In some scenarios, there can be a need that only the sink node must be able to wake up sensor nodes. In that case, the sink-initiated mode would be used, and only the sensor nodes must be accompanied with a WUR. Some application scenarios require both modes of the GWR-MAC protocol. In such cases, all the network nodes must be equipped with a WUR and data radio, as was illustrated in Figure 5.4. In recent years, WUR-based MAC solutions have gained researchers' attention and other similar type of solutions, which can be used in a hierarchical network context, and have been recently discussed and reviewed [29, 30].

5.3 Wake-Up Receiver Solutions

WUR designs have progressed substantially in the last decade, and they might be utilized in various IoT applications in the near future. The current designs consume power well below 100 μW, which is already over hundred times less than a typical commercial radio frequency (RF) transceiver designed for WSN consumption. Low power consumption enables a node to listen the channel constantly for years with a single coin size battery. But due to the tradeoff between power consumption and sensitivity, sensitivities of WURs are usually considerably worse compared with commercial RF transceivers. Consequently, to reach the WUR and the RF transceiver of the node, more transmit power is needed to transfer a WUS to the WUR than to transmit data to the RF transceiver.

Receiver design always includes tradeoffs regarding, e.g., data rate, power consumption, and sensitivity. The two main performance comparison metrics of the proposed solutions have been power consumption and sensitivity, followed by data rate. In some works, energy per bit is also used as a comparison metric. However, utilized wake-up packet lengths are usually only few bytes in size, so receiver's active power consumption becomes more important than energy per bit. Also due to the small size of the wake-up packet, it is transmitted reasonably fast even with low data rates. Later, we discuss in detail the different receiver architectures that can be used for WUS detection. Furthermore, future research directions will be outlined.

5.3.1 Wake-up receiver architectures

Data receivers typically use direct conversion architecture, but due to the challenges of flicker noise and DC offset, they are rarely used in WURs.

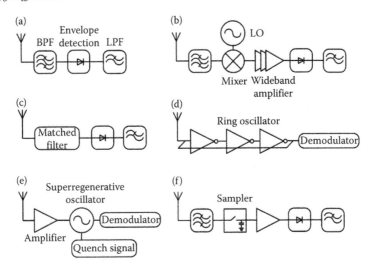

FIGURE 5.7
Typical wake-up receiver architectures: (a) RF envelope detection, (b) uncertain-IF, (c) matched filter, (d) injection-locking, (e) superregenerative oscillator, and (f) subsampling.

Also common superheterodyne receiver architecture has been proposed to be used as a WUR. However, in those solutions, duty-cycling is often employed, and active power consumptions of different superheterodyne solutions are significantly more than in WURs based on other architectures. Because of the demanding requirements of power consumption and sensitivity, receiver designers have proposed many different receiver architectures for WUR purposes. Figure 5.7 shows the block diagram of most common architectures, which are based on RF envelope detection (RFED), uncertain-intermediate frequency (IF), matched filter, injection-locking, superregenerative oscillator, and subsampling.

Being the most straightforward option for energy detection, the RFED architecture is the most commonly used solution. In the RFED-based receivers, local oscillator is not needed since the envelope of the RF signal is detected and the signal is therefore directly down converted to baseband. The envelope is detected from a wide bandwidth. Therefore, the incoming signal is band pass filtered to ensure low noise at the envelope detector input.

Accurate RF synthesizers typically consume too much power to be used in WURs. The uncertain-IF receiver architecture utilizes synthesizer that consumes low power at the cost of poor frequency stability. After the signal is down converted, the signal is located in a wide frequency range. Wideband amplification is typically used to improve signal-to-noise ratio (SNR) followed by envelope detection.

Majority of WURs use narrowband WUSs even though in-band interference is a major challenge for receiver architectures that are based on received signal power. With a spread spectrum technique, the signal is spread in frequency domain, which makes it more resistant against interference due to spreading gain that will be achieved by dispreading the signal at the receiver. To dispread the spread spectrum signal energy efficiently, e.g., passive surface acoustic wave matched filter can be used. Low-power consumption is achieved at a cost of sensitivity due to high insertion losses of the surface acoustic wave matched filter.

Receiver based on injection-locking architecture has an oscillator that locks to a received carrier frequency if it is close to the oscillator's natural oscillation frequency. If injection locking does not occur, but the carrier frequency disturbs the oscillator, it will be seen at the oscillator output. This is called injection polling. Hence, injection-locking receivers are usually used to detect frequency shift keying (FSK)-modulated signals.

The superregenerative receiver is based on power detection. Oscillation start-up time in a superregenerative oscillator correlates with the received signal strength. By detecting time difference between oscillation start-up times, a simple low-power receiver that has large gain can be built. Superregenerative oscillator architecture block diagram shown in Figure 5.7e has a low noise amplifier (LNA) at the front end, because it is needed to isolate feed through from oscillator to antenna, and it also amplifies the signal. It depends on superregenerative receiver characteristics that how much the oscillator emits to surroundings. If emission is larger than regulations allow, isolation must be implemented. LNA can be added at the front end of other architectures as well, which would give gain at the cost of power consumption. Typically, in WURs, amplification is rather done at IF than in RF because it is more power friendly.

Subsampling-based receiver architecture down converts the signal with a discrete-time sampler instead of with a continuous-time mixer. Input signal is mixed with harmonic components in the discrete-time sampler, producing replicas in multiple frequencies that gives more freedom to select an appropriate IF.

Different WUR designs are compared in Figure 5.8, where the trade-off between sensitivity and power consumption is clearly illustrated. At the moment, there is no superior architecture that would be clearly more suitable for WUR purposes in comparison with other solutions.

5.3.2 Future research directions

WUR sensitivities vary from -40 dBm to around -90 dBm in designs that have active power consumption less than 50 μW, whereas typical commercial short-range transceiver usually has a sensitivity of at least -95 dBm. Therefore, there is still work to be done to reach the gap between WURs and data transceivers. If the wake-up range is lower than the data radio range, the

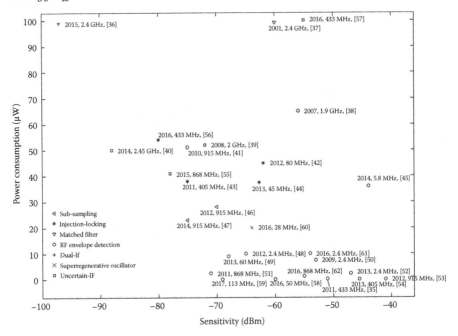

FIGURE 5.8
Comparison of wake-up receivers and their architectures.

usage of wake-up signaling will require higher node density in the network or higher transmit power. The gap becomes even larger issue if one would like to integrate WURs to low power wide area network (LPWAN) sensor nodes. In LPWAN, wireless communications range outdoors can be more than 10 km [67], and in indoors, one base station can cover a large real estate [68], whereas, in a traditional WSN communication, range of the sensor node is usually some tens of meters [69]. Figure 5.9 shows the main differences between the WSN and LPWAN. An application for LPWAN, where utilizing the WUR would be beneficial, could be, e.g., remote patient monitoring. Sensor nodes can be configured to report periodically patient's well-being directly via a remote base station, which enables that patient does not need to carry gateway device. In normal conditions, sensor nodes can be configured to transmit patient's vital data to the base station couple of times per day, but if the patient's physical condition has some alarming signs, and doctor needs data more frequently, the WUR usage enables that the reporting frequency could be changed with low latency. It is expected that this kind of situation occurs rarely; therefore, it is assumed to be more energy efficient to use nodes that are equipped with WUR instead of using duty-cycling based MAC protocol. To achieve 10 km range with transmit power according to regulations, commercial LPWAN transceivers are operating in sub 1 GHz bands, and they are designed to be

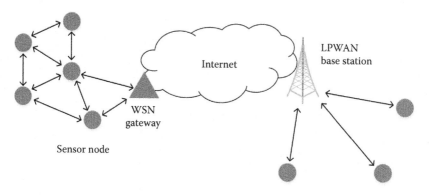

FIGURE 5.9
Main differences between the WSN and the LPWAN.

highly sensitive. For example, SX1272 transceiver by Semtech [70] achieves a sensitivity of −137 dBm at a data rate of 300 bps. Semtech is part of the LoRa alliance [71], which is targeting to standardize solution for the LPWAN. Another important player in the LPWAN field at the moment is Sigfox [72]. Sigfox's technology-compliant transceivers, such as the AX5043 by Axsem [73], have similar performance than the SX1272. Most of the WURs found from the literature have much higher data rates, from 100 to 350 kbps. One direction for future WUR research could therefore be the sensitivity improvement at the cost of data rate. Since the length of the wake-up packet is usually in the order of few bytes, the WUS can be transmitted sufficiently fast even with low data rates while maintaining energy efficiency.

The world is rapidly evolving into a networking society where more and more wireless devices communicate with each other. Gartner, a technology research and advisory firm, estimated in 2014 that there will be 26 billion IoT-based devices in 2020 [74]. ABI Research, a technology market intelligence firm, estimated the number of devices to be 30 billion [75] and Cisco estimate is that 50 billion devices and objects will be connected to IoT by 2020 [8]. These estimates show that the IoT market will be huge. Assuming that most of the devices will be wireless, the number of radios can be even higher since a device might be equipped with multiple radios. More wireless devices mean that there will be more interference, but on the other hand, new millimeter wave bands will be exploited that will ease the situation. More interference will be harmful, especially for the RFED architecture, since it is based on energy detection. This should be taken into account while planning future research activities related to WURs.

WUS generation and transmission are often neglected in the WUR design and energy efficiency performance evaluation. However, since OOK is the most common modulation method being used, it can be generated with most of the commercial transceivers. Also, data rate can be usually configured to match receiver's rate. In LPWANs, the base stations are already being installed all around the world by SigFox and Lora alliance; therefore, the hardware changes would be expensive. To make WURs as a widespread technology in the future, integrating WURs to existing and future IoT networks should get more attention in the research.

5.4 Energy-Efficient Target Scenarios

5.4.1 Application scenarios

The hierarchical architecture is scalable for various application scenarios that are deployed using heterogeneous devices. However, wake-up radio-based hierarchical architecture is targeted especially to applications that require communication rarely and in addition require a low latency reaction to events.

A very good example is an area surveillance network. In that case, the motion detection sensor (e.g., passive infrared) nodes are continuously monitoring the environment to detect intruder movements in the sensing area. Once the event is detected, sensor nodes must be able to report event rapidly. In the energy-efficient hierarchical architecture case, it means that the next layer in the hierarchy must be awakened with a very low delay. If the event is detected to be critical, the highest layer also will be awakened and an alarm will be generated to the user through the backbone. Another example is structural monitoring of, e.g., bridges or buildings. In that case, the condition of the monitored structure can be queried rarely. On the other hand, the sensor nodes can send alarm if some critical changes have been detected. Also in that case, the sensor nodes can perform continuous monitoring and wake up the higher layer node only when required to enable long lifetime for the network nodes. Hierarchical wake-up radio-based architecture can as well be utilized for WBAN and industrial applications. In WBAN case, the wake-up radios could be equipped, for instance, to implants and on-body nodes that must have very long lifetime once installed into human body. In industrial applications, WUR-based nodes can be installed to monitor, e.g., pipe valves or certain parts of engines. For military and critical infrastructure scenarios, there are also many use cases for hierarchical WUR-based network architecture, since there is a need for different types of longtime monitoring applications, e.g., for surveillance, reconnaissance, and security purposes. Figure 5.10 summarizes example applications and communication and sensor technologies that can be used in their implementation.

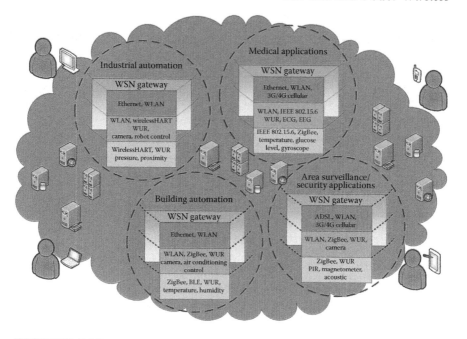

FIGURE 5.10
Examples of hierarchical WSN architecture application areas and techniques.

5.4.2   Energy efficiency comparison of wake-up mechanisms

This section introduces an energy efficiency comparison of WUR- and DCR-based wake-up mechanisms discussed in Section 5.2.2. The intelligent hierarchical operation is assumed for both WUR and DCR approaches, i.e., the lower-layer devices of the hierarchical architecture are performing continuous sensing and will wake up the higher layers when required. At first, the comparison is done with different duty-cycle and event-frequency values in a typical surveillance scenario for critical infrastructure protection, which corresponds with the case of subnetwork 2 in Figure 5.2. For both approaches, source-initiated communication is assumed, i.e., the focus is on the scenario where the sensor nodes have sensed something and trigger the wake up of the higher layer. The used analytical model can also be applied to other WSN scenarios with layered hierarchical architecture, but in that case detailed assumptions must be adjusted to match the particular application under evaluation. As an example of that, energy comparison results for a two-tier WBAN case using a sink-initiated GWR-MAC will also be introduced in this section. In addition, in the end of the section, the results of WUR design tradeoff effect to energy consumption will be discussed.

In the DCR approach, each layer's low-power transceivers need to wake up and sleep according to a predefined schedule. It is assumed that the nodes in each network layer have low-power radios (e.g., IEEE Std. 802.15.4-based) for duty cycling to enable the wake-up mechanism. In the DCR-based network case, the transmitter sends a DE message to inform the node at a higher layer that it should stay awake. If a higher-layer node receives a DE message during the duty cycle listening, it will stay on, send an ACK message, and wait for a data transmission.

In the WUR-based network, WUSs are used to activate the other layers when needed. The GWR-MAC protocol introduced in Section 5.2.3 is used in the WUR-based networks. In the surveillance network case, the lower layer transmits the WUS, and the higher-layer node sends back a beacon message, as described in the GWR-MAC protocol's source-initiated mode definition. Then the lower-layer nodes can send their data to the higher-layer node. The same GWR-MAC wake-up procedure is used for the EL to wake up the IL and in turn the IL to wake up the AL. In the WBAN example case, the sink node transmits a WUS to a sensor node when it needs to be awakened to receive a message. Once awakened, the sensor node will send ACK back to the sink. The sink node will then send data (or control) message back to the sensor node(s).

Authors have originally proposed an analytical model, which can be used to compare the energy efficiency of the GWR-MAC-based and conventional duty-cycle MAC (DCM)-based networks as a function of number of events in a hierarchical network for surveillance scenario [23]. The energy efficiency comparison takes into account the energy consumption of the node's core components: microcontroller unit, transceiver, and sensors. A low-power microcontroller typically has the active, standby, and sleep modes. The transceiver has the transmit (Tx), receive (Rx), idle, and sleep modes. The active Tx and Rx modes are the most energy-consuming parts. Therefore, it is important to put the nodes into sleep mode when possible. The sensing component consists of sensors and analog-to-digital (A/D) converters. The sensing component has different modes that affect energy consumption, e.g., sensor warm-up, active mode, and settle time of the A/D converter. The dominating energy consumption factors of each transceiver's components are taken into account: wake-up signaling, data transmission, and reception; MCU; and sensor active mode current consumption. The relevant energy consumption characteristics, affecting the WUR and DCR energy efficiency comparison, are then addressed. However, energy consumption during network initialization and run-time management (e.g., communication required for synchronization and routing) are not taken into account.

The energy efficiency equation for the WUR and DCR network comparison is defined as [7, 23]

$$\eta(\varepsilon, \lambda, t, \beta) = \frac{\min(E(\varepsilon, \Lambda, t, \beta))}{E(\varepsilon, \lambda, t, \beta)} \tag{5.1}$$

where E is the network energy consumption over time period t, ε is the number of events during t, λ is duty cycle, and β is bit error probability. In this case, the event means that nodes have a data packet to send for a hub about the sensed event. In eq. (5.1), the minimum of E is calculated over the duty-cycle value set $\Lambda = [0, 1]$. Note that $\Lambda = 1$ corresponds to the WUR case, since the receiver is listening the channel continuously. The metric introduced in eq. (5.1) defines the maximum energy efficiency to be one and enables comparison of the WUR- and DCR-based networks.

The total energy consumption during the operation time, t, as a function of number of events and bit error probability, for WUR-based network layers (EL, IL, and AL) can be calculated as

$$E_{\text{WUR}}^{\text{EL}}(\varepsilon, t, \beta) = E_s^{\text{EL}}(t) + E_{\text{MCU}}^{\text{EL}}(\varepsilon, t) + E_{\text{Tx,WUS}}(\varepsilon, t, \beta) + E_{\text{wait,BC}}(\varepsilon, t)$$

$$+ E_{\text{Rx,BC}}(\varepsilon, t) + E_{\text{C}}(t) + E_{\text{Tx,D}}^{\text{EL}}(\varepsilon, t, \beta) + E_{\text{clk}}(t)$$

$$E_{\text{WUR}}^{\text{IL}}(\varepsilon, t, \beta) = E_s^{\text{IL}}(t) + E_{\text{MCU}}^{\text{IL}}(\varepsilon, t) + E_{\text{Tx,WUS}}(\varepsilon, t, \beta) + E_{\text{Tx,BC}}(\varepsilon, t)$$

$$+ E_{\text{wait,BC}}(\varepsilon, t) + E_{\text{Rx,BC}}(\varepsilon, t) + E_{\text{Rx,WUS}}(\varepsilon, t, \beta) \quad (5.2)$$

$$+ E_{\text{C}}(t) + E_{\text{clk}}(t) + E_{\text{Tx,D}}^{\text{IL}}(\varepsilon, t, \beta) + E_{\text{Rx,D}}^{\text{IL}}(\varepsilon, t, \beta)$$

$$E_{\text{WUR}}^{\text{AL}}(\varepsilon, t, \beta) = E_{\text{MCU}}^{\text{AL}}(\varepsilon, t) + E_{\text{Rx,WUS}}(\varepsilon, t, \beta) + E_{\text{Tx,BC}}(\varepsilon, t) + E_{\text{C}}(t)$$

$$+ E_{\text{clk}}(t) + E_{\text{Tx,D}}^{\text{AL}}(\varepsilon, t, \beta) + E_{\text{Rx,D}}^{\text{AL}}(\varepsilon, t, \beta),$$

where $E_{\text{Tx,WUS}}$ is the energy consumption of WUS transmissions, $E_{\text{Rx,WUS}}$ is the energy consumption of WUS receptions, $E_{\text{Tx,BC}}$ is the energy consumption of beacon transmission, $E_{\text{wait,BC}}$ is the energy consumption of beacon listening, $E_{\text{Rx,BC}}$ is the energy consumption of beacon receptions, E_{C} is the constant energy consumption of WUR, and E_{clk} is the energy consumption of the clock needed to maintain the time synchronization. E_s^{x} is the energy consumption of sensing, $E_{\text{MCU}}^{\text{x}}$ is the energy consumption of MCU, $E_{\text{Tx,D}}^{\text{x}}$ and $E_{\text{Rx,D}}^{\text{x}}$ and are the energy consumption of data transmissions and receptions, respectively, calculated separately for each layer (i.e., x is EL, IL, or AL).

The total energy consumption during t, as a function of number of events, duty-cycle percentage and bit error probability, for DCR-based network layers (EL, IL, and AL) can be calculated as

$$E_{\text{DCR}}^{\text{EL}}(\varepsilon, \lambda, t, \beta) = E_s^{\text{EL}}(t) + E_{\text{MCU}}^{\text{EL}}(\varepsilon, t) + E_{\text{Rx,DC}}^{\text{EL}}(\lambda, t) + E_{\text{clk}}(t)$$

$$+ E_{\text{Tx,DE}}^{\text{EL}}(\varepsilon, t, \beta) + E_{\text{Rx,BC}}^{\text{EL}}(\varepsilon, t, \beta) + E_{\text{Tx,D}}^{\text{EL}}(\varepsilon, t, \beta)$$

$$E_{\text{DCR}}^{\text{IL}}(\varepsilon, \lambda, t, \beta) = E_s^{\text{IL}}(t) + E_{\text{MCU}}^{\text{IL}}(\varepsilon, t) + E_{\text{Rx,DC}}^{\text{IL}}(\lambda, t) + E_{\text{clk}}(t) \quad (5.3)$$

$$+ E_{\text{Rx,DE}}^{\text{IL}}(\varepsilon, t, \beta) + E_{\text{Tx,BC}}^{\text{IL}}(\varepsilon, t, \beta) + E_{\text{Tx,DE}}^{\text{IL}}(\varepsilon, t, \beta)$$

$$+ E_{\text{Rx,BC}}^{\text{IL}}(\varepsilon, t, \beta) + E_{\text{Tx,D}}^{\text{IL}}(\varepsilon, t, \beta) + E_{\text{Rx,D}}^{\text{IL}}(\varepsilon, t, \beta)$$

$$E_{\text{DCR}}^{\text{AL}}(\varepsilon, \lambda, t, \beta) = E_{\text{MCU}}^{\text{AL}}(\varepsilon, t) + E_{\text{Rx,DC}}^{\text{AL}}(\lambda, t) + E_{\text{clk}}(t) + E_{\text{Rx,DE}}^{\text{AL}}(\varepsilon, t, \beta)$$

$$+ E_{\text{Tx,BC}}^{\text{AL}}(\varepsilon, t, \beta) + E_{\text{Tx,D}}^{\text{AL}}(\varepsilon, t, \beta) + E_{\text{Rx,D}}^{\text{AL}}(\varepsilon, t, \beta)$$

where λ is the duty-cycle percentage, $E_{\text{Rx,DC}}^{x}$ is the energy consumption of channel listening according to the duty cycle, $E_{\text{Tx,DE}}^{x}$ and $E_{\text{Rx,DE}}^{x}$ and is the energy consumption of DE message transmission and reception, respectively, when x is EL, IL, or AL. Depending on the duty-cycling based MAC protocol features, the DE message can be replaced, e.g., by using a preamble before the data packet.

By multiplying the energy consumption of different layers nodes at WUR-based network (eq. 5.2) and DCR-based network (eq. 5.3) with the number of nodes at each layer, the network total energy consumption during operation time, t, can be easily derived. Interested reader can find more details about the network energy consumption derivation from Refs. [7, 23].

Figure 5.11 shows network total energy consumption comparison results, which are calculated using the parameters chosen to represent typical values for nodes equipped with WUR, IEEE Std. 802.15.4, and IEEE Std. 802.11b communication interfaces and sensors, as described for the surveillance scenario. The number of nodes at the EL is 100, and the number of nodes at the IL is 10, i.e., there is in average 10 sensor nodes assumed to be associated with one IL node, which acts as a coordinator node for the sensor nodes. It can be seen that, in the WUR-based network, energy consumption is drastically lower with the whole range of studied duty-cycle values when the event frequency is low. For the lowest number of events case, the energy consumption gain of the WUR approach is more than two orders of magnitude in comparison with a DCR with $\lambda = 5\%$. For the highest number of events per hour, DCR has energy consumption gain of 12% with the smallest duty cycle percentage value. More details about the analytical model, parameters, and results can be found in Ref. [23].

In Refs. [28, 76], authors have shown that hierarchical architecture and its analytical energy efficiency model introduced in Ref. [23] can also be applied for WBAN scenario. Further, in Ref. [69] the wake-up radio and DCR-based network energy consumption comparison is discussed in WBANs case. WUR and DCR energy efficiency comparison results, which are calculated using four different WUR parameters, are shown in Table 5.1. Typical state-of-the-art performance values are used in WUR1 and WUR2 for power consumption and sensitivity. WUR3 sensitivity is set to be the same as for DCR transceiver [77]. In the WUR4 case, the assumption is that sensitivity can be improved by using very low data rate while remaining at very low Rx mode power consumption. In the calculations, it has been assumed that data packet payload is 255 bytes and communication is error free ($\beta = 0$). More details about the used analytical model can be found in Ref. [28].

The energy efficiency comparison results for WUR- and DCR-based WBANs as a function of number of events per hour is presented in Figure 5.12.

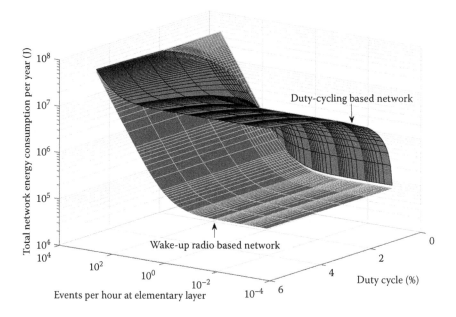

FIGURE 5.11
Network energy consumption comparison as a function of event per hour and duty cycle.

TABLE 5.1
Parameters Used for Different Radios in the Energy Efficiency Comparison

Radio	Sensitivity (dBm)	Tx Power (dBm)	Data Rate	Power Consumption	
				Tx Mode [77] (mW)	Rx Mode
WUR1	−70	0	200 kbps	52.2	5 μW
WUR2	−80	−10	200 kbps	33.9	10 μW
WUR3	−95	−25	200 kbps	25.5	50 μW
WUR4	−95	−25	300 bps	25.5	5 μW
DCR	−95	−25	971 kbps	25.5	56 mW [77]

It can be observed that the GWR-MAC-based network outperforms the DCM-based network's lowest duty cycle ($\lambda = 0.5\%$) case when the number of events is less than 12 per hour. When compared with DCM network with $\lambda = 3\%$, the GWR-MAC-based approach is more energy efficient if the number of events is below 60 per hour. The results for different WURs shows that WUR's Rx mode power consumption has remarkable effect to the total energy efficiency, since it is continuously listening the channel to detect WUSs. The WUR1-based network features lowest sensitivity for WUR. However, it is the most energy efficient when $\varepsilon < 12$, even though it requires highest power used by

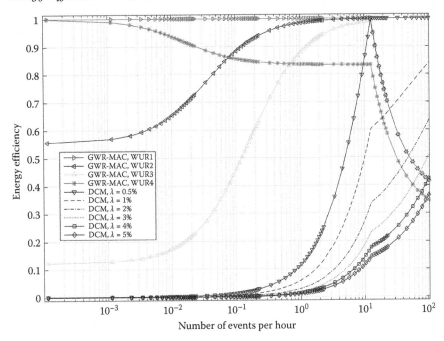

FIGURE 5.12
Energy efficiency comparison for WUR-based and DCM-based hierarchical network.

the transmitter. WUR3 case has the highest Rx mode power consumption, which leads to drastically lower energy efficiency if events occur rarely. This observation highlights the importance of constant mode power consumption (Rx mode) minimization for WURs. All the studied WUR cases lead to higher energy efficiency than DCR approach when the number of events is less than 12 per hour. When the number of events increases above 12 per hour, the energy consumption of data communication starts to dominate in the network overall energy consumption, and the difference between WUR's energy efficiency is not visible anymore. Furthermore, it can be observed that if sensitivity of WUR can be improved by decreasing the data rate that will lead to energy-efficient solution for very rare event cases. When the event frequency increases, the longer transmission and reception duration will cause lower energy efficiency in comparison to higher data rate WUR solutions. From the results of Figure 5.12, it can be concluded that the WUR-based approach is drastically more energy efficient than duty-cycle based approach when the event frequency is low.

In Ref. [76], a dual-radio solution has been proposed combining an ultra wideband (UWB) communication and WUR (UWB-WUR) for WBANs. A dual-radio approach for asymmetric communication links is based on WUR usage so that sensor nodes will include IR-UWB in transmit-only mode for data transmissions (uplink), and WUR is used for control message receptions from the hub node (downlink). Solution enables that idle listening will not occur at the sensor nodes, which presumably can improve energy efficiency, especially when the downlink traffic does not occur often. Therefore, the solution introduced in Ref. [76] takes advantage of sink-initiated wake-up procedure which was introduced in Figure 5.6. Energy consumption and energy efficiency of the UWB-WUR solution has been studied by the authors in Ref. [76]. Results will be shortly introduced and more details of the energy consumption calculation can be found in Ref. [76]. Figure 5.13 shows the energy consumption comparison for UWB-WUR approach and the DCR approach based on duty cycling in the WBAN case. Results are given for different duty cycle values between 0.3% and 2.5%, and error-free transmissions have been assumed for both approaches. It can be observed that UWB-WUR approach consumes less energy until the number of wake ups per year increases to approximately to three per minute. After that point, the DCR network with the lowest duty cycle consumes less energy.

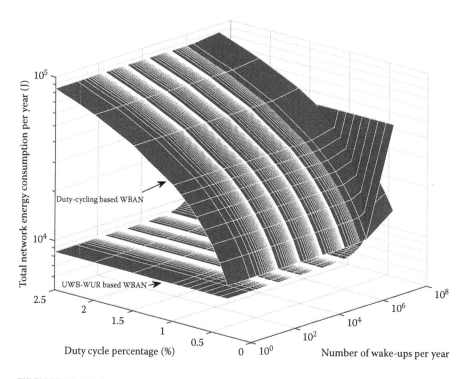

FIGURE 5.13
Energy consumption comparison of UWB-WUR based approach and duty-cycling based approach for WBAN.

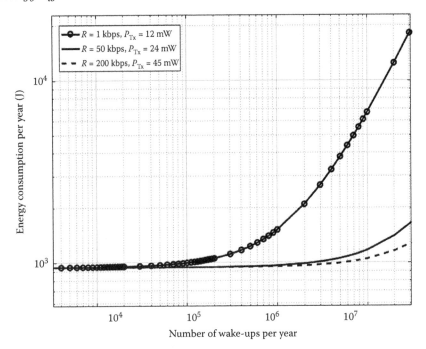

FIGURE 5.14

Energy consumption comparison for a Tx-Rx link when using different parameter setting for WUR.

Figure 5.14 shows energy consumption comparison for a link where Tx node sends a WUS to a receiver node (WUR). For the WUR, different parameter settings are used to find out how the energy savings that can be achieved by performing tradeoff between WUR data rate and sensitivity. Three different WUR settings are evaluated in the results of Figure 5.14: (1) WUR with data rate $R = 1$ kbps and required transmitter's power consumption $P_{Tx} = 12$ mW; (2) WUR with $R = 50$ kbps and $P_{Tx} = 24$ mW; and (3) WUR with data rate $R = 200$ kbps and $P_{Tx} = 45$ mW. The rationale behind these parameter values is that low data rate WUR can be designed to be more sensitive. Therefore, for more sensitive WUR solution, a lower transmit power is required to achieve successful WUS detection. For example, in Ref. [77], it is shown that the transmitter's power consumption is halved when output power is reduced by 25 dBm. In the aforementioned WUR settings, the sensitivities vary so that the setting (1) corresponds to most sensitive receiver and setting (3) to the least sensitive. The results of Figure 5.14 are calculated for a single link, taking into account Tx and Rx energy consumption. From the results of Figure 5.14, it can be observed that when the number of wake ups is more than 10 per hour, the energy consumption of wake-up link with lowest data rate performance starts to increase. The rationale is that, since the data rate is low, the WUS transmission and reception take more time, and their contribution to energy consumption starts to be remarkable, even the required transmission power is lowest for that WUR parameter setting. It can be observed that when the number of wake ups is high, the highest data rate wake-up link consumes least energy, even the required transmit power is highest for that WUR setting, because it has the lowest sensitivity. Therefore, the results of Figure 5.14 illustrate that it is important to take into account the WUR design factor to minimize the energy consumption that depends on how frequently the wake ups are needed in the particular network. Further details of the model that can be used to calculate the results of Figure 5.14 can be found in Ref. [76].

5.5   Summary

Hierarchical architecture for WSN and WBAN, which can be used to provide data to IoTs, was discussed in this section. Introduced architecture is designed to enable the deployment of multiple technologies in the same network. Therefore, it can be used for different types of monitoring scenarios, e.g., sensor networks for monitoring of environment and building structures, as well as for many other types of WSN and WBAN applications. Functionalities for different layers are designed so that the network can fulfill its requirements with low-power operation. Energy efficiency is achieved by utilizing a wake-up signaling that can be used to activate the layers only when required. A generic wake-up radio-based MAC protocol designed for hierarchical architecture was introduced. GWR-MAC protocol includes a bidirectional wake-up procedure and data transmission period for which channel access method can be selected depending on the application characteristics. Different WUR architectures were introduced and future research directions were outlined. Architecture's energy efficiency performance results were shown for a typical area surveillance scenario and WBAN case. Results show that WUR-based networks have a remarkable potential to improve energy efficiency, in comparison to traditional duty-cycle operation, in the case of low event rate applications. Comparison for different WUR parameter settings is also made in this section to illustrate the WUR's data rate and sensitivity tradeoff effect to energy consumption. The DCR approach was found to be more energy efficient only when sufficiently low duty cycle is combined with a high number of events. However, in many practical solutions, the duty cycle is fixed and must be large enough to handle the worst-case traffic. Duty cycle should be changed dynamically when the event frequency changes to save energy. Furthermore, the very low duty-cycle operation would require very strict synchronization to enable that transmissions would be done exactly at correct times according to the duty cycle. Strict synchronization maintenance will cause additional energy consumption. The wake-up radio research area is still quite unexplored and has gained researchers attention in recent years. Therefore, the purpose of the hierarchical architecture with GWR-MAC protocol is to enable more efficient usage of WURs in WSNs and WBANs, and to foster future research and development.

Bibliography

[1] Raghunathan V, Schurgers C, Park S & Srivastava M (2002) Energy-aware wireless microsensor networks. *IEEE Signal Processing Magazine* 19(2): 40–50.

[2] Ares BZ, Park PG, Fischione C, Speranzon A & Johansson KH (2007) On power control for wireless sensor networks: System model, middleware component and experimental evaluation. Proc. ECC, pp. 1–8.

[3] Hohlt B, Doherty L & Brewer E (2004) Flexible power scheduling for sensor networks. Proc. IPSN, pp. 205–214.

[4] Demirkol I, Ersoy C & Alagoz F (2006) MAC protocols for wireless sensor networks: A survey. *IEEE Communications Magazine* 44(4): 115–121.

[5] Anastasi G, Conti M, Francesco MD & Passarella A (2009) Energy conservation in wireless sensor networks: A survey. *Ad Hoc Networks* 7(3): 537–568.

[6] Nechibvute A, Chawanda A & Luhanga P (2012) Piezoelectric energy harvesting devices: An alternative energy source for wireless sensors. *Smart Materials Research* 2012: 13, Article ID 853481. doi: 10.1155/2012/853481.

[7] Karvonen H (2015) Energy efficiency improvements for wireless sensor networks by using cross-layer analysis. Ph.D. thesis, University of Oulu, Finland, March 2015.

[8] Evans D (2011) The internet of things: How the next evolution of the internet is changing everything. Cisco Internet Business Solutions Group (IBSG) white paper, April 2011.

[9] Markets and Markets (2014) Wireless devices market for medical by technology (BT/BLE, Wi-Fi, ZigBee, ANT+), component (sensors, ICs, processors), application (monitoring, medical therapeutics, diagnosis, fitness & wellness), and geography – global forecast to 2020. Market report, December 2014.

[10] Buratti C & Verdone R (2008) A hybrid hierarchical architecture: From a wireless sensor network to the fixed infrastructure. Proc. Wireless Conference, EW 2008.

[11] Khan Z, Catalot D & Thiriet J (2009) Hierarchical wireless network architecture for distributed applications. Proc. ICWMC '09, pp. 70–75.

[12] Yang J, Gao Y & Zhang Z (2011) Cluster-based routing protocols in wireless sensor networks: A survey. Proc. ICCSNT '11, volume 3, pp. 1659–1663.

[13] ZigBee Alliance. ZigBee alliance webpage. www.zigbee.org. Accessed 22.03.2018.

[14] Z-Wave Alliance. Z-Wave alliance webpage. www.z-wavealliance.org. Accessed 22.03.2018.

[15] Kumar D, Aserib TC & Patelc RB (2009) EEHC: Energy efficient heterogeneous clustered scheme for wireless sensor networks. *Journal of Computer Communications* 32(4): 662–667.

[16] Wu M & Collier M (2011) Extending the lifetime of heterogeneous sensor networks using a two-level topology. Proc. CIT, pp. 499–504.

[17] Slimane JB, Song YQ, Kouba A & Frikha M (2009) A three-tiered architecture for large scale wireless hospital sensor networks. Proc. MobiHealthInf, pp. 1–12.

[18] Zhang M, Song J & Zhang Y (2005) Three-tiered sensor networks architecture for traffic information monitoring and processing. Proc. IROS, pp. 2291–2296.

[19] Kulkarni P, Ganesan D, Shenoy P & Lu Q (2005) SensEye: A multitier camera sensor network. Proc. MM '05, pp. 229–238.

[20] Lopes CER, Linhares FD, Santos MM & Ruiz LB (2007) A multitier, multimodal wireless sensor network for environmental monitoring. *Springer Link Lecture Notes in Computer Science* 4611: 589–598.

[21] Zatout Y, Campo E & Llibre JF (2009) WSN-HM: Energy-efficient wireless sensor network for home monitoring. Proc. ISSNIP '09, pp. 367–372.

[22] Stefanov A & Stojanovic M (2010) Hierarchical underwater acoustic sensor networks. Proc. WUWNet '10, pp. 1–4.

[23] Karvonen H, Suhonen J, Petäjäjärvi J, Hämäläinen M, Hännikäinen M & Pouttu A (2014) Hierarchical architecture for multi-technology wireless sensor networks for critical infrastructure protection. Special Issue on Intelligent Infrastructures, *Wireless Personal Communications* 76(2): 209–229. doi: 10.1007/s11277-014-1686-2.

[24] Crossbow. TelosB datasheet. www.willow.co.uk/TelosB_Datasheet.pdf. Accessed 23.03.2018.

[25] FriendlyARM. FriendlyARM Mini6410 specification. www.friendlyarm. net/products/mini6410. Accessed 23.03.2018.

[26] Kuorilehto M, Kohvakka M, Suhonen J, Hämäläinen P, Hännikäinen M & Hämäläinen T (2007) *Ultra-Low Energy Wireless Sensor Networks in Practice: Theory, Realization and Deployment.* John Wiley & Sons, Hoboken, NJ.

[27] Karvonen H, Petäjäjärvi J, Iinatti J, Hämäläinen M & Pomalaza-Ráez C (2014) A generic wake-up radio based MAC protocol for energy efficient short range communication. IEEE PIMRC Workshop: The Convergence of Wireless Technologies for Personalized Healthcare, September 2–5, 2014, Washington, DC.

[28] Karvonen H, Petäjäjärvi J, Iinatti J & Hämäläinen M (2014) Energy efficient IR-UWB WBAN using a generic wake-up radio based MAC protocol. The Third Ultra Wideband for Body Area Networking Workshop (UWBAN-2014), Co-located with the 9th International Conference on Body Area Networks (BodyNets-2014), September 29–October 1, 2014, London, UK.

[29] Oller J, Demirkol I, Casademont J, Paradells J, Gamm GU & Reindl L (2016) Has time come to switch from duty-cycled MAC protocols to wake-up radio for wireless sensor networks?. *IEEE/ACM Transactions on Networking* 24(2): 674–687.

[30] Djiroun FZ & Djenouri D (2017) MAC protocols with wake-up radio for wireless sensor networks: A review. *IEEE Communications Surveys & Tutorials* 19(1): 587–618, Firstquarter 2017.

[31] Gu L & Stankovic JA (2005) Radio-triggered wake-up for wireless sensor networks. *Springer Journal on Real-Time Systems* 29(2): 157–182.

[32] Van der Doorn B, Kavelaars W & Langendoen K (2009) A prototype low-cost wakeup radio for the 868 MHz band. *International Journal of Sensor Networks* 5(1): 22–32.

[33] Ansari J, Pankin D & Mähnen P (2009) Radio-triggered wake-ups with addressing capabilities for extremely low power sensor network applications. *International Journal of Wireless Information Networks* 16(1): 118–130.

[34] Petäjäjärvi J, Karvonen H, Mikhaylov K, Pärssinen A, Hämäläinen M & Iinatti J (2015) WBAN energy efficiency and dependability improvement utilizing wake-up receiver. *Special Issue on Innovation of Medical Information and Communication Technology for Dependable Society, IEICE Transactions on Communications* E98-B(4): 535–542.

[35] Marinkovic SJ & Popovici EM (2011) Nano-power wireless wake-up receiver with serial peripheral interface. *IEEE Journal on Selected Areas in Communications* 29(8): 1641–1647.

[36] Salazar C, Kaiser A, Cathelin A & Rabaey J (2015) A -97dBm-sensitivity interferer-resilient 2.4GHz wake-up receiver using dual-IF multi-N-path architecture in 65nm CMOS. Proc. ISSCC, pp. 396–398.

[37] Tomabechi S, Komuro A, Konno T, Nakase H & Tsubouchi K (2001) Design and implementation of spread spectrum wireless switch with low power consumption. *IEICE TRANSACTIONS on Fundamentals of Electronics, Communications and Computer Sciences* E84-A(4): 971–973.

[38] Pletcher N, Gambini S & Rabaey J (2007) A 65 μW, 1.9 GHz RF to digital baseband wakeup receiver for wireless sensor nodes. Proc. CICC, pp. 539–542. doi: 10.1109/CICC.2007.4405789.

[39] Pletcher N, Gambini S & Rabaey J (2009) A 52 μW wake-up receiver with -72 dBm sensitivity using an uncertain-IF architecture. *IEEE Journal of Solid-State Circuits* 44(1): 269–280. doi: 10.1109/JSSC.2008.2007438.

[40] Bryant C & Sjoland C (2014) A 2.45GHz, 50μW wake-up receiver front-end with $-$88dBm sensitivity and 250kbps data rate. Proc. ESSCIRC, pp. 235–238. doi: 10.1109/ESSCIRC.2014.6942065.

[41] Xiongchuan H, Rampu S, Xiaoyan W, Dolmans G & de Groot H (2010) A 2.4 GHz/915 MHz 51 μW wake-up receiver with offset and noise suppression. Proc. ISSCC, pp. 222–223. doi: 10.1109/ISSCC.2010.5433958.

[42] Joonsung B & Hoi-Jun Y (2012) A 45μW injection-locked FSK wake-up receiver for crystal-less wireless body-area-network. Proc. A-SSCC, pp. 333–336. doi: 10.1109/IPEC.2012.6522693.

[43] Pandey J, Shi J & Otis B (2011) A 120 μW MICS/ISM-band FSK receiver with a 44 μW low-power mode based on injection-locking and 9x frequency multiplication. Proc. ISSCC, pp. 460–462. doi: 10.1109/ISSCC.2011.5746397.

[44] Hyunwoo C, Joonsung B & Hoi-Jun Y (2013) A 37.5 μW body channel communication wake-up receiver with injection-locking ring oscillator for wireless body area network. *IEEE Transactions on Circuits and Systems I: Regular Papers* 60(5): 1200–1208. doi: 10.1109/TCSI.2013.2249173.

[45] Choi J, Lee IY, Lee K, Yun SO, Kim J, Ko J, Yoon G, Lee SG (2014) A 5.8-GHz DSRC transceiver with a 10-μA interference-aware wake-up receiver for the Chinese ETCS. *IEEE Transactions on Microwave Theory and Techniques* 62(12): 3146–3160. doi: 10.1109/TMTT.2014.2362118.

[46] Moazzeni S, Cowan GER & Sawan M (2012) A 28 μW subsampling based wake-up receiver with -70 dBm sensitivity for 915 MHz ISM band applications. Proc. ISCAS, pp. 2797–2800. doi: 10.1109/ISCAS.2012.6271891.

[47] Moazzeni S, Sawan M & Cowan GER (2014) An ultra-low-power energy-efficient dual-mode wake-up receiver. *IEEE Transactions on Circuits and Systems I: Regular Papers* 62(2): 517–526. doi: 10.1109/TCSI.2014.2360336.

[48] Kuang-Wei C, Xin L & Minkyu J (2012) A 2.4/5.8 GHz 10 μW wake-up receiver with $-65/-50$ dBm sensitivity using direct active RF detection. Proc. A-SSCC, pp. 337–340. doi: 10.1109/IPEC.2012.6522694.

[49] Wada T, Ikebe M & Sano E (2013) 60-GHz, 9-μW wake-up receiver for short-range wireless communications. Proc. ESSCIRC, pp. 383–386. doi: 10.1109/ESSCIRC.2013.6649153.

[50] Durante MS & Mahlknecht S (2009) An ultra low power wakeup receiver for wireless sensor nodes. Proc. STA, pp. 167–170. doi: 10.1109/SENSORCOMM.2009.34.

[51] Hambeck C, Mahlknecht S & Herndl T (2011) A 2.4 μW wake-up receiver for wireless sensor nodes with -71dBm sensitivity. Proc. ISCAS, pp. 534–537. doi: 10.1109/ISCAS.2011.5937620.

[52] Nilsson E & Svensson C (2013) Ultra low power wake-up radio using envelope detector and transmission line voltage transformer. *IEEE Journal on Emerging and Selected Topics in Circuits and Systems* 3(1): 5–12. doi: 10.1109/JETCAS.2013.2242777.

[53] Roberts NE & Wentzloff DD (2012) A 98nW wake-up radio for wireless body area networks. Proc. RFIC, pp. 373–376. doi: 10.1109/RFIC.2012.6242302.

[54] Seunghyun O, Roberts NE & Wentzloff DD (2013) A 116 nW multi-band wake-up receiver with 31-bit correlator and interference rejection. Proc. CICC, pp. 1–4. doi: 10.1109/CICC.2013.6658500.

[55] Armas ET, Ramos-Valido D, Khemchandani SL & del Pino J (2015) A 40.9 μW high sensitivity wake-up radio for wireless sensor networks using uncertain-IF architecture. Proc. DCIS, pp. 1–6. doi: 10.1109/DCIS.2015.7388585.

[56] Chen SE & Cheng KW (2016) A 433 MHz 54 μW OOK/FSK/PSK compatible wake-up receiver with 11 μW low-power mode based on injection-locked oscillator. Proc. ESSCIRC, pp. 137–140. doi: 10.1109/ESSCIRC.2016.7598261.

[57] Thanh PN, Tuan KN & Dong XM (2016) A 100-μW wake-up receiver for UHF transceiver. Proc. ICICDT, pp. 1–4. doi: 10.1109/ICICDT.2016.7542049.

[58] Nikoofard A & Mandal S (2016) An 11.5 nW broadband wake-up RF receiver with −60 dBm sensitivity at 50 MHz. Proc. ISCAS, pp. 2787–2790. doi: 10.1109/ISCAS.2016.7539171.

[59] Jiang H et al. (2017) A 4.5 nW wake-up radio with -69 dBm sensitivity. Proc. ISSCC, pp. 416–417. doi: 10.1109/ISSCC.2017.7870438.

[60] Petäjäjärvi J, Mikhaylov K, Karvonen H, Vuohtoniemi R & Iinatti J (2016) Superregenerative wake-up receiver with 20 μW power consumption for human body communications. Proc. NTMS, pp. 1–5.

[61] Cheng KW, Lin JS & Chen SE (2016) Reference-less ultra-low-power wake-up receiver with noise suppression. Proc. URSI AP-RASC, pp. 994–997. doi: 10.1109/URSIAP-RASC.2016.7601309.

[62] Magno M, Jelicic V, Srbinovski B, Bilas V, Popovici E & Benini L (2016) Design, implementation, and performance evaluation of a flexible low-latency Nanowatt wake-up radio receiver. *IEEE Transactions on Industrial Informatics* 12(2): 633–644. doi: 10.1109/TII.2016.2524982.

[63] IEEE (2011) IEEE standard for local and metropolitan area networks - Part 15.4: Wireless medium access control (MAC) and physical layer (PHY) specifications for low-rate wireless personal area networks (LR-WPANs). Standard, The Institute of Electrical and Electronics Engineers, Inc. IEEE Std. 802.15.4-2011 (Revision of IEEE Std. 802.15.4-2006).

[64] IEEE standard for low-rate wireless networks. In: IEEE Std. 802.15.4-2015 (Revision of IEEE Std. 802.15.4-2011), pp. 1–709, April 22, 2016.

[65] IEEE Std. 802.15.6-2012, "IEEE standard for local and metropolitan area networks - Part 15.6: wireless body area networks," The Institute of Electrical and Electronics Engineers, Inc, Standard, 2012.

[66] ETSI TC SmartBAN, "Smart body area networks (SmartBAN): Low complexity medium access control (MAC)", TS DTS/SmartBAN(14) 006001r5, December 2014.

[67] Petäjäjärvi J, Mikhaylov K, Pettissalo M, Janhunen J & Iinatti J (2017) Performance of a low-power wide-area network based on LoRa technology: Doppler robustness, scalability, and coverage. *SAGE International Journal of Distributed Sensor Networks* 13(3): 1–16. doi: 10.1177/1550147717699412.

[68] Petäjäjärvi J, Mikhaylov K, Yasmin R, Hämäläinen M & Iinatti J (2017) Evaluation of LoRa LPWAN technology for indoor remote health and wellbeing monitoring. *Springer International Journal of Wireless Information Networks* 24(2): 1–13. doi: 10. 1007/s10776-017-0341-8.

[69] Petäjäjärvi J, Mikhaylov K, Vuohtoniemi R, Karvonen H & Iinatti J (2016) On the human body communications: Wake-up receiver design and channel characterization. *EURASIP Journal on Wireless Communications and Networking* 2016(179): 1–17doi: 10. 1186/s13638-016-0674-5.

[70] Semtech, SX1272 datasheet: SX1272/73 - 860 MHz to 1020 MHz low power long range transceiver.

[71] LoRa Alliance. www.lora-alliance.org/. Accessed 23.03.2018.

[72] Sigfox company webpage. www.sigfox.com/. Accessed 23.03.2018.

[73] Axsem, AX5043 datasheet: Advanced high performance ASK and FSK narrow-band transceiver for 70-1050 MHz range.

[74] Gartner, "Gartner says the internet of things installed base will grow to 26 billion units by 2020", Press release, January 2, 2014.

[75] ABI Research, "More than 30 billion devices will wirelessly connect to the internet of everything in 2020", Press release, May 2013.

[76] Karvonen H, Petäjäjärvi J, Niemela V, Hämäläinen M, Iinatti J & Kohno R (2017) Energy efficient UWB-WUR dual-radio solution for WBANs. 11th International Symposium on Medical Information and Communication Technology (ISMICT), Lisbon, pp. 64–68.

[77] Texas Instruments, CC2420 datasheet: 2.4 GHz IEEE 802.15.4/Zigbee-ready RF transceiver.

6

All-Digital Noise-Shaping Time-to-Digital Converters for Mixed-Mode Signal Processing

Fei Yuan

Ryerson University

CONTENTS

The advancement of Complementary Metal Oxide Semiconductor (CMOS) technology has resulted in a sharp increase in the time resolution of digital circuits and a rapid deterioration of the performance of mixed analog-digital circuits arising from a shrinking voltage headroom, worsening device match, deteriorating linearity, and disturbances coupled from neighboring digital blocks. Time-mode signal processing where information is represented by the difference between the occurrence of two digital signals offer a viable and technology-friendly means to combat technology scaling induced challenges encountered in the design of mixed analog-digital systems. Time-to-digital converters (TDCs) that map a time variable to a digital code are the most important building blocks for time-mode signal processing [1]. Although the deployment of TDCs in time-of-flight measurement dates back to 1970s [2, 3], the applications of TDCs in mixed-mode signal processing such as analog-to-digital converters (ADCs) [4–6] and phase-locked loops (PLLs) [7, 8] emerged recently. TDCs can be loosely classified into *sampling TDCs* such as delay-line TDCs [9–11] and *noise-shaping TDCs* such as gated ring oscillator (GRO) TDCs [5]. Sampling TDCs suffer from a low time resolution due to the absence of a noise-suppressing mechanism capable of lowering quantization noise to below gate delay. Although a high resolution can be obtained using vernier TDCs [9, 10, 12, 13] or pulse-shrinking TDCs [14, 15], both are at the expense of excessive silicon area and power consumption. Noise-shaping TDCs achieve a high time resolution by displacing a portion of in-band quantization noise to frequencies outside signal band. TDCs with a high time resolution are pivotal to applications such as all-digital PLLs (ADPLLs), where TDCs function as a phase detector. As the quantization noise of TDC phase detectors directly affects the overall phase noise of ADPLLs, minimizing the quantization noise of TDCs within the loop bandwidth of ADPLLs is critical.

This chapter deals with all-digital noise-shaping TDCs. Section 6.1 examines all-digital open-loop noise-shaping TDCs, including GRO TDCs, gated relaxation oscillator TDCs, and switching ring oscillator (SRO) TDCs. Section 6.2 studies all-digital closed-loop noise-shaping TDCs. Time registers including gated delay cell (GDC) time registers, switched delay unit (SDU) time registers, gated delay line (GDL) time registers, and gated discharge path (GDP) time registers are investigated first. It is followed by an in-depth investigation of all-digital time adders including GDC time adders, SDU time adders, GDP time adders, unidirectional GDL (UniGDL) time adders, and bidirectional GDL (BiGDL) time adders. All-digital time integrators that are built upon GDC time adders, SDU time adders and ring oscillators, GDP time adders and registers, and BiGDLs are investigated. Finally, the design and simulation results of an all-digital 1-1 Multi-Stage Noise Shaping (MASH) $\Delta\Sigma$ TDC utilizing differential GDP time integrators and an all-digital first-order $\Delta\Sigma$ TDC utilizing differential BiGDL time integrators are presented in Section 6.3. The chapter is concluded in Section 6.4.

6.1   All-Digital Open-Loop Noise-Shaping TDCs

Open-loop noise-shaping TDCs provide first-order noise shaping without employing a negative feedback loop. Three open-loop noise-shaping TDCs, namely GRO TDCs, gated relaxation oscillator TDCs, and switched ring oscillator TDCs, are investigated in this section.

6.1.1   Gated ring oscillator TDCs

A GRO TDC is a ring oscillator whose operation is gated by time variable T_{in} to be digitized, specifically the oscillator will oscillate when $T_{in} = 1$ and halt oscillation when $T_{in} = 0$ [5]. The GRO can be either a generic static CMOS inverter ring oscillator or a voltage-controlled ring oscillator. In both cases, T_{in} only affects the state of oscillation and has no impact on the frequency of the oscillator. Since the number of the oscillation cycles of the oscillator during $T_{in} = 1$ is proportional to the duration of T_{in}, it can be considered as the digital representation of T_{in} and hence performs time-to-digital conversion. Note that one can also keep T_{in} unchanged while letting the control voltage of the oscillator be the signal to be digitized. In this case, the GRO functions as an ADC. The number of the oscillation cycles during $T_{in} = 1$ can be recorded using a resettable counter, whose content is readout at the falling edge of T_{in} and reset during $T_{in} = 0$, as shown in Figure 6.1. The counter is essentially a phase quantizer with quantization error 2π. Increasing the frequency of the oscillator improves the resolution of the TDC. Both multipath [16] and active inductors [17] are effective in achieving this. The former speeds up oscillators by preskewing the load capacitor of the delay stages of the oscillators, whereas the latter shortens the charge/discharge processes of the load capacitor by means of resonance.

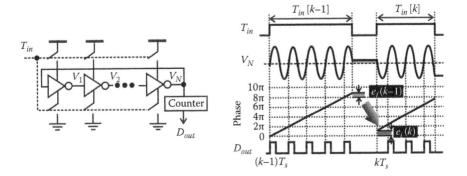

FIGURE 6.1
Gated ring oscillator TDCs with counter-based phase readout.

GRO-based TDCs possess a key characteristic of first-order noise shaping. To illustrate this, consider a GRO-based TDC. During $T_{in} = 0$, since the delay stages of the oscillator are isolated from both supply voltage and ground rails, the output voltage of the delay stages of the oscillator remains unchanged, provided that the effect of the leakage of pn-junctions at the output nodes, charge injection from gated transistors, and disturbances coupled with neighboring devices is negligible. The residual phase of $(k-1)$th sampling cycle, denoted by $e_f(k-1)$ and in the form of the charge of the load capacitor, is carried over in its entirety to kth sampling cycle and becomes the initial phase of kth sampling cycle, i.e., $e_i(k) = e_f(k-1)$, where subscripts i and f identify the initial and final conditions, respectively. The net phase accumulation in kth sampling cycle is composed of two parts: (i) $K_{vco}v_{in}$ that accounts for the number of oscillation cycles of the oscillator during $T_{in} = 1$, where K_{vco} is the voltage-to-phase gain of the oscillator and (ii) $e_f(k) - e_i(k)$ that accounts for the difference between the phase inherited from the previous sampling cycle and the residual phase of the current sampling cycle. Since $e_i(k) = e_f(k-1)$, we have

$$e_f(k) - e_i(k) = e_f(k) - e_f(k-1). \tag{6.1}$$

Thus

$$\phi(k) = K_{vco}v_{in} + [e_f(k) - e_f(k-1)] = K_{p,vco}v_{in} + [e(k) - e(k-1)]. \tag{6.2}$$

Note that we have dropped subscript f in (6.2) with the understanding that $e(k)$ is the quantization error of kt. It is seen from (6.2) that only the difference between two consecutive quantization errors affects the output phase. Quantization error is therefore first-order shaped. GRO-based TDCs perform noise shaping without a negative feedback loop, thereby enabling them to perform rapid digitization [18].

The resolution of GRO-based TDCs can be improved if the logic state of the stages of GROs is utilized. Although increasing the number of stages of GROs improves the resolution of GRO-based TDCs, the larger the number of the stages of GROs, the worse is the gate delay mismatch. Since gate delay mismatch manifests itself as phase uncertainty, it is subject to the constraint of phase continuity and is hence first-order shaped. GRO-based TDCs are vulnerable to the effect of leakage of pn-junctions at the output nodes of delay stages, charge injection from gated transistors, and disturbances coupled with neighboring devices. These effects constitute *skew errors* [19–21]. Among them, charge injection from gated transistors is most critical. This is because the duration of $T_{in} = 0$ is typically small due to oversampling. As a result, the effect of charge leakage and disturbances coupled with neighboring devices during $T_{in} = 0$ is rather insignificant. Reducing the dimension of gated transistors, though minimizing charge injection, slows down the oscillator and worsens time resolution.

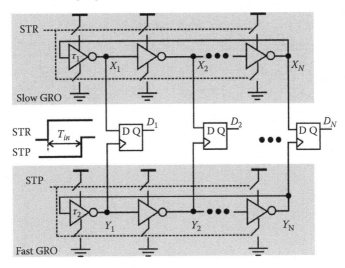

FIGURE 6.2
Vernier-gated ring oscillator TDCs.

To improve the resolution of GRO-based TDCs, vernier architectures popular in delay-line, TDCs can be used [22]. A vernier GRO TDC consists of two GROs with slightly different oscillation frequencies. The front edge of T_{in} [start (STR)] is applied to the slow GRO while the rear edge of T_{in} [stop (STP)] is routed to the fast GRO, as shown in Figure 6.2. When the front edge of T_{in} is asserted, the slow GRO will start to oscillate. The fast GRO will start to oscillate only when the rear edge of T_{in} arrives. Since x_j leads y_j initially, we have $D_j = 0$. Once the rising edge of the fast GRO catches up with that of the slow GRO, $D_j = 1$ will be set and time-to-digital conversion will end. To prevent D-type Flip Flop (DFFs) from entering a meta-state, each DFF has a built-in set-reset latch functioning as a time amplifier that stretches the time difference between the front and rear edges of T_{in}, such that no violation of the timing constraints of DFFs will occur [23].

The conversion time of GRO-based TDCs is bottlenecked by the latency of counter-based phase quantizers. To improve conversion time, the output of GROs, which is phase, can be converted to frequency using digital differentiators, as shown in Figure 6.3 [24]. For an N-stage GRO, the number of $0 \rightarrow 1$ transitions within one oscillation period of the oscillator is N. To capture all N $0 \rightarrow 1$ transitions, the period of sampling clock T_s should be set slightly smaller than $T_{GRO}/2$ where T_{GRO} is the period of the GRO. Note that the output of the TDC is represented by the number of $0 \rightarrow 1$ transitions of the output of the oscillator per sampling period. As phase quantization occurs before differentiation, quantization error is processed by the differentiators, leading to first-order noise shaping of the quantization error.

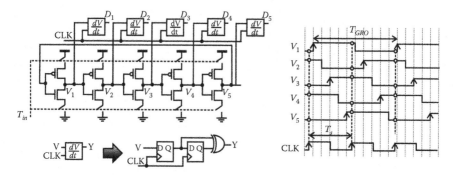

FIGURE 6.3
Gated ring oscillator TDCs with frequency readout.

6.1.2 Gated relaxation oscillator TDCs

Unlike ring oscillators that are sensitive to supply voltage fluctuation, relaxation oscillators exhibit a low sensitivity to supply voltage uncertainity as the charging currents of the capacitors of these oscillators are from current references [25]. First-order noise-shaping characteristics intrinsic to GROs are also possessed by gated relaxation oscillators [26]. We use the two-capacitor relaxation oscillator shown in Figure 6.4 to illustrate this. Whether the oscillator oscillates or not is dictated by the gating signal T_{in} to be digitized. Since the charge of capacitors C1 and C2 will be held unchanged when $T_{in} = 0$, the residual phase of the oscillator in $(k-1)$th sampling cycle is retained and carried over in its entirety to kth sampling cycle in a similar way as that in GRO-based TDCs, leading to first-order noise shaping. Unlike GRO-based TDCs where the load capacitance of delay stages is comparable with the capacitances of gated transistors, the capacitance of gated relaxation oscillators is typically much larger when compared with that of gated transistors. As a result, the effect of charge injection from gated transistors, the leakage of pn-junctions at the output nodes, and disturbances coupled with neighboring

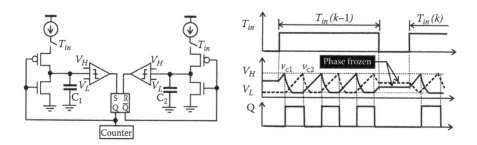

FIGURE 6.4
Gated relaxation oscillator TDCs with phase readout.

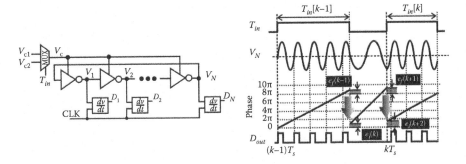

FIGURE 6.5
Switched ring oscillator TDC with frequency readout.

devices becomes negligible. Since the frequency of gated relaxation oscillators is typically much lower when compared with that of GROs, these TDCs are not particularly suitable for applications where oversampling ratio is high.

6.1.3 Switched ring oscillator TDCs

As pointed out earlier, GRO-based TDCs suffer from skew errors arising from the leakage of pn-junctions at output nodes of GRO-based TDCs, charge injection from gated transistors, and disturbances coupled with neighboring devices. Konishi *et al.* showed that if the OFF state of a GRO is replaced with another oscillation state where the output voltage of the oscillator is well defined, skew errors are minimized, as shown in Figure 6.5 [21, 27]. Because the charge of the load capacitor of the oscillator at the end of $(k-1)$th sampling cycle is carried over in its entirety to the next sampling cycle, first-order noise-shaping intrinsic to GROs is also possessed by SROs. Since SROs oscillate in both states, gating error arising from the inability of GROs to emerge from a holding state instantaneously is reduced. The price paid is additional power consumption as the oscillator oscillates in both $T_{in} = 1$ and $T_{in} = 0$ states.

6.2 All-Digital Closed-Loop Noise-Shaping TDCs

The absence of a negative feedback mechanism in the preceding open-loop noise-shaping TDCs makes them less resilient against the effect of process, voltage, and temperature uncertainty. Further, first-order noise shaping is often inadequate for applications where a large signal-to-noise ratio is required. It is well understood that $\Delta\Sigma$ operations offer desirable characteristics such

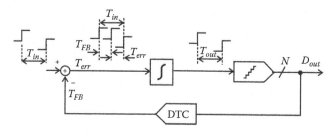

FIGURE 6.6
All-digital $\Delta\Sigma$ time-to-digital converters.

as less sensitive to the effect of process, voltage, and temperature uncertainty and high-order noise-shaping capability. TDCs utilizing $\Delta\Sigma$ operations so as to possess these attractive characteristics are desirable. An all-digital time integrator whose input and output are both time variables is needed, as shown in Figure 6.6. Integrating a time variable T_{in} performs

$$\int T_{in}(\tau)d\tau \approx \sum_{n=1}^{N} T_{in}[n] = T_{in}[N] + \sum_{n=1}^{N-1} T_{in}[n]. \tag{6.3}$$

It is seen that time integration consists of two separate operations, namely time addition that performs $\sum_{j=1}^{k} T_{in}[j] = T_{in}[k] + \sum_{j=1}^{k-1} T_{in}[j]$ and time registration that stores $\sum_{j=1}^{k} T_{in}[j]$ and releases it upon a read request. In this section, we study all-digital time registers. It is followed with an in-depth investigation of all-digital time adders. All-digital time integrators utilizing time adders and time registers are explored. Finally, the design and simulation results in two all-digital $\Delta\Sigma$ TDCs.

6.2.1   All-digital time registers

Time registers store a time variable and release it upon a read request. In this section, we study four all-digital time registers, namely GDC time registers, SDU time registers, GDL time registers, and GDP time registers.

6.2.1.1   GDC time register

Figure 6.7 shows the simplified schematic of the time register utilizing GDCs [28]. The time register consists of two identical GDCs and two static inverters. It has two control commands : H (hold) that withholds time input T_{in} and RD (read) that readouts the stored time variable. Each GDC is composed of a 2-to-1 multiplexer, a DFF, and a gated inverter. The GDC has two control

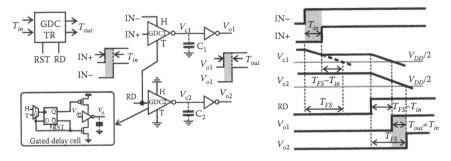

FIGURE 6.7
Time register using gated delay cells.

signals: T (trigger) and H (hold). The operation of the GDC is briefly depicted here: Upon the reset of the DFF (RST = 1), the gated inverter is enabled. If H = 0, the output of the inverter is set by its input. Otherwise, the inverter is disabled and the output of the inverter is held.

Let us now examine the operation of the GDC time register. Let T_{FS} be the amount of the time for the voltage of the capacitors to drop from V_{DD} to $V_{DD}/2$. When $T_{in} = 1(\text{IN}+ = 1/\text{IN}- = 0)$, since the read command RD = 0, GDC will be enabled. C1 starts to discharge and v_{c1} drops with time. When IN$- = 1$, GDC1 will enter a holding state and v_{c1} will remain unchanged. For GDC2, since H = T = 0 and RD = 0, we have C2 fully charged. When RD = 1, for GDC1, with T = 1 we will have Q = 0. As a result, GDC1 will resume discharge and v_{c1} will continue to drop. When v_{c1} drops to $V_{DD}/2$, $v_{o1} = 1$ will be set. For GDC2, C2 starts to discharge and v_{c2} drops. When it dips below $V_{DD}/2$, $v_{o2} = 1$ will be set. It can be shown from the timing diagram that $T_{out} = T_{in}$. Time variable T_{in} is thus stored indefinitely and will be readout when read command RD is asserted.

6.2.1.2 Switched delay unit time register

Bakhshian and Roberts showed that the SDU shown in Figure 6.8 can be used as a time register [29 30]. To store time variable T_{in}, Reset command RST = 1 is asserted before the arrival of $T_{in} = 1$. The capacitors of SDUs are charged to V_{DD}. During $T_{in} = 1$ (IN+ = 1 and IN$- = 0$), the capacitor of SDU1 discharges while that of SDU2 remains charged. When IN$- = 1$ arrives, the discharge of the capacitor of SDU1 will be halted. To readout the stored data, RD = 0 is asserted, forcing SDU1 to resume its discharge process and SDU2 to start its discharge process. When the voltage of the capacitors drops below $V_{DD}/2$, OUT+ = 1 and OUT$- = 1$ will be set. For SDU1, we have $T_{in} + x = T_{FS}$ while for SDU2, we have $x + T_{out} = T_{FS}$, where x is defined explicitly in the figure. It follows that $T_{out} = T_{in}$. Time variable T_{in} is stored and will be readout when read command RD is asserted.

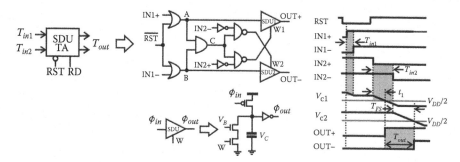

FIGURE 6.8

Time register using switched delay units.

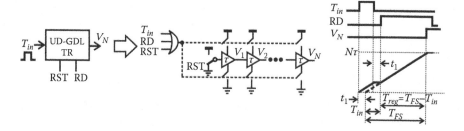

FIGURE 6.9

Time register using a unidirectional gated delay line. When RST $= 1$ is asserted, $v_{1,...,N} = 0$V.

6.2.1.3 UniGDL time register

Figure 6.9 shows the simplified schematic of UniGDL time registers proposed by Kim et al. [31, 32]. The time variable T_{in} to be stored is the gating signal of the UniGDL that is reset initially, e.g. $v_{1,...,N} = 0$ V. When $T_{in} = 1$, "1" will propagate from the beginning of the GDL for duration $T_{in} = 1$. When $T_{in} = 0$, the voltage of the load capacitor of the delay stages will remain unchanged. To readout the stored time variable, read command RD is asserted, resuming the propagation of "1" until it reaches the end of the delay line. The duration from the assertion of RD to the time instant at which "1" reaches the end of the delay line, denoted by T_{reg}, is given by $T_{reg} = T_{FS} - T_{in}$, where T_{FS} is the amount of the time for "1" to propagate from the beginning to the end of the GDL when the gating signal is present all the time. When compared with GDC time registers and SDU time registers, UniGDL time registers feature a simple configuration, subsequently low power consumption and a small silicon area. It should be noted that the output of UniGDL time registers is the complementary of T_{in} rather than T_{in} itself. To extract T_{in}, prior knowledge of T_{FS} is needed.

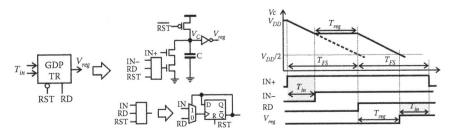

FIGURE 6.10
Time register using a gated discharge path.

6.2.1.4 Gated discharge path time register

Figure 6.10 shows the simplified schematic of a time register with a GDP [33]. The capacitor is fully charged initially by reset command RST. The time variable to be stored is the gating signal of the discharge path. The capacitor will discharge when $T_{in} = 1$ (IN+ = 1/IN− = 0) and discharge will be halted when IN− = 1. To readout the stored time variable, read command RD is asserted. Note that IN+ = 1 is required during the readout operation. When the voltage of the capacitor drops below $V_{DD}/2$, $V_{reg} = 1$ will be set. Let T_{FS} be the time for v_c to drop from V_{DD} to $V_{DD}/2$ with the discharge path activated all the time and T_{reg} be the time from the assertion of read command RD to the time instant at which $V_{reg} = 1$ is set. It can be shown from the timing diagram that $T_{reg} = T_{FS} - T_{in}$. Similar to UniGDL time registers, the output of GDP time registers is the complementary of T_{in} rather than T_{in} itself. To extract T_{in}, prior knowledge of T_{FS} is needed. One convenient way is to set the duration of IN+ and IN− to $2T_{FS}$, as shown in the figure. In this case, the stored T_{in} is given by the time variable bordered by the rising edge of v_{reg} and the falling edge of IN+.

6.2.2 All-digital time adders

Time adders perform the addition of two time variables. In this section, we study five all-digital time adders, namely GDC time adders, SDU time adders, GDP time adders, UniGDL time adders, and BiGDL time adders.

6.2.2.1 GDC time adder

The GDC time register studied earlier can be used to construct a time adder, as shown in Figure 6.11 [28]. The time adder consists of two identical GDCs and two static inverters. The operation of the time adder is depicted using the timing diagram shown in the figure. The load capacitors of the two GDCs are precharged to V_{DD} before the start of an addition operation. When the front edges of T_{in1} and T_{in2} arrive, both capacitors will start to discharge and

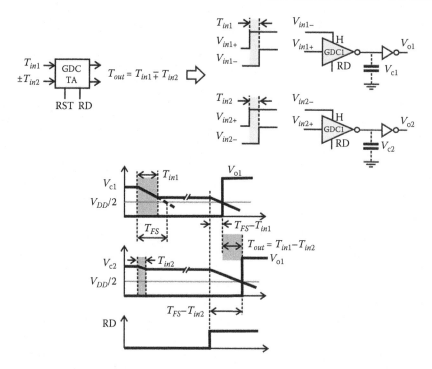

FIGURE 6.11
Time adder using gated delay cells.

the voltages of the capacitors will drop with time. Discharge processes will be halted when the rear edges of T_{in1} and T_{in2} arrive. To readout the result of the addition of T_{in1} and T_{in2}, read command RD is asserted and the discharge processes of the capacitors resume. When the voltage of the capacitors drops below $V_{DD}/2$, $v_{o1} = 1$ and $v_{o2} = 1$ will be set. The output of the addition is given by the time interval bordered by the rising edge of v_{o1} and that of v_{o2}. In order for the time adder to function properly, $T_{in1}, T_{in2} \leq T_{FS}$ is required, where T_{FS} is the time for the voltage of the capacitor to discharge from V_{DD} to $V_{DD}/2$.

6.2.2.2 SDU time adder

Time addition can be performed using the SDU time adder shown in Figure 6.12 [29, 30]. To add time variables T_{in1} and T_{in2} that are nonoverlapping, reset command RS = 0 is asserted before the arrival of time variables, allowing the capacitors of both SDUs to be charged fully. When $T_{in1} = 1$ arrives (IN1+ = 1 and IN1− = 0), we will have A = 1 and B,C = 0. Since $T_{in2} = 1$ has not arrived yet, W1,2 = 1. The capacitor of SDU1 starts to discharge while that of SDU2 remains charged. When IN1− = 1, A,B,C = 1 will be set. The discharge of the capacitor of SDU1 will be halted. When $T_{in2} = 1$

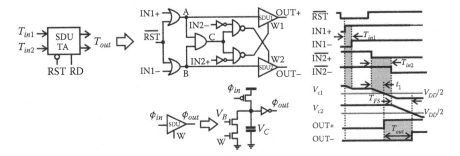

FIGURE 6.12
Time adder using a switched delay unit.

(IN2+ = 1 and IN2− = 0), we will have W1 = 1 and W2 = 0. SDU1 will resume its discharge process while the capacitor of SDU2 will remain charged. When IN2− = 1, the capacitor of SDU2 will start to discharge. When the voltage of both capacitors drops below $V_{DD}/2$, OUT+ = 1 and OUT− = 1 will be set. For SDU1, we have $T_{in1} + t_1 = T_{FS}$ where T_{FS} is the time for the capacitor to discharge from V_{DD} to $V_{DD}/2$. One can show from the timing diagram that $T_{out} = T_{in1} + T_{in2}$.

6.2.2.3 GDP time adder

Figure 6.13 shows a time adder using GDPs [33]. The time adder consists of two identical GDPs that discharge the same capacitor. The gating signals of the GDPs are the time variables to be added. The capacitor is fully charged by reset command RST before addition. The capacitor will discharge when $T_{in1} = 1$ or $T_{in2} = 1$ and retain its charge when the gating signals are absent. When the voltage of the capacitor v_c drops below $V_{DD}/2$, $V_{sum} = 1$ will be set, marking the end of the addition operation. Let T_{FS} be the time for v_c to drop from V_{DD} to $V_{DD}/2$ with only one of the two gated discharge paths

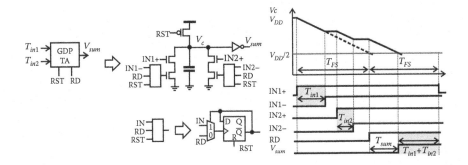

FIGURE 6.13
Time adder using gated discharge paths.

activated and T_{sum} be the time from the assertion of read command RD to the time at which $V_{sum} = 1$. It can be shown from the timing diagram that $T_{in1} + T_{in2} = T_{FS} - T_{sum}$. To extract T_{sum}, the knowledge of T_{FS} is needed. If the width of IN1+ = 1 is set to $2T_{FS}$, $T_{in1} + T_{in2}$ will be the time from the rising edge of V_{sum} to the falling edge of IN1+. It should be noted that only positive time variables are permitted. This is because IN1,2+ must arrive before IN1,2- in order for the circuit to function properly.

6.2.2.4   UniGDL time adders

The uniGDL shown in Figure 6.9 can function as a time adder that performs the addition of two nonoverlapping positive time variables T_{in1} and T_{in2}, as shown in Figure 6.14 [31, 32]. Its operation is briefly depicted here: When $T_{in1} = 1$, "1" will propagate rightwards for duration T_{in1}. During $T_{in1} = 0$, the logic state and voltage of the output of the delay stages remain unchanged. When $T_{in2} = 1$, "1" will continue to propagate for duration T_{in2}. Propagation will be halted when $T_{in2} = 0$. To readout the sum of the addition, read command RD is asserted and the propagation of "1" resumes. The addition operation will end when "1" reaches the end of the delay line. Let T_{FS} be the time for "1" to propagate from the beginning to the end of the delay line with the gating signal present all the time. Since $x = T_{in1} + T_{in2}$, the duration from the assertion of RD to the time instant at which "1" reaches the end of the delay line is given by $T_{FS} - (T_{in1} + T_{in2})$. If cascading two identical time adders as shown in Figure 6.9, we will have $T_{out} = T_{FS} - [T_{FS} - (T_{in1} + T_{in2})] = T_{in1} + T_{in2}$.

6.2.2.5   BiGDL time adders

UniGDL time adders can only perform the addition of two nonoverlapping positive time variables. To perform the addition of two nonoverlapping time variables whose polarity is arbitrary, let us consider the BiGDL line shown in Figure 6.15. Each stage consists of two identical GDCs and is capable of

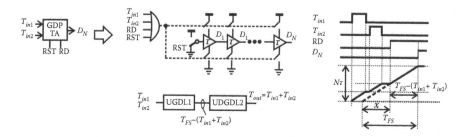

FIGURE 6.14

Time adder using a unidirectional gated delay line. When RST = 1 is asserted, $D_{1,2,...,N} = 0$.

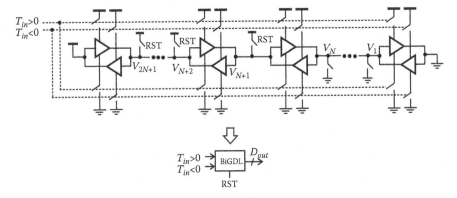

FIGURE 6.15
Bidirectional gated delay line.

propagating signals in both directions with the same gate delay, depending upon the polarity of T_{in} [34, 35]. A total of $2N+1$ identical biGDCs are needed for an output range $-N\tau \leq T_{out} \leq N\tau$, where τ is the gate delay. The voltage of nodes $N+1$, ..., $2N+1$ is initialized to V_{DD} while that of nodes $1, \ldots, N$ is initialized to 0 V. As a result, $T_{out} = 0$ corresponds to $v_1 = \cdots v_N = 0$V and $v_{N+1} = \cdots = v_{2N+1} = V_{DD}$. In addition, the leftmost node is connected to V_{DD} while the rightmost node is grounded, ensuring "1" will propagate rightwards when $T_{in} > 0$ and "0" will propagate leftwards when $T_{in} < 0$.

Figure 6.16 shows the simplified schematic of a BiGDL time adder and its timing diagram. The operation of the time adder is briefly depicted here : Time variables T_{in1} and T_{in2} to be added are nonoverlapping and their polarity is arbitrary. Assume T_{in1} arrives first. The sign block exacts both the value and sign of T_{in1}, denoted by $|T_{in1}|$ and $\text{Sign}(T_{in1})$, respectively, and route them to a 1-to-2 demultiplexer with $\text{Sign}(T_{in1})$ as the demultiplexing signal and $|T_{in1}|$ as the demultiplexed. If $T_{in1} > 0$, "1" will propagate rightwards for duration $|T_{in1}|$, otherwise, "0" will propagate leftwards for duration $|T_{in1}|$. When T_{in2} arrives, its value and sign will be determined by the sign block. "1" will continue to propagate rightwards for duration $|T_{in2}|$ if $T_{in2} > 0$ or leftwards for duration $|T_{in2}|$ is $T_{in2} < 0$. To readout the result of the addition operation, the polarity of the result of the addition T_{out} needs to be determined first. If $T_{out} > 0$, the rightmost "1" will be located in the right half of the gated delay line, otherwise, it will be located in the left half of the gated delay line. Since $v_1 = \ldots = v_N = 0$ and $v_{N+1} = \ldots = V_{2N+1} = 1$ initially, we will have $v_N = 1$ if $T_{out} > 0$, $v_{N+2} = 1$ if $v_{out} < 0$, and $v_N = 0$ & $v_{N+1} = 1$ if $T_{out} = 0$. Once the polarity of T_{out} is determined, the following operations are carried out:

1. If $T_{out} > 0$ (Figure 6.16a), we will have $\text{Sign}(T_{out}) = 1$. RD is asserted and "1" propagates rightwards until $v_1 = 1$. Let T_{FS} be the time for

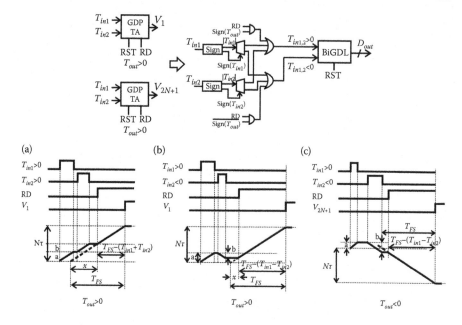

FIGURE 6.16

Bidirectional gated delay line time adder. (a) $T_{in1}, T_{in2} > 0$. (b) $T_{in1} > 0$, $T_{in2} < 0$, and $|T_{in1}| > |T_{in2}|$, (c) $T_{in1} > 0$, $T_{in2} < 0$, and $|T_{in1}| < |T_{in2}|$.

$v_{N+1} = 1$ to propagates rightwards until $v_1 = 1$. Since $a = kT_{in1}$, $b = kT_{in2}$ where k is the slope of the voltage of the load capacitor, we have $a+b = k(T_{in1}+T_{in2})$. As a result, $x = (a+b)/k = T_{in1}+T_{in2}$. The interval from the assertion of RD to $v_1 = 1$ yields $T_{FS} - (T_{in1} + T_{in2})$.

2. If $T_{out} > 0$ (Figure 6.16b), we will have Sign(T_{out})=1. RD is asserted and "1" propagates rightwards until $v_1 = 1$. Since $a = kT_{in1}$, $b = -kT_{in2}$, we have $a + b = k(T_{in1} - T_{in2})$. As a result, $x = (a + b)/k = T_{in1} - T_{in2}$. The interval from the assertion of RD to $v_1 = 1$ yields $T_{FS} - (T_{in1} - T_{in2})$.

3. If $T_{out} < 0$ (Figure 6.16c), we will have Sign $(T_{out}) = 0$. RD is asserted and "0" propagates leftwards until $v_{2N+1} = 0$. Since $a = kT_{in1}$, $b = -kT_{in2}$, we have $a + b = k(T_{in1} - T_{in2})$. As a result, $x = (a + b)/k = T_{in1} - T_{in2}$. The interval from the assertion of RD to $v_1 = 1$ yields $T_{FS} - (T_{in1} - T_{in2})$.

6.2.3 Time integrators

Integrating a time variable accumulates its value recursively. The accumulation process consists of two separate operations : time addition that performs

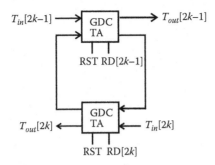

FIGURE 6.17
Time integrator utilizing GDC time adders.

$$\sum_{j=1}^{k} T_{in}[j] = T_{in}[k] + \sum_{j=1}^{k-1} T_{in}[j]$$ and time registration that stores $\sum_{j=1}^{k} T_{in}[j]$ and releases it upon a read request. This section studies all-digital time integrators realized using GDCl time adders, SDUs and ring oscillators, GDP time adders and registers, and BiGDLs.

6.2.3.1 Time integrator using GDCs

The preceding GDC time adder and register can be utilized to construct an all-digital time integrator. Figure 6.17 shows the simplified schematic of such a time integrator [8]. The time integrator consists of two back-to-back connected time adders that are operated in a time-interleaved manner to perform time accumulation or integration.

6.2.3.2 Time integrator using SDUs

If an SDU is placed in a ring oscillator, as shown in Figure 6.18, the period of the oscillator can be altered by changing the duration of the absence of the write command of the SDU [30]. To illustrate this, let the delay of the inverters be τ_{inv} and the high-to-low propagation delay of the SDU be τ_{SDU}. The low-to-high delay of the SDU is assumed to be the same as that of the inverters. Consider the discharge process of the capacitor. When v_c drops below $V_{DD}/2$, v_{11} will start to rise. When v_{11} exceeds $V_{DD}/2$, v_{21} will start to drop. Continuing this process, we arrive at the time at which v_{41} exceeds $V_{DD}/2$. If the write command of the SDU is present, v_{c1} will drop, otherwise, v_{c1} will remain unchanged and start to drop only after the duration of the absence of the write command.

Let us consider two identical ring oscillators, each consists of an SDU and three static inverters. The output of the time integrator T_{out} is the time difference between the rising edges of the output of the oscillators. An SDU time adder is employed to add the time variable T_{in} to be integrated and the output of the time integrator. A logic circuit (not shown) is used to ensure that

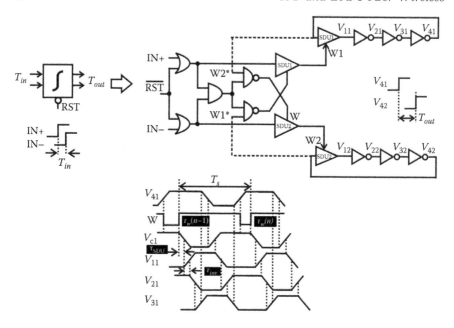

FIGURE 6.18
Time integrator using an SDU and a pair of SDU-embedded ring oscillators.

the time difference between W1* and W2* is $T_{out}(n)$. As a result, $T_{out}[k+1] = T_{out}[k] + T_{in}[k]$. Time integration is therefore performed. Since T_{out} is defined as the time difference between the rising edges of the two oscillators, the range of the time integrator is upper bound by the period of the oscillators. Note that not only mismatches between the oscillators give rise to timing errors, phase accumulation intrinsic to oscillators is another source of timing errors. The need for four SDUs and two ring oscillators makes the time integrator less attractive for low-power applications.

6.2.3.3 Time integrator using GDP time adder and register

Time integration can also be performed using one GDP time adder and one GDP time register, as shown in Figure 6.19. We briefly depict its operation here : At the assertion of reset command RST, the output of the time adder V_{sum} and that of the time register V_{reg} are set to zero. When $T_{in}[1]$ arrives, the load capacitor of the time adder will start to discharge. Since read command RD1 of the time adder is asserted T_{FS} after the rising edge of IN1+ and the time duration bordered by the rising edge of RD1 and that of V_{sum} is the complementary of $T_{in}[1]$, e.g., $T_{FS} - T_{in}[1]$, the time interval bordered by the rising edge of V_{sum} and that of $\overline{IN1+}$ is $T_{in}[1]$. For input port 2 of the time adder, since V_{reg} is 0 for the most part of the integration cycle, the discharge path associated with input port 2 of the time adder is disabled. As

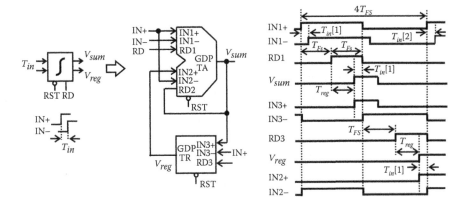

FIGURE 6.19
Time integrator using a gated discharge path time adder and a gated discharge path time register.

a result, towards the end of the first integration cycle, the output of the time integrator is given by $T_{out}[1] = T_{in}[1]$. Note that T_{out} is the time interval bordered by the rising edge of V_{sum} and that of $\overline{IN1}+$. In the second half of the integration cycle where IN1+ = 0, $T_{in}[1]$ is fed to the time register. With RD3 asserted T_{FS} after the rising edge of IN3−, the time interval bordered by rising edge of V_{reg} and that of the second integration cycle of IN1+ is $T_{in}[1]$. It is evident that the time register not only stores $T_{in}[1]$ but also releases it to the input port 2 of the time adder just before the arrival of the rising edge of the second integration cycle. In the second integration cycle, $T_{in}[1]$ is fed to the input port 2 of the time adder. It is followed by $T_{in}[2]$ fed to the input port 1 of the time adder. The time adder performs $T_{out}[2] = T_{in}[1] + T_{in}[2]$. The result is the time interval bordered by the rising edge of V_{sum} and that of $\overline{IN1}+$. It is fed to the time register in the second half of the integration cycle. This process continues and integrating $T_{in}[k]$ is performed. Since it is based on the GDP time adder and register, only positive time variables can be integrated. The need for explicit addition and registration operations per integration increases power consumption and integration time.

6.2.3.4 Time integrator using BiGDL

The BiGDL studied earlier can be used as a time integrator with no constraint on the polarity of the time variable to be integrated. Time variable T_{in} to be integrated is the gating signal of the BiGDL. Integrating T_{in} is performed via the propagation of "1" in the BiGDL. The output of the time integrator is a thermometer code $1 \ldots 10 \ldots 0$ with the location of the rightmost "1" set by the output of integration. If the rightmost "1" is on the right side of the middle

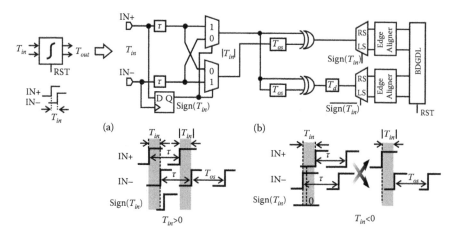

FIGURE 6.20
Bidirectional gated delay line time integrator.

node (node N+1), $T_{out} > 0$, otherwise, $T_{out} < 0$. To integrate T_{in}, an absolute-value block that generates both the value and polarity of T_{in}, denoted by $|T_{in}|$ and $\text{Sign}(T_{in})$, respectively, is employed, as shown in Figure 6.20. Both IN+ and IN− are delayed by τ, where $\tau \geq |T_{in}| + \tau_{DFF}$. Such a delay is necessary as the sign of T_{in} is determined by sampling IN+ using IN−. If $T_{in} > 0$ (IN+ leads IN− by T_{in}), by the time that the sign signal is generated, the signals to be multiplexed are already available. However, if $T_{in} < 0$ (IN− leads IN+ by T_{in}), the sign signal is generated even though the rising edge of IN+ might have not arrived yet. The polarity of T_{in} sets the direction of the propagation of "1" while its value sets the distance of propagation. If T_{in} is smaller than the minimum width of the gating signal T_{gate}, which is the minimum amount of time required to switch on the gated transistors, the GDL will not function. A constant time offset T_{os} is therefore added to T_{in} before a gating operation with $T_{os} \geq T_{gate}$. To integrate T_{in}, "1" propagates rightwards for duration $T_{in} + T_{os}$ first performing the integration of $T_{in} + T_{os}$ and then leftwards for duration T_{os} performing the integration of $-T_{os}$, thereby yielding the result of the integration of T_{in}. Since the left-shift operation can only be carried out after the right-shift operation is completed, the left-shift operation is delayed by T_d, where $T_d \geq T_{os} + |T_{in,max}|$. The edge aligners generate nonoverlapping commands RS and $\overline{RS}$ for right-shift and LS and $\overline{LS}$ for left-shift.

The output range of the time integrator is the time delay of half the BiGDL and can be made large once the delay line is sufficiently long. In reality, it is constrained by power consumption, silicon area, and gate-delay mismatch induced nonlinearity. The time integrator enjoys rapid integration and low power consumption accredited to its simple configuration.

6.3 All-Digital $\Delta\Sigma$ TDCs

In this section, an all-digital 1-1 MASH $\Delta\Sigma$ TDC utilizing GDP time integrators and registers and an all-digital first-order $\Delta\Sigma$ TDC utilizing differential BiGDL time integrators are presented.

6.3.1 1-1 MASH $\Delta\Sigma$ TDC using GDP time integrator and register

Figure 6.21 shows the configuration of a differential GDP time integrator. It consists of two single-ended GDP time integrators. Note the polarity of the time variable to be integrated in the two single-ended time integrators differs. The 1-1 MASH $\Delta\Sigma$ TDC shown in Figure 6.21 consists of two identical $\Delta\Sigma$ TDCs. The quantization error of the coarse TDC is extracted from the output of the differential time integrator of the coarse TDC. The value of T_{FB2} is determined based on the maximum output of the time integrator of the coarse TDC.

A differential sinusoidal time input of frequency 415 kHz and amplitude 50 ps is fed to the TDC. The sampling frequency is 25 MHz. The TDC was designed in an IBM 130 nm 1.2 V CMOS technology and analyzed using Spectre from Cadence Design Systems with BSIM4 device models. The spectrum of the TDC is plotted in Figure 6.22, together with that from behavioral analysis. It is seen that the TDC exhibits second-order noise shaping. Also observed is that the in-band noise floor of the TDC is approximately 20 dB higher when compared with that of the ideal TDC, arising from the imperfections of the TDC. The signal-to-noise ratio of the TDC is 38.66 dB over $48-415$ kHz. The Effective Number of Bits (ENOB) of the TDC is 6.13 and its resolution is 1.45 ps.

6.3.2 $\Delta\Sigma$ TDC with differential BiGDL time integrator

Figure 6.23 shows the simplified schematic of a $\Delta\Sigma$ TDC utilizing differential BiGDL time integrators. The gating signals are $\hat{T}_{err}^+ = T_{os} + |T_{err}|$ and $\hat{T}_{err}^- = T_{os} - |T_{err}|$. To integrate T_{err}^+, "1" propagates rightwards to integrate $T_{os} + |T_{err}|$ and leftwards to remove the result of the integration of T_{os}. T_d is required for reasons given earlier. The shift block generates right-shift commands (RS and $\overline{RS}$) and left-shift commands (LS and $\overline{LS}$). Similar operations are carried out for integrating T_{err}^- as well. The quantizer will output 1 if T_{out}^+ leads T_{out}^- and 0 otherwise. Each BiGDL has 32 stages. Since the leftmost stage is connected to 1.2 V and the rightmost stage is connected to 0 V, only 15 stages on each side of the middle node are used for time integration. The location of the rightmost "1" is determined using the logic preceding the DTCs. Once it is determined, T_{out} is obtained using DTCs for time quantization. The $\Delta\Sigma$ TDC was designed in an IBM 130 nm 1.2 V CMOS technology. The input of

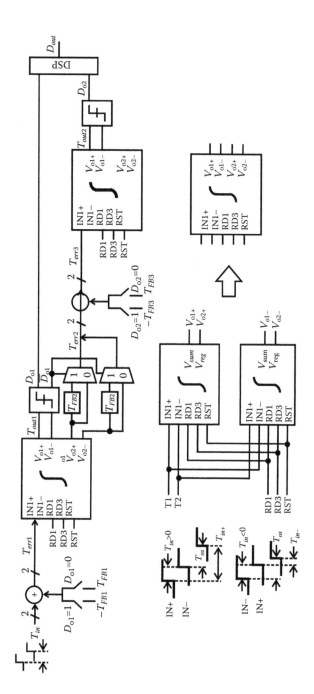

FIGURE 6.21

1-1 MASH time-mode $\Delta\Sigma$ TDC utilizing differential GDP time integrators.

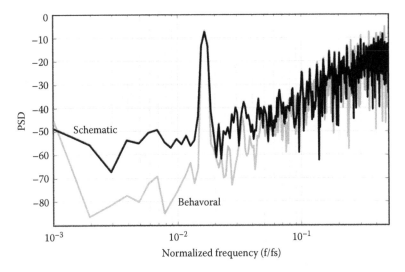

FIGURE 6.22
Spectrum of 1-1 MASH time-mode $\Delta\Sigma$ TDC with GDP time integrators and registers. 1024 samples with Hanning window (Copyright ©IEEE).

the modulator is a sinusoidal time signal of frequency 231 kHz and amplitude 430 ps. The sampling clock is 25 MHz. Figure 6.24 shows the spectrum of the TDC. For comparison purpose, the spectrum of the same TDC with a single-ended BiGDL time integrator is also shown. First-order noise shaping is evident. Also observed is a second-order harmonic tone that is greatly reduced with the differential BiGDL time integrator. The Signal to Noise + Distortion Ratio (SNDR) of the TDC is 34.6 dB over frequency range from the corner frequency of flicker noise to the second harmonic.

6.4 Conclusions

This chapter began with a close examination of open-loop noise-shaping TDCs. GRO TDCs, gated relaxation oscillator TDCs, and switched ring oscillator TDCs were studied. GRO-based TDCs suffer from inherent skew errors caused by clock feedthrough of gating signals, charge injection from gated transistors, and disturbances coupled with neighboring devices and gating errors arising from their inability to emerge from a holding state instantaneously. SRO-based TDCs are largely immune from skew errors and less affected by gating errors at the expense of more power consumption. Although gated relaxation oscillator TDCs are less prone to skew errors, they are unsuitable

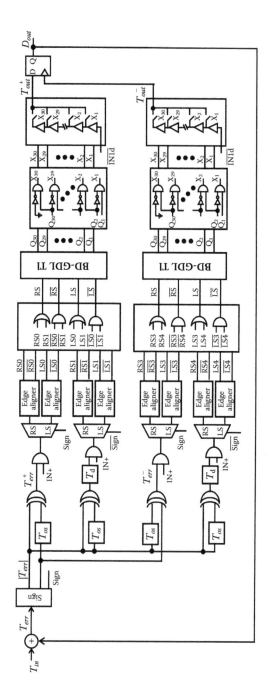

FIGURE 6.23
First-order $\Delta\Sigma$ TDC with differential BiGDL time integrators.

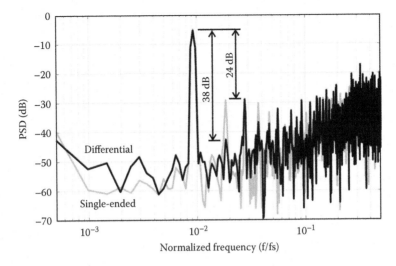

FIGURE 6.24
Spectrum of first-order $\Delta\Sigma$ TDC with differential BiGDL time integrators.
2048 samples with Hanning window (Copyright ©IEEE).

for applications constrained by a large oversampling ratio. The second part
of the chapter focused all-digital closed-loop noise-shaping TDCs. All-digital
time registers including GDC time registers, SDU time registers, UniGDL time
registers, and GDP time registers were investigated. Mismatches existing in
GDC and SDU time registers affect performance. UniGDL time registers feature a low silicon area and a virtually unlimited range. GDP time registers
can only handle positive time variables. All-digital time adders including GDC
time adders, SDU time adders, GDP time adders, UniGDL time adders, and
BiGDL time adders were investigated. The characteristics of these time adders
are similar to those of the corresponding time registers. GDP time adders only
operate on positive time variables. The same hold for UniGDL adders. Over-
lapping between time variables is not allowed in UniGDL and BiGDL time
adders. BiGDL time adders permit both positive and negative time variables.
All-digital time integrators realized using GDC time adders, SDUs and ring
oscillators, GDP time adders and registers, and BiGDLs were investigated.
SDU time integrators are power hungry as each SDU time integrator is made
of four SDUs and two ring oscillators. Mismatches between the ring oscillators
also affect performance. Further, the output range of SDU time integrators
is limited by the period of the oscillators. GDP time integrators do not permit negative time variables. Since one time adder and one time register are
needed, these time integrators are also power hungry. BiGDL time integrators
consume the least amount of power and silicon area and are capable of performing rapid time integration. Finally, the design and simulation results of
an all-digital 1-1 MASH $\Delta\Sigma$ TDC with differential GDP time integrators and

an all-digital first-order $\Delta\Sigma$ TDC with differential BiGDL time integrators were presented.

Acknowledgement - Financial support from Natural Science and Engineering Research Council of Canada and computer-aided design tools from CMC Microsystems, Kingston, ON, Canada are gratefully acknowledged by the author.

Bibliography

[1] F. Yuan, Ed., *CMOS time-mode circuits and systems: Fundamentals and applications.* New York: CRC Press, 2015.

[2] T. Yoshiaki and A. Takeshi, "Simple voltage-to-time converter with high linearity," *IEEE Trans. Instrument. Meas.*, vol. 20, no. 2, pp. 120–122, May 1971.

[3] D. Porat, "Review of sub-nanosecond time-interval measurements," *IEEE Trans. Nuclear Sci.*, vol. NS-20, pp. 36–51, Sept. 1973.

[4] M. Straayer and M. Perrott, "A 12-bit, 10-MHz bandwidth, continuous-time $\Delta\Sigma$ ADC with a 5-bit, 950-MS/s VCO-based quantizer," *IEEE J. Solid-State Circuits*, vol. 43, no. 4, pp. 805–814, Apr. 2008.

[5] ——, "A multi-path gated ring oscillator TDC with first-order noise shaping," *IEEE J. Solid-State Circuits*, vol. 44, no. 4, pp. 1089–1098, Apr. 2009.

[6] M. Park and M. Perrott, "A single-slope 80 Ms/s ADC using two-step time-to-digital conversion," in *IEEE International Symposium on Circuits and System*, pp. 1125–1128, Taipei, Taiwan, May 2009.

[7] T. Tokairin, M. Okada, M. Kitsunezuka, T. Maeda, and M. Fukaishi, "A 2.1-to-2.8-GHz low-phase-noise all-digital frequency synthesizer with a time-windowed time-to-digital converter," *IEEE J. Solid-State Circuits*, vol. 45, no. 12, pp. 2582–2590, Dec. 2010.

[8] J. Hong, S. Kim, J. Liu *et al.*, "A 0.004 mm^2 250μW $\Delta\Sigma$ TDC with time-difference accumulator and a 0.012 mm^2 2.5 mW bang-bang digital PLL using PRNG for low-power SoC applications," in *IEEE International Conference Solid-State Circuits Digest of Technical Papers*, pp. 240–242, San Francisco, CA, Feb. 2012.

[9] T. Rahkonen, J. Kostamovaara, and S. Saynajakangas, "Time interval measurements using integrated tapped CMOS delay lines," in *Proceedings IEEE Mid-West Symposium on Circuits and System*, pp. 201–205, Champaign, IL, Aug. 1990.

[10] Y. Aria, T. Matsumura, and K. Endo, "A CMOS four-channel × 1K memory LSI with 1-ns/b resolution," *IEEE Trans. Circuits Syst. II*, vol. 27, no. 3, pp. 359–364, Mar. 1992.

[11] T. Rahkonen and J. Kostamovaara, "The use of stabilized CMOS delay lines in the digitization of short time intervals," *IEEE J. Solid-State Circuits*, vol. 28, no. 8, pp. 887–894, Aug. 1993.

[12] C. Gray, W. Liu, W. Noije, T. Hughes, and R. Cavin, "A sampling technique and its CMOS implementation with 1 Gb/s bandwidth and 25 ps resolution," *IEEE J. Solid-State Circuits*, vol. 29, pp. 340–349, Mar. 1994.

[13] C. Ljuslin, J. Christiansen, A. Marchioro, and O. Klingsheim, "An integrated 16-channel CMOS time to digital converter," *IEEE Trans. Nuclear Sci.*, vol. 41, no. 4, pp. 1104–1108, Aug. 1994.

[14] Y. Park and F. Yuan, "Two-step pulse-shrinking time-to-digital converter," *Microelectron. J.*, vol. 60, pp. 45–54, Feb. 2017.

[15] Y. Park and F. Yuan, "Time-interleaved pulse-shrinking time-to-digital converter with reduced conversion time," *Analog Integr. Circuits Signal Process.*, vol. 91, no. 3, pp. 385–398, June 2017.

[16] S. Lee, B. Kim, and K. Lee, "A novel high-speed ring oscillator for multiphase clock generation using negative skewed delay scheme," *IEEE J. Solid-State Circuits*, vol. 32, no. 2, pp. 289–291, Feb. 1997.

[17] F. Yuan, *CMOS current-mode circuits for data communications*. New York: Springer, 2006.

[18] G. Taylor and I. Galton, "A mostly-digital variable-rate continuous-time delta-sigma modulator ADC," *IEEE J. Solid-State Circuits*, vol. 45, no. 12, pp. 2634–2646, Dec. 2010.

[19] K. Hwang and L. Kim, "An area efficient asynchronous gated ring oscillator TDC with minimum GRO stages," in *Proceedings of the IEEE International Symposium on Circuits and System*, pp. 3973–3976, Paris, France, May 2010.

[20] A. Elshazly, S. Rao, B. Young, and P. Hanumolu, "A 13b 315 $f_{s,rms}$ 2 mW 500 MS/s 1 MHz bandwidth highly digital time-to-digital converter using switched ring oscillators," in *International Solid-State Circuits Conference Digest of Technical Papers*, pp. 464–465, San Francisco, CA, Feb. 2012.

[21] T. Konishi, K. Okumo, S. Izumi, M. Yoshimoto, and H. Kawaguchi, "A 61 dB SNDR 700μm second-order all-digital TDC with low-jitter frequency shift oscillator and dynamic flipflops," in *Symposium on VLSI Circuits Digest of Technical Papers*, pp. 190–191, Honolulu, Hawaii, June 2012.

[22] P. Lu, Y. Wu, and P. Andreani, "A 90 nm CMOS digital PLL based on vernier-gated-ring-oscillator time-to-digital converter," in *Proceedings of the IEEE International Symposium on Circuits and System*, pp. 2593–2596, Seoul, South Korea, May 2012.

[23] M. Lee and A. Abidi, "A 9B, 1.25 ps resolution coarse-fine time-to-digital converter in 90 nm CMOS that amplifies a time residue," *IEEE J. Solid-State Circuits*, vol. 43, no. 4, pp. 769–777, Apr. 2008.

[24] M. Hovin, A. Olsen, T. Lande, and C. Toumazou, "Delta-sigma modulators using frequency-modulated intermediate values," *IEEE J. Solid-State Circuits*, vol. 32, no. 1, pp. 13–22, Jan. 1997.

[25] F. Yuan and N. Soltani, "A low-voltage low vdd sensitivity relaxation oscillator for passive wireless microsystems," *IET Electronics Letters*, vol. 45, no. 21, pp. 1057–1058, Oct. 2009.

[26] Y. Cao, P. Leroux, W. D. Cock, and M. Steyaert, "A 1.7 mW 11b 1-1-1 MASH $\Delta\Sigma$ time-to-digital converter," in *IEEE International Solid-State Circuits Conference Digest of Technical Papers*, pp. 480–481, San Francisco, CA, Feb. 2011.

[27] T. Konishi, K. Okumo, S. Izumi, M. Yoshimoto, and H. Kawaguchi, "A 51 dB SNDR DCO-based TDC using two-stage second-order noise shaping," in *Proceedings of IEEE International Symposium on Circuits and System*, pp. 3170–3173, Seoul, South Korea, May 2012.

[28] S. Kim, *Time domain algebraic operation circuits for high performance mixed-mode system.* MS Thesis, Korean Advanced Institute of Science and Technology, 2010.

[29] M. Ali-Bakhshian and G. Roberts, "Digital storage, addition and subtraction of time-mode variables," *IET Electron. Lett.*, vol. 47, no. 16, pp. 910–911, Aug. 2011.

[30] M. Ali-Bakhshian and G. Roberts, "A digital implementation of a dual-path time-to-time integrator," *IEEE Trans. Circuits Syst. I*, vol. 59, no. 11, pp. 2578–2591, Nov. 2012.

[31] K. Kim, Y. Kim, W. Yu, and S. Cho, "A 7b 3.75 ps resolution two-step time-to-digital converter in 65 nm CMOS using pulse-train time amplifier," *IEEE J. Solid-State Circuits*, vol. 48, no. 4, pp. 1009–1017, Apr. 2013.

[32] K. Kim, W. Yu, and S. Cho, "A 9 bit, 1.12 ps resolution 2.5 b/stage pipelined time-to-digital converter in 65nm CMOS using time-register," *IEEE J. Solid-State Circuits*, vol. 49, no. 4, pp. 1007–1016, Apr. 2014.

[33] Y. Park, D. Amor, and F. Yuan, "Time integrator for mixed-mode signal processing," in *Proceeding of the IEEE International Symposium on Circuits and System*, pp. 826–829, Motreal, Quebec, May 2016.

[34] Y. Park and F. Yuan, "Low-power all-digital delta-sigma TDC with bi-directional gated delay line time integrator," in *IEEE Mid-West Symposium on Circuits and Systems*, Accepted. pp. 679–682, Boston, MA, Aug. 2017.

[35] Y. Park and F. Yuan, "All-digital delta-sigma TDC with differential bi-directional gated-delay-line time integrator," in *IEEE Mid-West Symposium on Circuits and Systems*, Accepted. pp. 1513–1516, Boston, MA, Aug. 2017.

7

Power-Efficient CMOS Power Amplifiers for Wireless Applications

Haoyu Qian

Qualcomm Technologies Inc.

Suraj Prakash and Jose Silva-Martinez

Texas A&M University

CONTENTS

7.1 Introduction to CMOS Power Amplifiers for Wireless Applications

The continuous advancement of semiconductor technologies, especially complimentary metal-oxide-semiconductor (CMOS) technology, has enabled an exponential growth of the wireless communication industry. Future generations of wireless communication call for even further levels of integration, and for now, the only circuit block that is rarely integrated in CMOS along with other parts of the system is the RF power amplifier (PA). RF PA design involves handling large voltages and large amounts of current at radio frequencies, which in the present wireless communication standards are in the gigahertz range.

In this chapter, nonlinear effects of the PA as well as power-efficient CMOS PA are systematically presented. An analysis of nonlinear effects proposes a formal mathematical description of multitone nonlinearity and its relationship with two-tone tests. A thorough analysis of a current mode digitally predistorted RF PA for a wideband code-division multiple access (WCDMA) application standard is described and implemented in CMOS. The digital (gain)–analog (current reduction) technique resolves the main tradeoff between power efficiency and linearity.

7.1.1 Overview

Because of the exponential increase in the wireless communication market, the trend has been towards lower cost and power radio frequency (RF) transceivers. A typical transmitter is displayed in Figure 7.1.

PA is one of the major power consumers in the RF transceiver [1], and the design and implementation of highly efficient CMOS. PA has been an active research and development area during the last few years [2–4]. The 3–5 G communication standards use a high data rate and bandwidth-efficient modulations that result in a high peak-to-average power ratio (PAPR); thus, the use of linear transmitters is highly recommended. As a result of the high PAPR in such modulations during orthogonal frequency-division multiplexing (OFDM), the probability density function (PDF) of the transmitted power will peak in the power back-off (PBO) region. However, the power efficiency of linear PAs reaches its maximum performance at the peak output power and drops drastically in the PBO region. The PA usually operates in the low-efficiency region most of the time.

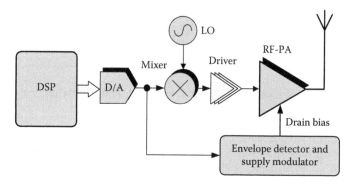

FIGURE 7.1
Simplified schematic of a typical direct conversion transmitter.

Envelope tracking [3–6] and PA segmentation [2, 7–15] are two efficient enhancement techniques that have gained much interest recently. However, the envelope tracking system is becoming less effective in advanced CMOS technologies as the power supply scales down, which is shown in Figure 7.1. The relatively low drain-source voltage allowed by the technology limits the benefits of this approach. Additionally, wide bandwidth standards require wideband agile switching regulators, which serve as a tradeoff between regulator power efficiency, switch size, component rating, and switching ripple [5].

On the other hand, the use of on-chip transformers in segmented PAs can result in unreliable impedance matching. These segmentations must be accompanied by a tunable impedance matching network that makes these solutions sensitive to process–voltage–temperature variations [11, 12]. The PA based on digital-to-analog converter (DAC) switching used in polar PAs is an interesting approach to be further exploited in this chapter's design.

This chapter describes the design of a 1.9 GHz linear-segmented PA. To improve the efficiency in the PBO region, a combination of PA segmentation and digital signal processing are employed. The PA sections are directly connected to the output impedance matching network equipped with class AB common-mode feedback (CMFB) mechanism to reduce common-mode variations when (de)activating the PA segments. The digital prewarping scheme improves the signal-to-noise ratio of the solution under PBO conditions.

This chapter is organized as follows. Section 7.2 provides a thorough analysis of the nonlinear effects of PAs, especially a closed-form analysis of the relationship between multitone adjacent-channel leakage ratio (ACLR) and two-tone intermodulation distortions. In Section 7.3, the digitally aided architecture is described in detail, and an in-depth analysis of the impact of linearity due to timing mismatch is carried out. The design of the PA building blocks is presented in Section 7.4, and the measurement results and discussions are presented in Section 7.5. Finally, the conclusions are presented in Section 7.6.

7.2 Nonlinear Effects of Power Amplifier

7.2.1 Linearity limitations

Conventional modulations, such as frequency modulation, frequency-shift keying, and Gaussian minimum-shift keying have their information stored in the frequency and/or phase variations—not in the envelope amplitude, and thus do not require linear power amplification. If a signal's envelope amplitude varies then linear amplification is required [16]. Multicarrier modulations such as OFDM present amplitude variations that need to be preserved throughout the transceiver chain; therefore, linear PAs are required. OFDM has found its popularity in wideband digital communications such as the wireless local access network, digital television, and fourth generation long-term evolution, due to its multicarrier nature that enables more reliable transmission and reception compared with single-carrier schemes [17]. Lack of PA linearity would result in distortion of the amplified signal, which results in unwanted components in frequencies apart from the desired designated frequency bands.

7.2.2 Multitone nonlinearity

Linearity is one of the major concerns in current and near future wireless communication systems. The requirements on spectral leakage, often in the form of an ACLR, are usually the most demanding specifications for RF-integrated circuit design. Because of the nonlinearity of the RF front-end, the shape of the input signal is not preserved, which means the spectrum is not limited to the desired bandwidth. This effect is termed spectral regrowth [18, 19] and is quantified by ACLR, which is defined as the ratio of the integrated power in the adjacent channel to the power in the transmitted channel [20].

Because of the quasirandom nature of signals used in communications, it is not easy to analytically relate ACLR to PA design parameters. Multitone signals are similar to the band-limited signals used in communication systems in the frequency domain, but its simulation is still relatively expensive. Depending on the number of tones at the input source, it may not be well-supported by the simulator [21]. This subsection presents an analysis that relates a multitone ACLR to two-tone, third-order intermodulation distortion (IMD3) of a nonlinear system, which enables a quick estimate of the PA spectral regrowth using a fast two-tone simulation in the early design phase.

Consider an input signal that consists of N identical tones evenly spaced in the frequency domain:

$$v_{\text{in}} = v_a \sum_{n=1}^{N} \cos\left[\omega_0 + (n-1)\,\Delta\omega\right] t \tag{7.1}$$

where $\Delta\omega$ denotes the angular frequency spacing. Rearranging (7.1) reveals more clearly that the normalized input resembles a modulated signal:

$$\frac{v_{\text{in}}}{v_a} = \frac{1}{2}\left(\sum_{n=1}^{N} \exp\left\{j\left[\omega_0 + (n-1)\,\Delta\omega\right]t\right\} + \sum_{n=1}^{N} \exp\left\{j\left[-\omega_0 - (n-1)\,\Delta\omega\right]t\right\}\right)$$

$$= \frac{1}{2}\left(e^{j\omega_0 t}\frac{1 - e^{jN\Delta\omega t}}{1 - e^{j\Delta\omega t}} + e^{-j\omega_0 t}\frac{1 - e^{-jN\Delta\omega t}}{1 - e^{-j\Delta\omega t}}\right)$$

$$= \frac{1}{2}\left(e^{j\omega_0 t}\frac{e^{jN\frac{\Delta\omega}{2}t}}{e^{j\frac{\Delta\omega}{2}t}}\frac{e^{-jN\frac{\Delta\omega}{2}t} - e^{jN\frac{\Delta\omega}{2}t}}{e^{-j\frac{\Delta\omega}{2}t} - e^{j\frac{\Delta\omega}{2}t}}\right.$$

$$+ e^{-j\omega_0 t} \frac{e^{-jN\frac{\Delta\omega}{2}t}}{e^{-j\frac{\Delta\omega}{2}t}} \frac{e^{jN\frac{\Delta\omega}{2}t} - e^{-jN\frac{\Delta\omega}{2}t}}{e^{j\frac{\Delta\omega}{2}t} - e^{-j\frac{\Delta\omega}{2}t}} \Bigg) \qquad (7.2)$$

$$= \frac{\sin\left(N\frac{\Delta\omega}{2}t\right)}{\sin\left(\frac{\Delta\omega}{2}t\right)} \cos\left(\omega_0 + \frac{N-1}{2}\Delta\omega\right)t$$

Clearly $\left(\omega_0 + \frac{N-1}{2}\Delta\omega\right)$ is the center frequency of the frequency band from ω_0 to $(\omega_0 + (N-1)\Delta\omega)$, which can be viewed as the carrier frequency; hence, the N-tone signal would have an "envelope" of

$$v_{\text{env}} = v_a \frac{\sin\left(N\frac{\Delta\omega}{2}t\right)}{\sin\left(\frac{\Delta\omega}{2}t\right)} \qquad (7.3)$$

Using the L'Hopital's rule, it can be shown that

$$v_{\text{env,max}} = \lim_{t\to 0} v_a \frac{\sin\left(N\frac{\Delta\omega}{2}t\right)}{\sin\left(\frac{\Delta\omega}{2}t\right)} = N v_a \qquad (7.4)$$

To calculate the root-mean-square (RMS) value of v_{in} composed of N equal-magnitude tones, Parseval's identity is used, leading to

$$v_{\text{in,rms}} = \sqrt{2 v_a^2 \sum_{n=1}^{N} \left(\frac{1}{2}\right)^2} = \sqrt{\frac{N}{2}} v_a \qquad (7.5)$$

Combining the results from (7.4) to (7.5), the PAPR of an N-tone signal is computed as

$$\text{PAPR} = \frac{P_{\text{max}}}{P_{\text{avg}}} = \frac{v_{\text{env, max}}^2/2}{v_{\text{in,rms}}^2} = N \qquad (7.6)$$

where a normalized resistance of 1 Ω is assumed; thus, a modulated signal with a PAPR of 10 dB can be emulated by a 10-tone signal. Although the frequency spectrum of a multitone signal may have a similar profile to that of a modulated signal with similar PAPR, it is still different from the continuous band-limited signal's spectrum due to the discrete nature of its spectrum. This is evident when the signals go through a nonlinear system, and a spectral regrowth is observed. Even when two signals show similar ACLR, the spectra can still have a different shape.

The second-order distortion would not contribute to tones in the adjacent channels for high-IF narrow-band signals; therefore, only third-order distortions are considered in analyzing multitone adjacent channel powers (ACP). The third-order distortion output is

$$i_{\text{out3}} = g_3 v_{\text{in}}^3 = g_3 v_a^3 \left(\sum_{k=1}^{N} \cos\omega_k t\right) \left(\sum_{l=1}^{N} \cos\omega_l t\right) \left(\sum_{m=1}^{N} \cos\omega_m t\right) \qquad (7.7)$$

Next, the amplitude of each tone in the adjacent channel is calculated, and the ACP is the sum of their output power. The analysis of this function is cumbersome, and the details can be found in Ref. [22]; This section is restricted to only the most relevant results. If a two-tone signal with equal amplitude of v_a is applied to the PA, the IMD3 is expressed in decibels as

$$\text{IMD3}|_{\text{dB}} = 10 \log \left(\frac{3}{4} \left| \frac{g_3}{g_1} \right| v_a^2 \right)^2 \tag{7.8}$$

It can be shown [22] that the ACLR of an N-tone signal that is applied to the PA can be related to a two-tone IMD3 described by (7.8) as:

$$\text{ACLR}_{\text{max}} = 10 \log \frac{P_{\text{adj}}}{P_{\text{ch}}} = \text{IMD3} - 10 \log N + 10 \log F(N) \tag{7.9}$$

where

$$F(N) = \frac{1}{4} \left[\frac{1}{5} (N-1)^5 + (N-1)^4 + \frac{5}{3} (N-1)^3 + (N-1)^2 + \frac{2}{15} (N-1) \right]$$

Analysis in Ref. [23] assumes that the phases of all distortion components are random. Such an assumption provides a lower limit of ACLR:

$$\text{ACLR}_{\text{min}} = \text{IMD3} - 10 \log N + 10 \log G(N) \tag{7.10}$$

where

$$G(N) = \frac{1}{12} \left[4N \left(N^2 - 1 \right) - 3 \left(N^2 - N \bmod 2 \right) \right]$$

Therefore, in the design phase, a simple two-tone simulation can be used to get IMD3, and from (7.9) to (7.10), one can obtain a rough idea of the multitone ACLR range, which indicates what could happen in the corresponding ACLR as a result of a modulated input signal that has a similar PAPR.

To verify the analysis, an RF PA was designed and fabricated in Taiwan Semiconductor Manufacturing Company (TSMC) 40 nm CMOS. The linear PA operates at 1.9 GHz, with a measured continuous-wave (CW) saturated output power of $P_{\text{SAT}} \approx 35$ dBm and a power gain of 38 dB. For more details on the design and implementation of PA, readers are referred to Section 7.4 and Refs. [14, 24].

Based on the two-tone test and thus the corresponding IMD3, (7.9) can be computed and compared with simulation results. The results are shown in Figure 7.2.

Note that, as a reference, the ACLR predicted by (7.10) is also included in the plot. As expected, the simulation results are better than the results of (7.9) but worse than that predicted by (7.10). As previously mentioned, the purpose of this analysis is to provide a quick estimate of the ACLR at the early design phase; therefore, a prediction of the ACLR range should suffice.

As a second testbed, an OFDM signal with a PAPR of about 9 dB is applied to the PA and the output power spectrum is measured. The normalized output spectrum is calculated in Ref. [22] and the corresponding simulation

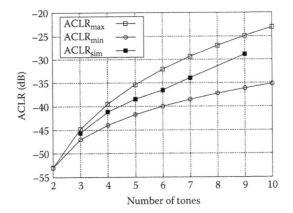

FIGURE 7.2
Simulated and predicted multitone adjacent channel leakage ratio (ACLR) as a function of number of in-band tones.

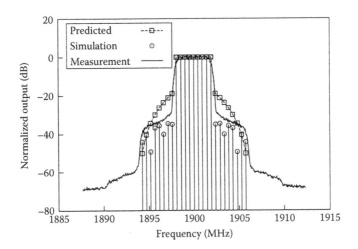

FIGURE 7.3
Multitone output spectrum: theoretical, simulated, and experimental results.

results are compared with the measured PA output spectrum, as shown in Figure 7.3.

As expected, the theoretical calculations from (7.9) represent the worst case, and the simulation results give a better nonlinearity performance. The ACLR turns out to be similar, which again validates the proposed methodology. The simulated nine-tone ACLR is about -30 dBc, whereas the measurement result is -33 dBc ACLR for the OFDM signal.

7.3 Power Amplifier Architecture

Since most communication systems in 3G and onward have a Gaussian distribution power transmission PDF as a function of output power in dBm, the architecture targeted in such communication systems partition the signal in a linear-in-dB manner to maximize its effectiveness. The best power efficiency in current PAs is utilized for large signals. Thus, the aim of the proposed approach is to maintain the PA input signal's large size; which requires digital prewarping techniques. The incoming signal is segmented into four regions with adjacent regions, which differ in maximum voltage by 6 dB as shown in Figure 7.4.

The four regions are distinguished by the values of the control phases $\phi_1 - \phi_3$ These control bits correspond to the two most significant bits (MSBs) of the baseband signal; thus, baseband signal power is identified in the digital signal process. The control phases control the segments of the PA, and then

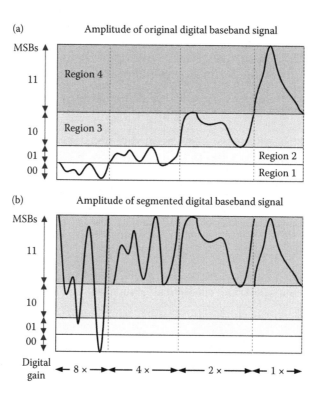

FIGURE 7.4

Correlation between control phases and baseband signal amplitude: (a) input signal and (b) digitally segmented signal.

correlate the PA current consumption and gain with the signal containing the MSBs.

Figure 7.5 shows the conceptual schematic of the proposed system. Ignoring the sign bit, the MSBs of the digital representation of the baseband signal magnitude manage the segments while the least significant are converted into analog format and then up-converted by the mixer. The PA and its driver are divided into four sections in a binary fashion. The control bits $\phi_1 - \phi_3$ drive the PA sections through the drivers. If the signal strength falls in region ϕ_0; for instance, the control phases $\phi_1 - \phi_3$ are zero, and then only the unswitchable section manages the signal $S_{in}(t)$. To minimize the switches in the signal path, the drivers are turned off by switching the connection to V_{DD}. DC coupling is used to drive the PA sections, thereby avoiding the use of large capacitors that introduce a significant delay in the signal path. The architecture is designed such that when the drivers are turned off, the PA sections also shut off. As a result, the drivers and PA sections are dynamically correlated with signal power providing further power savings.

Because of the manipulation of segments, the PA power gain follows this pattern, which makes the PA gain signal dependent on linear amplifiers.

An elegant yet efficient solution is to use digital gain equalization to overcome this shortcoming. The signal strength is evaluated and amplified accordingly in the digital domain such that the digital gain and gain attenuation due to PA switching compensate each other, leading to a constant power gain factor across all operating conditions. A unique property of this approach is that small signals are noise-free amplified in the digital domain, making them more

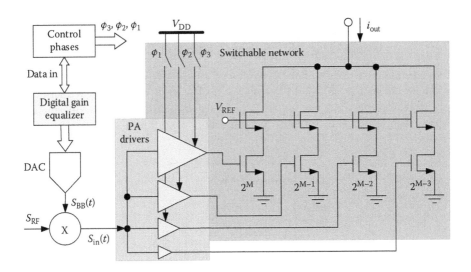

FIGURE 7.5
Simplified schematic of the proposed architecture employing three binary weighted switchable arrays.

tolerant to thermal noise due to the mixer, PA drivers, and PA sections. The digital gain by multiples of two is an easy and cheap operation, since it only requires a bit-shift to the left in the digital domain. If the digital gain equalized signal reaching PA input is fully synchronized with the manipulation of the PA sections, the PA output signal is smooth when transitioning across different segments.

7.3.1 Timing mismatch analysis

One concern is the timing alignment of the RF signal path and the digital control phase path. A simplified model of the system shown in Figure 7.6 is used to capture the essence of the timing mismatch.

Let us consider the case of only one-bit control, ϕ_3. Suppose that there is a timing delay of τ seconds between the RF signal path and the control phase, i.e., the control signal arrives at the switch before the corresponding RF signal reaches the PA cells. Assume a modulated input signal $s_{\text{in}}(t) = s_{\text{BB}}(t) s_{\text{RF}}(t) = \cos(\omega_{\text{BB}} t) \cos(\omega_{\text{RF}} t)$, where ω_{BB} and ω_{RF} represent the baseband and RF angular frequencies, respectively. For simplicity, the amplitude of the input tone and gain of the mixer are chosen to be unity. If all PA sections are active, then the output power is then described as $s_{\text{out}-N}(t) = A_{\text{VPA}} s_{\text{in}}(t)$, where A_{VPA} is the PA gain. When the PA is partitioned at its halfway point, the PA output is $s_{\text{out}-N}(t) = 0.5 A_{\text{VPA}} s_{\text{in}}(t)$. However, the baseband equalizer recognizes that the signal power is small and amplifies it by 6 dB; $s_{\text{in}}(t)$ is then a pre-equalized version of the original baseband input signal and can be expressed as follows:

$$s_{\text{in}}(t) = \begin{cases} 2 s_{\text{BB}}(t) s_{\text{RF}}(t) & \text{if} -0.5 \leq S_{\text{BB}}(t) \leq 0.5 \\ s_{\text{BB}}(t) s_{\text{RF}}(t) & \text{if} -0.5 > S_{\text{BB}}(t) \text{ or } S_{\text{BB}}(t) > 0.5 \end{cases} \quad (7.11)$$

If the timing is perfectly aligned, while the magnitude of the baseband signal is smaller than the threshold voltages, the PA gain reduces by a factor of two. At the same time, the signal is digitally amplified by two, while in this region, the overall gain remains constant since the digital amplification and

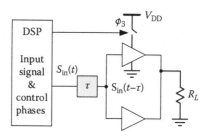

FIGURE 7.6
Simplified model for timing mismatch analysis.

PA attenuation are fully synchronized. On the other hand, if there is a timing mismatch of τ s between the time we manipulate the PA segments and the signal traveling through the up-convertor and amplification chain, then the operations are misaligned resulting in an error (glitch like) at the PA output. The delay occurs when the signal travels through the DAC, the mixer, drivers, and PA sections. If the PA sections are turned off earlier, the PA gain drops by 6 dB and stays in this condition until the equalized signal reaches the gate of the PA. This scenario is illustrated in Figure 7.7a, and the resulting error signal is plotted in Figure 7.7b.

The RF component is not shown to simplify the plot. For the sake of simplicity, let us denote $\theta = \omega_{\mathrm{BB}}t$; then, the third Fourier coefficient of the error signal can be calculated as follows:

$$a_3 = \left(\frac{G_{m-\mathrm{PA}}}{\pi}\right)\left(-\frac{1}{2}\int_{\theta_{1,5}}^{\theta_{1,5}+\theta_\tau}\cos\theta\cos3\theta d\theta + \int_{\theta_{3,7}}^{\theta_{3,7}+\theta_\tau}\cos\theta\cos3\theta d\theta\right) \quad (7.12)$$

where $\theta_i = \omega_{\mathrm{BB}}t_i$, $i = 1, 2, 3, 4$ and $\theta_\tau = \omega_{\mathrm{BB}}\tau$. Calculation of the integrations and then rearranging the expanded terms, and noting from Figure 7.7 that $\theta_1 = \frac{\pi}{3}, \theta_3 = \frac{2\pi}{3}, \theta_5 = \frac{4\pi}{3}, \theta_7 = \frac{5\pi}{3}$ would lead to

$$a_3 = \left(\frac{\sqrt{7}G_{m-\mathrm{PA}}}{2\pi}\right)\left(\frac{1}{2}\sin2\theta_\tau\sin(2\theta_\tau - \phi) - \sin\theta_\tau\sin(\theta_\tau + \phi)\right) \quad (7.13)$$

where $\phi = \tan^{-1}\left(\frac{1}{3\sqrt{3}}\right) = 0.19$ rad. If we assume that $\theta_\tau \ll \phi$, then (7.13) reduces to the simpler yet intuitive result

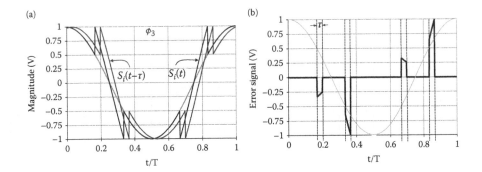

FIGURE 7.7
PA output waveforms (RF component is not shown for simplicity): (a) pre-warped signal with and without timing delay and (b) error waveform due to timing mismatch between ϕ_3 and $S_i(t - \tau)$.

$$|a_3| \approx \left(\frac{\tau}{T_{\text{BB}}} \right) G_{m-\text{PA}} \tag{7.14}$$

where T_{BB} is the baseband signal period. Since $a_1 \approx G_{m-\text{PA}} S_{\text{BB-pk}}$ in this simplified analysis, the third-order intermodulation distortion due to the timing mismatch IMD3 is proportional to $\frac{3\tau}{4T_{\text{BB}}}$. For a baseband signal of 10 MHz ($T_{\text{BB}} = 10^{-7}$ s), the delay error τ must be under 1.3×10^{-9} s to maintain IM3 under -40 dB. Timing mismatches in the other three PA segments add similar effects and increase the PA sensitivity to time delay mismatches.

To reduce the nonlinearity caused by the timing mismatch, a delay cell is added to the system to reduce the timing mismatch. The delay cell includes a replica of the preamp, but it acts as a digital driver. Extensive simulations in a WCDMA system, where the channel bandwidth is 3.84 MHz, showed that timing mismatch of 100 ps would result in a PA neighbor channel leakage power under -40 dB. Thus, it is highly recommended to use 100 ps delay as a cap for delay mismatch.

7.4 Power Amplifier System Design

The critical design of the proposed system is the switching scheme, which is applied to both the PA sections and their drivers. The PA design details are described in this section.

7.4.1 Output stage design

Figure 7.8 shows the schematic of the PA stage. Cascode configuration is used to improve its reliability and stability. The common-source transistors are standard thin oxide transistors that have lower input capacitance and higher transconductance; the common-gate transistors have thick oxide to withstand larger voltage swing. The transistors are optimized for linearity, and their sizes are also included in Figure 7.8.

The maximum RF current is expected at this stage; thus, extra care is needed in the design layout. Multiple pads for the output and ground nodes are used, and the ground pads of this stage are not shared with the remaining parts of the chip. The transistors are organized in clusters employing common-centroid techniques to facilitate the connectivity and to minimize transistor mismatches. The PA transistors are DC connected to the PA drivers; thus, no additional switches are required to enable or disable these sections. When M_1 transistors are switched on/off, a significant common-mode step in current occurs that may produce significant common-mode ringing. To alleviate this issue, a fast class AB CMFB circuit (not shown in Figure 7.8) is allocated at the PA output.

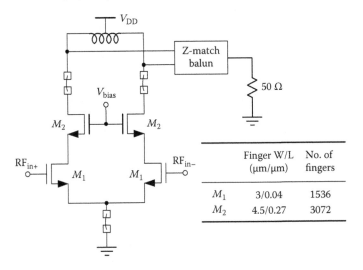

FIGURE 7.8
Schematic of the PA output stage; the core consists of 1536 replicas.

7.4.2   PA drivers

The schematic of the driver stage is shown in Figure 7.9. It consists of a differential pair with resistive load, a switch controlled by the control code, and a CMFB loop. Direct coupling between the driver and the PA stages reduces the switching delay. When the switch is opened, the driver's output common-mode voltage moves down very quickly, putting the differential pair

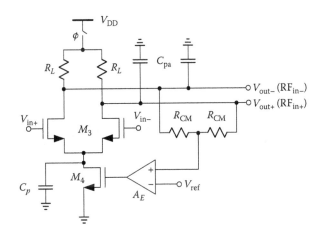

FIGURE 7.9
Conceptual schematic of the driver stage.

transistors in the triode region. The common-mode voltage drops and then breaks the loop during this condition, which helps to quickly turn down the preamplifier outputs. When the switch is closed again, the output voltage of the driver moves toward V_{DD} and is only limited by the time constant $R_L C_{PA}$.

Since the load resistors R_L are small, the time constant is small, and fast low-to-high transition is achieved. As soon as the common-mode level exceeds the reference voltage, the loop tries to reach its steady state; then, settling time of the CMFB is a function of the loop properties. Therefore, the use of fast CMFB is a must. Open-loop gain, closed-loop bandwidth, and stability are all important parameters to be considered when designing the CMFB loop [14].

7.4.3   Output impedance matching network

For the output impedance matching circuit, a multisection network was implemented. Figure 7.10 shows a half representation of the matching network. C_D and L_{bnd} stand for the drain capacitance and the bond wire inductance, respectively. The parasitic capacitance at the package on the PCB is accounted for in C_1. The transmission line with a length d and characteristic impedance Z_0 is formed by a microstrip line consisting of the PCB trace and the ground plane underneath. R_L represents the input impedance of a balun, which is 25 Ω for a half circuit. The optimal impedance R_T is determined by maximum linear output power design specification.

The load–pull simulations further allow us to optimize the choice of R_T. The component values can be determined by hand calculations, the Smith chart, or existing software packages. The summary of the component values is given in Table 7.1.

Two important design specifications for the output matching network are bandwidth and insertion loss. The matching circuit used in the proposed system is effectively a multisection design, and its bandwidth is sufficient for

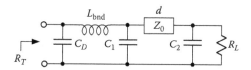

FIGURE 7.10
Two-section impedance matching network.

TABLE 7.1
Impedance Matching Network Component Values

C_D (pF)	C_1 (pF)	C_2 (pF)	L_{bnd} (pH)	Transmission Line W (mil)	d (mill)
10	15.9	3.27	400	150	244

WCDMA applications. To ensure robustness, the insertion loss of the output matching is simulated under process variations, as shown in Figure 7.11.

The worst case of the simulated insertion loss is around 1 dB when all component values are shrunk by 30%. However, this is the less likely case, since nonidealities usually result in additional parasitic components, making the effective component values larger. If all component values increase by 30%, the insertion loss is simulated to be only 0.2 dB.

The Q of the bond wire inductance affects the output matching network. This effect manifests itself in higher insertion loss at the frequency of interest. As shown in Figure 7.11, the insertion loss at 1.9 GHz is simulated with various Q and L values of the bond wire inductance across all four modes of operations. As can be seen, only at extreme conditions, i.e., inductance increased by 30% and $Q = 30$, does the insertion loss exceed 1 dB. Even in this case, since the output match adopts a multistage topology, there are many other components that can help optimize matching.

As the PA switches on and off different transistor sections, the output impedance of the transistor changes. However, the transistor is in either an active region or a cutoff region, and the drain capacitance is mainly due to the depletion region capacitance between the drain and the substrate plus the gate-drain overlap capacitance. A 2 MHz sinusoidal baseband signal modulated to the carrier frequency is input to the system; the transient waveforms are shown in Figure 7.12. The top plot shows the original sinusoidal baseband signal, along with a predistorted baseband that is to be input to the PA. The peak voltage amplitude before and after impedance matching network are 3.1 V and 10.7 V, respectively. The voltage transformation ratio of 3.45 thus implies an impedance transformation ratio of 11.9, as desired ($Z_\text{L}/Z_\text{T} = 50/4.5 = 11.1$). Notice that if the mismatches in segmented PA are small, the AC current delivered to the matching network is smooth for the entire power range. In practice, some glitches are present when switching

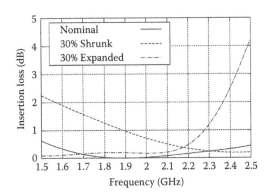

FIGURE 7.11
Insertion loss simulation with process variations.

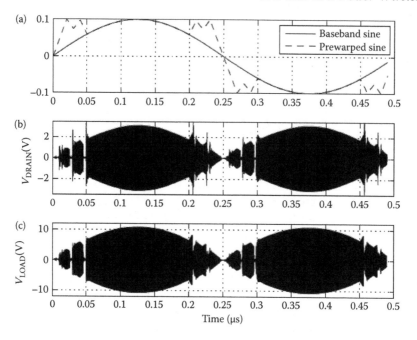

FIGURE 7.12

Transient simulation results: (a) input signal before and after digital prewarping (top trace), (b) output signal at drain voltage (middle trace), and (c) output signal after impedance matching network.

between segments mainly due to the unavoidable parasitic capacitors and timing offsets.

Also, the S_{22} of the PA in each mode of operation is simulated and measured. These can be found in Ref. [14]. Although there is some mismatch due to nonidealities, it is manageable, and can be optimized by tweaking the output matching network component values. Note that both simulation and measurement show that S_{22} does not vary much across different modes of operations of the PA. This is because, due to the cascode topology of the output stage of the PA, the output resistance of the PA is kept large when compared with the transformed R_T; therefore, the variation in PA output resistance is "absorbed" by the output matching network.

7.5 Measurement Results

The PA was fabricated in a TSMC 40 nm CMOS process, and Figure 7.13 shows the microphotograph of the chip.

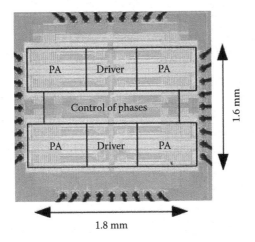

FIGURE 7.13
Microphotograph of the chip.

The chip area is approximately 2.88 mm². A single-tone CW signal of 1.9 GHz was applied to characterize the PA in all four operation modes. Figure 7.14 shows the measured gain, output power (P_{OUT}), and power-added efficiency (PAE) as a function of the input power (P_{IN}). The PCB and cable losses are de-embedded in the performance. The PA's output, P_{1dB} and P_{SAT}, are measured as 31 dBm and 35 dBm, respectively. The average power gain is 38 dB, and the PAE at P_{1dB} and P_{SAT} are 28.8% and 44.9%, respectively. As a

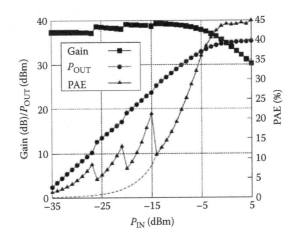

FIGURE 7.14
Measured gain, output power, and PAE as a function of input at 1.9 GHz.

comparison, the PAE of PA without the proposed power efficiency improvement techniques was measured, and is presented in Figure 7.14 as the dashed curve. At 7 dB backoff from P_{1dB}, the PAEs of the PA with and without segmentation are 21.3% and 8.1%, respectively.

A WCDMA baseband signal that is compliant with the 3rd Generation Partnership Project (3GPP) standard [20] was generated, preprocessed in digital, and up-converted to 1.9 GHZ with a bandwidth of 3.84 MHZ. The baseband signal that is prewarped is generated in MATLAB and input to the signal generator. Along with the baseband signal, the control phases are sent out to decoders and digital-to-analog converters implemented on the PCB in the measurement setup. According to [20], the ACLR at ±5 MHZ should be kept below −33 dBc for cellular handsets to comply with the standards. The measured output power spectrum is shown in Figure 7.15, with the PA under test transmitting a maximum linear power of 31 dBm. Data analysis from the spectrum analyzer shows the ACLR at a maximum power of 31 dBm or −35.8 dBc.

The ACLR as a function of maximum output power was measured, and the result is shown in Figure 7.16.

The linearity of the system was compromised in the PBO region, while still meeting the specifications. Another linearity figure of merit is the error

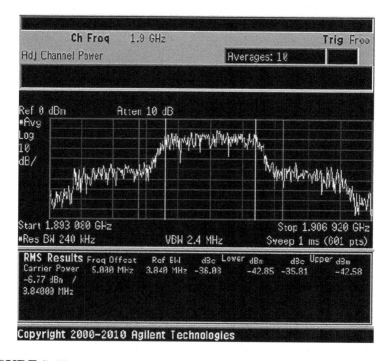

FIGURE 7.15
ACLR measured at maximum output power of 31 dBm.

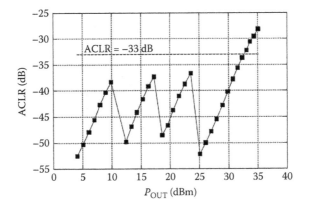

FIGURE 7.16
ACLR as a function of maximum output power.

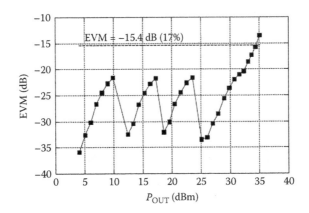

FIGURE 7.17
EVM as a function of maximum output power.

vector magnitude (EVM). For 3G WCDMA, the specification for EVM is less than −15 dB (17%). The EVM as a function of maximum output power of the PA is shown in Figure 7.17. At a maximum output power of 31 dBm, the EVM is −21 dB (8.95%).

The phase error is measured as an indication of the PA's AM/PM nonlinearity. The measured phase error can be under 2.5% up to 35 dBm output power. Recently, reported linear PAs with segmentation technique to improve PAE were compared with the proposed PA as shown in Table 7.2. The proposed PA achieved a remarkable peak PAE as well as outstanding marks at PBO regions.

TABLE 7.2
Comparison with Recent Publications

Reference	Frequency (GHz)	P_{SAT}/P_{AE} (dBm/%)	V_{DD} (V)	CMOS (nm)	Size (mm^2)	Number of Modes	PAE Increase at PBO		
							7 dB	10 dB	15 dB
[8]	2.4	23.1/42	1.5	130	5.48	2	3.3	4.3	3.6
[9]	2.4	27/32	1.2	130	2	2	5.4	4	3.4
[11]	2.4	23.1/42	3.3	180	0.88	2	8.7	8.7	7.2
[12]	2	23/38	2.5	250	2.48	3	10	N/A	N/A
[13]	2.45	31.5/25	3.3	65	2.7	3	10	5	N/A
[2]	2.45	26.3/33	2	90	1.88	2	9	7	3
[15]	2.2	43	1.2	65	6.25	2	6.2	4	2.1
This work	1.9	35.3/44.9	2.5	40	4	4	13	7.36	9

7.6 Conclusion

A 1.9 GHz segmented linear PA was designed and implemented in 40 nm CMOS technology. The input signal is segmented and strategically amplified in the digital domain while the PA is segmented, and its segments are properly manipulated to maintain its power gain invariant with voltage while achieving significant power savings. The PA achieved a saturated/maximum linear output power of 35/31 dBm with the corresponding peak PAEs of 44.9% and 28.8%, respectively. A fast yet efficient switching scheme that employs direct coupling between PA sections and drivers was demonstrated, which enabled the PA to improve the efficiency in the PBO region within a wideband communication standard.

Bibliography

[1] L. Larson, "RF and microwave hardware challenges for future radio spectrum access," *Proceedings of the IEEE*, vol. 102, no. 3, pp. 321–333, Mar. 2014.

[2] E. Kaymaksut and P. Reynaert, "Transformer-based uneven Doherty power amplifier in 90 nm CMOS for WLAN applications," *IEEE Journal of Solid-State Circuits*, vol. 47, no. 7, pp. 1659–1671, Oct. 2012.

[3] H. Son, J. Kim, J. Jang, I. Oh, C. Park and W. Y. Kim, "A CMOS envelope-tracking transmitter with an on-chip common-gate voltage modulation linearizer," *Microwave and Wireless Components Letters, IEEE*, vol. 24, no. 6, pp. 406–408, Jun. 2014.

[4] K. Oishi, E. Yoshida, Y. Sakai, H. Takauchi, Y. Kawano, N. Shirai, H. Kano, M. Kudo, T. Murakami, T. Tamura, S. Kawai, S. Yamaura, K. Suto, H. Yamazaki and T. Mori, "A 1.95 GHz fully integrated envelope elimination and restoration CMOS power amplifier with envelope/phase generator and timing aligner for WCDMA and LTE," *Solid-State Circuits Conference Digest of Technical Papers (ISSCC), 2014 IEEE International*, pp. 60–61, Feb. 2014.

[5] B. Sahu and G. Rincon-Mora, "A high-efficiency linear RF power amplifier with a power-tracking dynamically adaptive buck-boost supply," *IEEE Transactions on Microwave Theory and Techniques*, vol. 52, no. 1, pp. 112–120, Jan. 2004.

[6] F. Wang, A. Yang, D. Kimball, L. Larson and P. Asbeck, "Design of wideband envelope-tracking power amplifiers for OFDM applications,"

IEEE Transactions on Microwave Theory and Techniques, vol. 53, no. 4, pp. 1244–1255, Apr. 2005.

[7] A. Shirvani, D. Su and B. Wooley, "A CMOS RF power amplifier with parallel amplification for efficient power control," *IEEE Journal of Solid-State Circuits*, vol. 37, pp. 684–793, Jun. 2002.

[8] P. Reynaert and M. S. Steyaert, "A 2.54-GHz 0.13-μm CMOS PA with parallel amplification," *IEEE Journal of Solid-State Circuits*, vol. 42, no. 3, pp. 551–562, Mar. 2007.

[9] G. Liu, P. Haldi, T. J. K. Liu and A. Niknejad, "Fully integrated CMOS power amplifier with efficiency enhancement at power back-off," *IEEE Journal of Solid-State Circuits*, vol. 43, no. 3, pp. 600–609, Mar. 2008.

[10] J. Kim, Y. Yoon, H. Kim, K. H. An, W. Kim, H. W. Kim, C. H. Lee and K. Kornegay, "A linear multi-mode CMOS power amplifier with discrete resizing and concurrent power combining structure," *IEEE Journal of Solid-State Circuits*, vol. 46, no. 5, pp. 1034–1048, May 2011.

[11] Y. Yoon, J. Kim, H. Kim, K. A. An, O. Lee, C. H. Lee and J. Kenney, "A dual-mode CMOS RF power amplifier with integrated tunable matching network," *IEEE Transactions on Microwave Theory and Techniques*, vol. 60, no. 1, pp. 77–88, Jan. 2012.

[12] H. Hedayati, M. Mobarak, G. Varin, P. Meunier, P. Gamand, E. Sanchez-Sinencio and K. Entesari, "A 2-GHz highly linear efficient dual-mode BiCMOS power amplifier using a reconfigurable matching network," *IEEE Journal of Solid-State Circuits*, vol. 47, no. 10, pp. 2385–2404, Oct. 2012.

[13] A. Afsahi and L. Larson, "Monolithic power-combining techniques for watt-level 2.4-GHz CMOS power amplifiers for WLAN applications," *IEEE Transactions on Microwave Theory and Techniques*, vol. 61, no. 3, pp. 1247–1260, Mar. 2013.

[14] H. Qian, Q. Liu, J. Silva-Martinez and S. Hoyos, "A 35 dBm output power and 38 dB linear gain PA with 44.9% peak PAE at 1.9 GHz in 40 nm CMOS," *IEEE Journal of Solid-State Circuits*, vol. 51, no. 3, pp. 587–597, Mar. 2016.

[15] L. Ye, J Chen, L. Kong, P. Cathelin, E. Alon, and A. Niknejad, "A digitally modulated 2.4 GHz WLAN transmitter with integrated phase path and dynamic load modulation in 65 nm CMOS," *International Solid-State Circuits Conference Digest of Technical Papers (ISSCC'13)*, pp. 330–331, Feb. 2013.

[16] F. Raab, P. Asbeck, S. Cripps, P. Kenington, Z. Popovic, N. Pothecary, J. Sevic and N. Sokal, "Power amplifiers and transmitters for RF and

microwave," *IEEE Transactions on Microwave Theory and Techniques*, vol. 50, no. 3, pp. 814–826, Mar. 2002.

[17] J. G. Proakis and M. Salehi, *Digital Communications*, 5th ed., New York, NY: McGraw-Hill, 2007.

[18] S. C. Cripps, *RF Power Amplifiers for Wireless Communications*, 2nd ed., Norwood, MA: Artech House, Inc., 2006.

[19] B. Razavi, *RF Microelectronics*, 2nd ed., Upper Saddle River, NJ: Prentice-Hall, Inc., 2011.

[20] "3GPP TS 25.101 Technical Specification Rev. 12.3.0," 3rd Generation Partnership Project, Mar. 2014. Available: http://www.3gpp.org.

[21] T. Quarles, D. Pederson, R. Newton and A. Sangiovanni-Vincentell, "SPICE User Guide," EECS Department of the University of California at Berkeley. Available: http://bwrcs.eecs.berkeley.edu/Classes/IcBook/ SPICE/.

[22] H. Qian and J. Silva-Martinez, "Multitone ACLR and Its Applications to Linear PA Design," *IEEE Transactions on Circuits and Systems II: Express Briefs*, vol. 64, no. 10, pp. 1177–1181, 2017.

[23] J. C. Pedro and N. B. de Carvalho, "On the use of multitone techniques for assessing RF components' intermodulation distortion," *IEEE Transactions on Microwave Theory and Techniques*, vol. 47, no. 12, pp. 2393–2402, Dec. 1999.

[24] H. Qian and J. Silva-Martinez, "A 44.9% PAE digitally-assisted linear power amplifier in 40 nm CMOS," *Solid-State Circuits Conference (A-SSCC), 2014 IEEE Asian*, pp. 349–362, Nov. 2014.

8

Injection-Locking Techniques in Low-Power Wireless Systems

Yushi Zhou

Lakehead University

Fei Yuan

Ryerson University

CONTENTS

8.1   Injection Locking

In the middle of the 17th century, a Dutch physicist, Christiaan Huygens, the inventor of the pendulum clock, was first noticing an "odd kind of sympathy" from two pendulum clocks on the wall, which swing in a synchronized form no matter how these clocks began. This is what we called mechanic "injection locking," as sound pulses, generated from the movement of the pendulums, travel through the wall from each other, and eventually force them to synchronize [1]. A similar phenomenon can be observed in an electronic oscillator when an incident signal, having different frequencies, is injected into the oscillator. The process of the oscillator changing from the perturbation to the steady-state condition is called injection locking, and it is graphically interpreted in Figure 8.1. As we can see, Figure 8.1a is the spectrum of a free-running harmonic oscillator, assuming a purely sinusoidal signal at the output; Figure 8.1b exhibits an extra frequency component, ω_{inj}, far from the locking range, ω_L, which is the maximum frequency distance from the free-running frequency, ω_o, that the injection locking holds; Figure 8.1c shows that the ω_o starts shifting towards the injection signal; and Figure 8.1d is the spectrum when the oscillator is locked. The first examination of the injection locking was provided by Adler [2] in 1946, where the locking range, Eq. (8.1), and injection phase were detailed.

$$\frac{E_1}{E} = 2Q\frac{\Delta\omega}{\omega_0}, \tag{8.1}$$

where E_1 and E are the imposed voltage and the output voltage signal of the oscillator, respectively, Q denotes the quality factor of the LC circuit, and

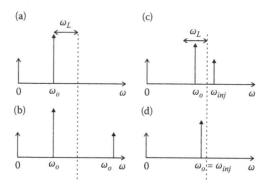

FIGURE 8.1
Injection-locked procedure represented in the simplified spectrum diagram. (a) Free-running, (b) ω_{inj} deviates far from the locking range ω_L, (c) under perturbation, and (d) locked.

ω_0 and $\Delta\omega$ are free-running frequency and the maximum frequency between the injection frequency and the free-running frequency over which a lock state can be established. Injection-locking techniques have been widely appreciated in the design of wireless and wireline data communication systems, and have drawn great attention. Literature [3–12] further explores the insights of injection locking and subsequently their uses in clock generation. For instance, the phasor domain treatment of the injection-locking and pulling of oscillators by Razavi provides the much needed insight of the locking and pulling processes of oscillators [13]. The one-side locking range is shown in Eq. (8.2)

$$\Delta\omega = \frac{\omega_o}{2Q} \frac{I_{inj}}{I_o} \frac{1}{\sqrt{1 - \left(\frac{I_{inj}}{I_o}\right)^2}}, \tag{8.2}$$

where I_{inj} and I_o are the current of the incident signal and that of the LC tank, respectively. If $I_{inj} \ll I_o$, Eq. (8.2) will be simplified to

$$\Delta\omega \approx \frac{\omega_o}{2Q} \frac{I_{inj}}{I_o}. \tag{8.3}$$

From Eq. (8.2), notice that the locking range is determined by the injection ratio $\frac{I_{inj}}{I_o}$ and the quality factor of the oscillator. In addition, the difference between the phase of the injection signal and the output of the oscillating under injection is of great importance in the determination of locking range. It was shown that the maximum locking range in the weak injection in which $I_{inj} \ll I_o$ is attained at 90° phase difference, leading to the zero crossing of the incident signal reaches the peaks of the output.

8.2 Linear-Feedback Approach for Harmonic Oscillators

In this section, a linear-feedback approach is presented for injection-locking techniques in harmonic oscillators.

8.2.1 Linear-feedback model of harmonic oscillators

Consider a typical negative feedback system with β setting to unity illustrated in Figure 8.2, where

$$\frac{V_{out}}{V_{in}}(j\omega) = \frac{H_o(j\omega)}{1 + H_o(j\omega)}. \tag{8.4}$$

Unlike an amplifier design, which is often given a sufficient phase margin at the unity-gain frequency to avoid undesired overshoot at the output port, in the self-sustaining autonomous system, an unstable feedback is expected. In

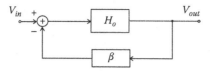

FIGURE 8.2
A negative feedback system.

other words, making the circuit perfectly to oscillate is the design goal. Provided a noise component at ω_o is injected to the circuit in which $H_o(j\omega_o) = -1$, resulting in an infinite loop gain. The noise experiences a total gain of unity and a phase-shift, $180°$, returning to the subtractor as a reversed input signal. As a result, the output of the subtractor is the amplified noise component, which continues to grow. Therefore, an oscillating signal is attained at V_{out}. If a negative-feedback system with its loop gain satisfies Barkhausen criteria, $|H_o(j\omega_o)| \geq 1$ and $\angle H_o(j\omega_o) = 180°$, the feedback system will oscillate at ω_o.

8.2.2   Linear-feedback model of injection-locked harmonic oscillators

The most widely recognized oscillators in the wireless communication systems are inductor-based oscillators due to their superior phase noise performance when compared with other types of oscillators such as ring oscillators. In this section, we focus on the locking range of inductor-based oscillators.

Figure 8.3a shows the simplified schematic of an injection-locked cross-coupled oscillator. The inverting buffer is inserted such that Barkhausen criteria are satisfied. Let I_{inj} denote the incident current and Z_T denote the load impedance of the cross-coupled oscillator. Applying Kirchhoff's Current Law (KCL)at node 1, we have $I_o + I_T = I_{inj}$ and $I_o = -g_m(Z_T I_T)$, where g_m is the transconductance of the transistor. The injection-locked oscillator can be represented by the block diagram shown in Figure 8.3b with I_{inj} the input and I_T the output. Note that Figure 8.3b is a linear system when a lock state is established as only one frequency, e.g. the frequency of the input exists in the system. When the injection signal is absent, the oscillator is an autonomous system that oscillates at ω_o, the self-resonance frequency of the LC tank.

Let us consider Figure 8.3b. By the definition of injection locking, if the injection frequency is $\omega_{inj} = \omega_o + \Delta\omega$, and the oscillator has a steady-state output in the locking state, the frequency of the oscillator is changed to ω_{inj}, and

$$I_o(\omega_o + \Delta\omega) = \frac{H_o(\omega_o + \Delta\omega)}{1 + H_o(\omega_o + \Delta\omega)} I_{inj}(\omega_o + \Delta\omega). \tag{8.5}$$

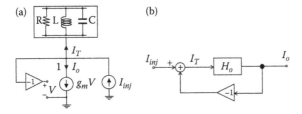

FIGURE 8.3
(a) Injection-locked oscillators and (b) block diagram of injection-locked oscillators.

In the case where $\Delta\omega \ll \omega_o$, and applying Tylor's expansion, it follows that

$$I_o(\omega_o + \Delta\omega) \approx \frac{H_o(\omega_o + \Delta\omega)}{1 + H_o(\omega_o) + \left[\dfrac{\partial H_o(\omega)}{\partial\omega}\right]_{\omega_o}\Delta\omega} I_{inj}(\omega_o + \Delta\omega). \qquad (8.6)$$

Provided $H_o(\omega_o) = -1$, after a few steps manipulation, we have the magnitude of $I_o(\omega_o + \Delta\omega)$ as

$$|I_o(\omega_o + \Delta\omega)| \approx \frac{|H_o(\omega_o + \Delta\omega)|}{\left|\dfrac{\partial H_o(\omega)}{\partial\omega}\right|_{\omega_o}|\Delta\omega|} |I_{inj}(\omega_o + \Delta\omega)|. \qquad (8.7)$$

Utilize the definition of quality factor given in ref. [14], and we know $H_o(\omega) = Ae^{j\phi}$. Thus,

$$|I_o(\omega_o + \Delta\omega)| \approx \frac{|H_o(\omega_o + \Delta\omega)|}{\dfrac{2Q}{\omega_o}|\Delta\omega|} |I_{inj}(\omega_o + \Delta\omega)|. \qquad (8.8)$$

From Figure 8.3b, it is clear that

$$|I_o(\omega_o + \Delta\omega)| = |H_o(\omega_o + \Delta\omega)||I_T(\omega_o + \Delta\omega)|. \qquad (8.9)$$

It follows from Eqs. (8.8) and (8.9) that

$$|\Delta\omega| = \frac{\omega_o}{2Q}\left|\frac{I_{inj}(\omega_o + \Delta\omega)}{I_T(\omega_o + \Delta\omega)}\right|. \qquad (8.10)$$

From preceding analysis, we consider harmonic oscillators as linear negative-feedback systems that are perturbed by the "external noise", the injection signal, which is an alternative perspective view of injection-locked harmonic oscillators. It provides an additional insight. More details can be found in ref. [15]. As we can see, if the injection is weak, i.e., $I_{inj} \ll I_o$, $I_o \approx I_T$ follows. Equation (8.10) is analogous to (8.1) and (8.3) for the maximum value of $\Delta\omega$. The way of increasing the locking range is either to make the injection signal stronger or to lower the quality factor of the LC tank.

8.3 Linear-Feedback Approach for Nonharmonic Oscillators

This section investigates the locking range of nonharmonic injection-locked oscillators such as ring oscillators or relaxation oscillators whose output is typically nonsinusoidal, revealing the existence of harsh nonlinearities in the systems.

8.3.1 Linear-feedback model of nonharmonic oscillators

Before we get into the details of injection-locked nonharmonic oscillators, we first develop a representation of nonharmonic oscillators. The output of an LC oscillator is a sinusoidal signal, whereas that of nonharmonic oscillators is typically a square wave-like signal. Let the fundamental frequency of the output of nonharmmonic oscillator be $\omega_o = 2\pi/T$. We know the purely sinusoidal signal has only one frequency component in the positive spectrum while a purely square wave shown in Figure 8.4 can be represented by an infinite number of frequency components, containing the fundamental and odd harmonics.

The output can be represented by

$$I_o = \sum_{n=1,3,\ldots}^{\infty} I_{o,n}(n\omega_o)\delta(\omega - n\omega_o), \tag{8.11}$$

where $\delta(.)$ is the discrete-time impulse also known as the unit impulse [16]. The preceding description shows that the spectrum of the square waveform, consisting of a train of impulses at ω_o, $3\omega_o$, $5\omega_o$, ... $n\omega_o$, leads to a critical observation: In the steady-state, the nonharmonic oscillator can be considered as an assembly of a set of harmonic oscillators, having free-running frequency at ω_o, $3\omega_o$, $5\omega_o$, ..., thereby greatly simplifying the analysis of injection-locked nonharmonic oscillators, as to be seen shortly.

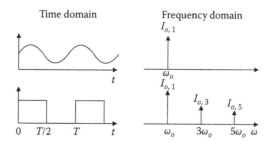

FIGURE 8.4

Waveform and spectrum of harmonic and nonharmonic oscillators.

8.3.2 Linear-feedback model of injection-locked nonharmonic oscillators

What happens if the nonharmonic oscillator is injection-locked to a sinusoidal signal? Figure 8.5 shows a cadence spectre simulation of a nonharmonic oscillator, e.g. a relaxation oscillator, locked to an external signal.

As expected, a sinusoidal, given sufficient injection strength, is injected to a nonharmonic oscillator with its frequency in the locking range, the oscillator frequency is shifted to the incident frequency. Further shown in Figures 8.6 and 8.7, not only is the fundamental harmonic locked but also the other harmonics of the oscillator are changed to $3\omega_{inj}$, $5\omega_{inj}$,..., accordingly.

This observation reveals that the single-tone input of the nonharmonic oscillator will affect the input of the representing harmonic oscillators. As explained in Section 8.2.1, injection locking for a harmonic oscillator can be modeled as a linear-feedback system. The injection-locked nonharmonic oscillator with an incident signal at ω_{inj}, therefore, can be represented by a set of injection-locked harmonic oscillators, each with a single-tone injection at ω_{inj}, $3\omega_{inj}$, $5\omega_{inj}$,..., as shown in Figure 8.8.

We note that this representation is analogous to Volterra series based analysis of nonlinear circuits. It has been demonstrated [17–21] that a periodically time-varying nonlinear circuit can be represented by a set of periodically time-varying linear circuits of the same topology but different inputs set by the nonlinearities of the circuits. Nonharmonic oscillators fall into the category of periodically time-varying nonlinear circuits, such that they can be analyzed using Volterra series approaches.

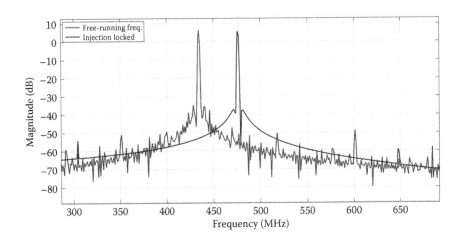

FIGURE 8.5
Spectrum of free-running frequency and injection locked frequency for the 1st-harmonic.

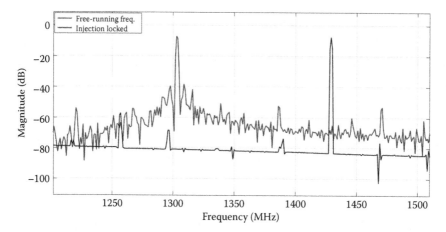

FIGURE 8.6

Spectrum of free-running frequency and injection locked frequency for the 3rd-harmonic.

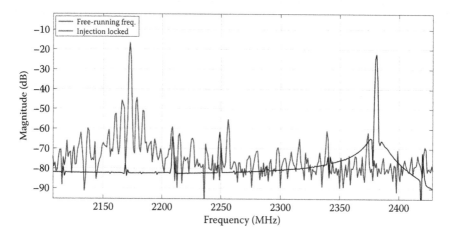

FIGURE 8.7

Spectrum of free-running frequency and injection locked frequency for the 5th-harmonic.

The foregoing discussion proves the feasibility of using a set of injection-locked harmonic oscillators to represent an injection-locked nonharmonic oscillator, which is only valid when the oscillator is locked. We introduce the open-loop gain, $H_{o,n}(n\omega_o)$, of the harmonics of the oscillators, and make use of a similar approach that is used to obtain the locking range of a harmonic oscillator. We have

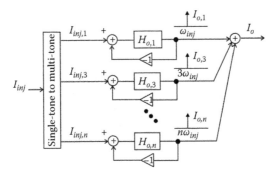

FIGURE 8.8
Representation of injection-locked nonharmonic oscillators with a single-tone
injection.

$$I_o \approx \frac{-I_{inj1,1}}{\left(\dfrac{\partial H_{o1}}{\partial \omega}\right)_{\omega_o} \Delta \omega} \delta(\omega - \omega_{inj}) + \frac{-I_{inj1,3}}{\left(\dfrac{\partial H_{o3}}{\partial \omega}\right)_{3\omega_o} 3\Delta \omega} \delta(\omega - 3\omega_{inj}) + \dots$$

$$(8.12)$$

After a few steps of algebraic manipulation, we can get the output power of
the oscillator

$$I_o^2 = I_{o,1}^2 + I_{o,3}^2 + I_{o,5}^2 + \dots, \tag{8.13}$$

where

$$I_{o,1}^2 = \left(\frac{\omega_o}{2Q_1}\right)^2 \frac{I_{inj,1}^2}{(\Delta \omega)^2}$$

$$I_{o,3}^2 = \left(\frac{\omega_o}{2Q_3}\right)^2 \frac{I_{inj,3}^2}{(\Delta \omega)^2} \dots$$

The summation of the expressions in Eq. (8.13) yields

$$(\Delta \omega)^2 = \left[\left(\frac{\omega_o}{2Q_1}\right)^2 \frac{I_{inj,1}^2}{I_o^2} + \left(\frac{\omega_o}{2Q_3}\right)^2 \frac{I_{inj,3}^2}{I_o^2} + \dots\right]. \tag{8.14}$$

Define

$$|\Delta \omega_n| = \left(\frac{\omega_o}{2Q_n}\right) \frac{I_{inj,n}}{I_o}. \tag{8.15}$$

Equation (8.15) shows that I_o is the total output current of the harmonic oscillators. Equation(8.15) thus quantifies the contribution of nth-order harmonic
oscillator $I_{inj,n}$ to the overall locking range of the nonharmonic oscillator.
Equation(8.14) can be written as

$$|\Delta \omega| = \sqrt{\sum_{n=1,3,\dots}^{\infty} (\Delta \omega_n)^2}, \tag{8.16}$$

where $\Delta\omega$ is the single side locking range of the injection-locked nonharmonic oscillators with a single-tone injection. Equation (8.16) reveals the relation between the locking range of the harmonic oscillators and that of the nonharmonic oscillators. If the harmonic oscillator is injection locked to an external signal, only fundamental frequency will be taken into account. And Eq. (8.16) is simplified to Eq. (8.10) for weak injection e.g. $I_{inj} \ll I_o$. Readers are referred to [22–24] for more details.

8.4 Injection-Locked Frequency Divider

Frequency dividers are common building blocks in the transceivers of wireless and wireline data communication. For instance, Figure 8.9 illustrates a sliding intermediate frequency Radio Frequency (RF)receiver in which a second intermediate frequency is employed. A way of generating a divided-by-2 clock is to use frequency division. In a phase-locked loop (PLL), consisting of a phase-frequency detector, a charge pump, a voltage-controlled oscillator (VCO), and a frequency-divider, depicted in Figure 8.10, the frequency divider is a crucial component with respect to power consumption. Frequency dividers can be implemented using static/dynamic logic [25,26], current-mode logic (CML) [27,28], Miller dividers [29,30], and injection-locked oscillators, including ring oscillators and resonant-based oscillators [4,31,5,32–48]. Power dissipation and maximum operation frequency dictate the choice of a frequency divider topology. In this section, a number injection-locked frequency dividers are examined.

8.4.1 Frequency divider

The rapid down scaling of the footprint of Complementary metaloxidesemiconductor (CMOS)transistors, and the aggressive reduction of the supply voltage

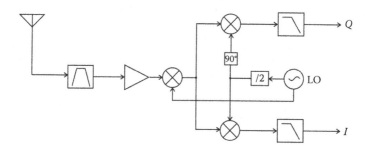

FIGURE 8.9
Divide-by-2 wireless receiver.

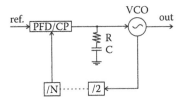

FIGURE 8.10
A phase-locked loop.

have greatly impacted on the performance of clock circuits. The attributes of flip-flop or latch logic lend themselves to be the most widely adopted frequency dividers in PLLs, as shown in Figure 8.11. They are particularly suitable for the applications at low-to-intermediate frequencies. As they are edge-triggered, altering two states through the negative feedback, they have a wide range of operation frequency, covering from near DC to circuits' bandwidth. One of the critical drawbacks of this type of frequency dividers, however, is their high power consumption due to the need for complete charging and discharging in every clock cycle.

Current-mode circuits, also named as "current-steering" circuits, where information is represented by the branch currents of the differential pair, offer high speed, low noise sensitivity from power supply, and inherently differential outputs. Therefore, CML is an attractive candidate of implementing frequency dividers. A CML D-latch circuit is shown in Figure 8.12a. The circuit senses the complementary inputs through a differential pair, which is controlled by a pair of clocks. The outputs are latched through a regenerative pair. It basically has two modes, namely *sensing mode* and *latched mode*. In the sensing mode: CK is active, and the tail current, I_t is steered to the transistors, M_1 and M_2. The voltage signals at node X and Y are amplified differentially. Once the CK is inactive, and $\overline{CK}$ goes up, transistors M_1 and M_2 are turning off but M_4 and M_5 are turning on. The CML circuits enters the latched mode in which the positive feedback loop formed by M_4 and M_5 regenerates the

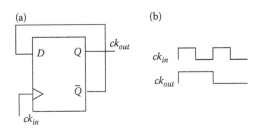

FIGURE 8.11
(a) D flip-flop and (b) timing diagram.

(a)   (b)

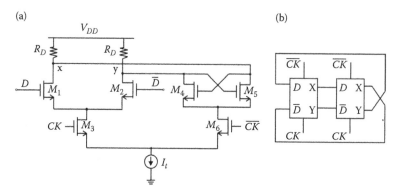

FIGURE 8.12
(a) CML latch and (b) divide-by-2 circuit.

difference between voltages at X and Y until one transistor turns off, and the other transistor take the whole I_t. Two CML latches form a negative feedback loop such that a divide-by-2 is realized, as shown in Figure 8.12b.

There are two concerns for CML frequency dividers in low-voltage high-speed applications: (i) The stacking of three transistors, M_1, M_3, and the tail transistor is a major hindrance in preventing CML from being used in applications where supply voltage is low. (ii) Operation frequency is severely limited by the capacitance at the drain of the differential pair and the latch. To alleviate the voltage headroom problem, the tail current transistor can be removed, and the transistor pair, which is handling CK and $\overline{CK}$, can be used as a pair of tail current to provide the constant current. To increase operation frequency, shunt peaking (inductive peaking) that offers larger bandwidth through resonating out the parasitic capacitance can be used.

Apart from conventional RS latches, true single-phase clocking (TSPC) logic is another logic topology of edge-triggered sequential circuits [49], as shown in Figure 8.13. Once the clock, CK, is high, M_2 shorts M_1 and M_3.

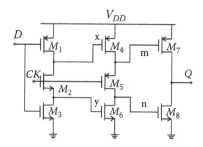

FIGURE 8.13
TSPC flip-flop.

As a result, the first stage becomes an inverter, delivering $\overline{D}$ to nodes x and y, while the second stage is disabled because M_5 is a P-type metal-oxide-semiconductor (PMOS). When CK is low, the second stage transfers the stored value to nodes m and n, and the first stage is disabled. A divide-by-2 is obtained using a negative feedback loop between the input and the output. In comparison with CML logic, TSPC has two attributes: (i) it does not consume static power and (ii) it operates at high frequencies without dissipating an excessive amount of power. TSPC logic, however, might fail at low frequencies due to charge leakage.

Miller dividers consist of a mixer, a feedback loop, and a low-pass filter (LPF), which absorbs the parasitic capacitance, resulting in relatively high operation frequency, depicted in Figure 8.14a. The development of Miller dividers is to use a Gilbert multiplier to implement frequency component generation. Figure 8.14b depicts the feedback model of Miller dividers, where the mixer is referred to the Gilbert cell and the LPF is associated with the load of the circuit. Provided CK operates at ω_{in}, and the output frequency is divided by 2, e.g. $\omega_{in}/2$. The output of the mixer includes $\omega_{in}/2$, $3\omega_{in}/2$, etc. With a properly defined cut-off frequency, the LPF suppresses frequency components higher than $\omega_{in}/2$, only picking up $\omega_{in}/2$ component at the output, which is sent back to the input of the mixer. Notice that the circuit does not oscillate, which is the key difference between Miller dividers and injection-locked frequency dividers.

8.4.2 High-speed low-power injection-locked frequency divider

Increasing demand for high-performance, low-cost, and low-power wireless transceivers has been the driving force for injection-locked prescalars in the design of frequency synthesizers. In Section 8.1, we saw that the oscillators are able to be locked to a frequency that is close to the free-running frequency of the oscillator. Oscillators can also lock to an incident whose subharmonic

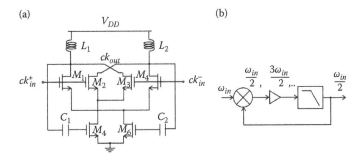

FIGURE 8.14
(a) Miller divider and (b) model.

is close to the free-running frequency of the oscillators, leading to injection-locked frequency dividers. Table 8.1 tabulates some injection-locked frequency dividers emerged recently.

A simple LC-based frequency divider is illustrated in Figure 8.15, consisting of a cross-coupled transistors, a tail-current transistor, and an inductor. In the absence of the injection signal, the oscillator operates at ω_o (free-running frequency). For frequency division of 2, we expect that the oscillator is locked to the injection frequency, $\omega_{inj} = 2\omega_o$. The question is which node should be used as the injection node. Due to the high impedance of the gate, the injection signal can be applied to the gate of M_3. In addition, the signal at node P also exhibits frequency $2\omega_o$. Considering M_1 and M_2 forms a single-balanced mixer, switching ON/OFF with respect to ω_o, the output of the mixers shows the frequency component at $\omega_o \pm 2\omega_o$. Because the LC circuit is designed to select ω_o, $\omega_o + 2\omega_o$ is therefore suppressed depending on the selectivity of the load [7]. The topology was demonstrated in a 0.5 μm CMOS process with supply voltage, 1.5 V. The oscillator was operating at 1.6 GHz. The measured locking range, centered at 3.2 GHz, was more than 190 MHz with the power dissipation 0.45 mW. It was increased to 370 MHz by increasing the tail current to 800 μA [4].

Wu and colleagues [32,51] incorporated shunt-peaking technique to improve the locking range without consuming more power. Equation (8.3) shows that to increase the locking range, one can either lower the quality factor of the oscillator or increase injection ratio, e.g. the ratio of injection signal I_{inj}

TABLE 8.1
Comparison of injection-locked frequency dividers.

Ref.	Year	Tech. (nm)	Div. ratio	Freq. (GHz)	Locking range (GHz)	Power (mW)
[36]	2007	180	8	6.5–18	3.055	3.6
[37]	2008	90	2	55	19.2	0.8
[38]	2009	90	2/4	51–74/82.5–89	–	3
[39]	2011	90	5	59.4–63.5	4.1	3.75
[40]	2010	180	5	39.36–42.1	0.74	18
[41]	2016	180	3	17–19.1	–	18
[50]	2016	180	3	4.88–11.9	4.1	6.67
[42]	2015	180	3	8–13	1.6	4.9
[43]	2015	180	3	20.4–23.8	3.4	3.9
[44]	2013	65	4	58.5–72.9	14.4/21.9	2.2
[45]	2017	180	4	8.8–12.7	3.9	3.6
[46]	2011	130	4	13.5–30.5	–	7.3
[47]	2011	350	4	2.4	0.41	3.63
[48]	2016	180	2	2.6–8.37	–	6.57

to oscillation signal I_{osc}. The former results in degrading the phase noise of the frequency divider and subsequently the overall phase noise of the PLL while the latter consumes more power. Wu et al. showed that the lock range is determined by the strength of the internal injection signal power rather than the external injection signal power. The larger the internal injection signal, the larger is the locking range. The parasitic capacitance at a drain of M_3, however, attenuates the injection signal at this node, subsequently reducing the locking range. Shunt-peaking technique used in amplifier design to boost internal injection signal is widely adopted, shown in Figure 8.16. A shunt inductor L_o is utilized to resonate out the capacitance at the drain of M_1, including C_{gd}, C_{db}, and C_{series} such that a wider signal bandwidth can be achieved.

In analogy with the design of a tuned amplifier, a variant frequency divider, derived from Figure 8.15, is depicted in Figure 8.17. Reference [32], on the top right, included an inductor, L_O to resonate out the parasitic capacitance, C_t, which was the total capacitance of $C_{gd3,db3,gs1,gs2}$, at ω_{inj}. In other words, L_o provided short circuit to the other harmonics of ω_o, except for the second harmonic of the ω_o, the same frequency as the injection frequency. This divide-by-2 circuit was verified in 0.35 μm CMOS process. Wafer measurements proved that with the shunt-peaking inductor, the locking range was 1.49 GHz with 9 GHz injection signal. The power dissipation was 1.12 mW with the supply voltage of 1.2 V. On the other hand, without L_o, the locking

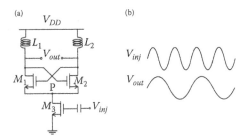

FIGURE 8.15
A differential LC-based frequency divider: (a) schematic and (b) signal at V_{out} and V_P.

FIGURE 8.16
A tuned amplifier.

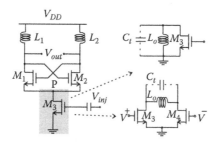

FIGURE 8.17
Enhanced Locking range topology. Top right: [32], bottom right: [51].

range was 0.84 GHz, 0.65 GHz narrower than the former. The circuit topology
was further demonstrated in ref. [51], a divide-by-3 frequency divider, where a
differential pair was incorporated to couple the injection signals to the source
of $M_{1,2}$. We can model a mild memoryless nonlinear system as

$$y(t) = \alpha_1 x(t) + \alpha_2 x(t)^2 + \alpha_3 x(t)^3 + \dots \qquad (8.17)$$

Equation (8.17) implies that the frequency component, $2\omega_o$, from the second
term, which multiplied by $3\omega_o$, results in $2\omega_o \pm 3\omega_o$. The output is conditioned
by the selection circuit so as to keep ω_o at the output. If we set the injection
frequency, ω_{inj}, as $3\omega_o$, since the output is operating at ω_o, divide-by-3 can
be attained. In Wu's design, a shunt inductor, L_o, was employed to increase
the injection power at $3\omega_o$ to have wider locking range. Measurement results
around 1 GHz locking range, centered at 16 GHz, was achieved.

We have seen that the injection signal in the divide-by-2 circuit can be
applied to the gate of the tail transistor. The inefficient injection attributed to
the relatively large tail transistor and subsequently the large parasitic capacitance results in a narrow locking range. A direct injection, where the injection
signal is applied to the drain of a cross-coupled transistor, either in series
or in parallel, features more efficient injection, shown in Figure 8.18. Consider M_{inj} as a mixer, switching ON/OFF at $2\omega_o$, and if the oscillator runs
at ω_o, the production of the two inputs of the mixer leads to $\omega_o \pm 2\omega_o$ at
the differential outputs of the oscillator, where the high-frequency component is filtered out by the LC circuit. Tiebout [33] demonstrated an improved
injection approach with a differential injections in the current reuse LC oscillator in which a symmetric cross-coupled P-type and N-type metal-oxide-
semiconductor (P/NMOS) pair was performed. The design was based upon
0.13 μm CMOS process with 3 mW power consumption, achieving divide-by-
2 at 15 GHz, 40 and 50 GHz. Using a similar circuit as Figure 8.18, where
the PMOS cross-coupled transistors were substituted by a single PMOS to
minimize the parasitic capacitance due to the large size of PMOS, Wu and
Yu [52] provided a design guidance in terms of how to choose a proper overdrive voltage of the single injection transistor and how to tradeoff amplitude

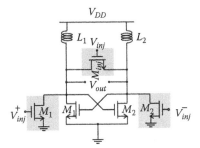

FIGURE 8.18
Direct injection-locked frequency divider.

of output signals for a wider locking range. They stated that maximizing the overdrive voltage of the injection transistor, M_{inj}, gave rise to a wider locking range. The divide-by-2 circuit was verified by simulation and measurement. A locking range from 66.4 to 76 GHz was achieved at relatively low current level, 4.4 mA.

Higher division of the prescaler is more preferable in PLL to lower the dynamic power consumption, proportional to the operation frequency. Figure 8.18 with single injection configuration, divide-by-4 was proposed by Yamamoto and Fujishima [53]. Observe that M_{inj} exhibits a nonlinear relation between I_{ds} and V_{gs}, and it can be formalized by Eq. (8.17), up to the third-order. It is well known that gain compression of the transistor is attributed to the negative α_3 in the strong inversion. This nonlinearity is the basis of implementation of divide-by-4 function. ω_o at one port became $3\omega_o$, multiplied by $\omega_{inj} = 4\omega_o$. Higher harmonics were mostly filtered out by the LC tank, and only the fundamental tone, ω_o, at the other port was extracted. To make stronger nonlinearity, M_{inj} was biased in the weak inversion. Simulation showed the 3rd-order harmonic increases with V_{ds}. The circuit was fabricated in CMOS 90 nm process. 12% locking range was measured with maximum input frequency, 71.6 GHz. The power consumption is 2.75 mW, comparable to [33] but with higher input frequency.

An improvement, employing a 4th-order LC, was made by Wu and Luong [44], to boost the harmonic component of the mixer so as to enhance the lock range, which is critical for the first stage of the dividers to robustly track the output of the VCO. Figure 8.19 is the simplified schematic. The cross-coupled transistors, $M_{1,2}$, provided current to sustain the oscillation at ω_o. One injection transistor, M_{inj}, in a same fashion as [53], operating at the subthreshold region, converted the injection signal at $4\omega_o$ to the corresponding current. The mixing product from M_{inj} consisted of fundamental and 3rd-order harmonic components of ω_o, associated with the 3rd-harmonic mixing and fundamental mixing, respectively. The novelty of this design was the top part of the oscillator, which was a 2nd-order LC tank in Figure 8.18, but was substituted by a 4th-order LC filter. The frequency response of this filter

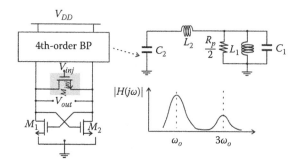

FIGURE 8.19
Direct injection-locked frequency divider with boosted 3rd-order harmonic.

showed intended two peaks at ω_o and $3\omega_o$. The inductance of $L_{1,2}$ and $C_{1,2}$ was determined based on the peak ratio of $Z(j\omega_o)/Z(j3\omega_o)$, setting to close to two-third to avoid detrimental perturbation on ω_o from $3\omega_o$. By properly sizing the transistors, the locking range was improved significantly. The design was implemented in the 65 nm CMOS process. The measurement showed that the lock range was 21.9%, from 58.53 to 72.92 GHz while consuming 2.2 mW.

Another promising divide-by-2 frequency divider shown in Figure 8.20 was demonstrated by Luo and Chen [37]. The authors proposed a dual-direct injection to increase locking range. Comparedwith conventional direct injection frequency dividers, this frequency divider routes the injection signal to the node where $2\omega_o$ can be observed through a coupling capacitor. In particular, a 2nd-order LC tank of self-resonant frequency at $2\omega_o$ was connected to the node. Therefore, a double injection was obtained without consuming more power. Theoretically, the locking range of double-direct injection is twofold that of

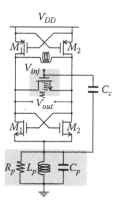

FIGURE 8.20
Dual-direct injection-locked frequency divider.

single-direct injection. The frequency divider was designed in 90 nm CMOS process and achieves ≃42.3% locking range at 45 GHz with only 0.8 mW power dissipation.

Other topologies were also proposed to increase the locking range. For instance, Chen and Tsao [38] utilized double mixers to demonstrate both divide-by-2 and divide-by-4, leading to 37% lock range with 3 mW power consumption. Kuo et al. [46] proposed a modified CML+direct injection cascode divide-by-4 structure, taking advantages of both designs to obtain 77% lock range from 13.5 to 30.5 GHz with 7.3 mW power consumption. In this design, since the first stage was the direct injection divide-by-2 circuit, therefore, the 2nd-order harmonic was used to mix with the injection frequency at $4\omega_o$. Through the second divide-by-2 stage, the modified CML divider, the function of divide-by-4 was performed. It is important to note that the power consumption is 2–3 times higher than that of other frequency dividers due to CML frequency-divider (FD) stacked on injection-locked frequency divider (ILFD).

Table 8.1 shows that odd-division frequency dividers are of great interest when compared with their even-division counterparts. Odd number modulus prescalers consume more power and necessitate more design effort when compared with even number modulus prescalars. For example, to form a divide-by-3 FD, at least two D flip-flops with some extra logic gates are needed whereas one D flip-flop has the full function of divide-by-2. In Figure 8.21, the common-source node of the injection pair exhibits even harmonics of oscillation frequency, i.e. $2\omega_o$, $4\omega_o$.... The mixing product of injection frequency and the frequency at node p gives rise to the desired frequency, ω_o.

Hsieh et al. [54] presented a differential injection to implement division of 3. The simplified circuit is shown in Figure 8.21a. The design was completed in a 65 nm CMOS process with a supply voltage of 1 V. The injection frequency was measured from 58.6 to 67.2 GHz, a divide-by-3 signal at the output was observed. Figure 8.21b shows two improved designs. High division

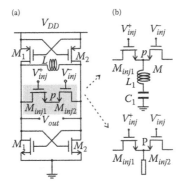

FIGURE 8.21
Direct injection for odd division: (a) [54] (b) [40] and [39].

can be obtained by making use of harmonic boosting technique at the common node of the injection transistors. A dedicated series LC tank in ref. [40] tends to shunt 2nd-order harmonic to the ground such that only strong 4th-order harmonic was left. As a result, divide-by-5 with 735 MHz locking range can be observed. The other design [39] was to use microstrips to create a $\lambda/4$ and $\lambda/2$ open stub at the common node of $M_{inj1,2}$. The transmission line termination seems more efficient than that of LC tank. Drastic improvement with regard to lock range, 4.1 GHz, was obtained. The power consumption was in an acceptable range, 3.75 mW.

8.5   Injection Locking in System Clock

To ensure a reliable communication between base stations and mobile devices, there has been a stringent constraint on the frequency of the timing reference. PLL is a dominant choice in the wireless communication systems because of its robustness. It is, however, generally preferred from a low-power point of view to obtain the system clock locally using an on-chip oscillators. In particular, the bulky and power-hungry crystal, the source of the reference clock in PLL, is necessitated to be obviated in ultra-low-power applications. Much simpler injection-lock based approaches to generate timing references, thereby fulfilling the low-power goal, are preferred [55,11,56–58]

Kocer and Flynn [55] implemented a fully integrated transponder for long-range telemetry applications, where a subharmonic injection-locking technique was adopted. An incident frequency of 450 MHz was injected to the transceiver in which a cross-coupled LC oscillator with cascaded PMOS transistors to provide current sources was running at 900 MHz as a timing reference. Since the design was a power-harvest system, the supply voltage was more prone to be unstable and varies with the time. It is therefore of great interest to stabilize the system clock. Because of the property of the injection-locked oscillator, the phase noise dropped to -113 dBc/Hz from -94 dBc/Hz when it was locked to the "clean" external RF signal. The design was based upon 0.25 μm CMOS process, achieving 7 Mbps data rate.

Bae et al. [57] specialized in body channel communication (BCC), using vertical electrodes through human bodies, having a high conductivity than that of air, to transmit data between Tx and Rx instead of antennas or inductors. BCC outperforms narrowband and ultra-wide band schemes with respect to energy efficiency attributed to the high conductivity of human body. Measurement results of BCC showed that S_{21} performed similar to that of bandpass filters covering frequency band 40–120 MHz for which signal transfer experiences less attenuation. Due to less restrictions on the frequency stability of medical sensor devices in temperature range between 25^{circ} and 45^{circ}, external crystal oscillators were eliminated and subsequently replaced with an open-loop digital controlled oscillator. The major drawback, however, are the low frequency stability of the oscillator, showing more than 10,000 ppm, and frequency variation over time without calibration. Drawing the phase noise close to the "clean" signal source is the unique characteristic of injection-locked oscillators. Injection locking, therefore, was utilized. The RF signal is provided by the base station and conveys the calibrating signal periodically to stabilize the system clock of the transceiver. Digital calibration with 100 kHz/code resolution was performed. The design was implemented using 180 nm CMOS process. With double frequency-shift-keying (FSK) modulation scheme, the data rate was tested at 1 kbps–10 Mbps.

8.6   Injection Locking in Ultra-Low-Power Receivers

Emerging wireless sensor networks (WSNs), a component of Internet of Things, bring autonomous and intelligent environmental monitoring to industrial operations, the natural environment, smart cities, smart homes, and medical healthcare, to name just a few. IEEE 802.15 group provides a number of short-range, low-power wireless scenarios, e.g. wireless body sensor network, Bluetooth, and ZigBee. Unlike other wireless communications, these applications are specialized in short-range, low-date rate, ultra-low-power consumption, and small footprint. For instance, operating at subgigahertz, 402–405 MHz, medical implantable communication service has been proved by Federal Communications Commission for years. The demand for date rate ranges from tens of kb/s to Mb/s with significantly less restriction of sensitivity. Thus, some techniques that are not applicable to high data rate access, e.g. 802.11, WCDMA, etc., become feasible and have enjoyed popularities in the past decades. It is well known that modulation schemes have a significant impact on transceiver architectures and transceiver architectures trade energy efficiency for spectral efficiency. In our discussion, the former is more in favor attributed to the aggressive demand for the expansion of the lifetime of devices. There have been three modulation schemes widely appreciated in ultra-low-power transceivers, namely on–off-keying (OOK), FSK, and phase-shift-keying (PSK). Noncoherent architectures outperform coherent demodulation as less power-hungry blocks are employed. Among noncoherent architectures, frequency/phase-to-amplitude conversion cascaded with an envelop detector approach is attractive due to low complexity and low power consumption. Superregenerative receivers and injection-locking based receivers have drawn attention from industry and academia. Since the core of demodulation is an oscillator and because of their similarity of operation mechanism, superregenerative and injection-locking techniques are discussed in this section briefly. Table 8.2 summarizes recently developed receivers. It is noticed that amplitude modulation is widely adopted in superregenerative

TABLE 8.2

The state-of-the-art ultra-low-power receiver. (SR: Superregenerative; IL: Injection locking)

Ref.	Year	Process (nm)	Mode	Freq. band (GHz)	Data rate (Mbps)	Power Cons. (μW)	Sens. (dBm)	Mod.
[61]	2005	–	SR	1.9	0.005	450	−100.5	OOK
[59]	2007	130	SR	2.4	0.5	2800	−90	OOK
[62]	2009	90	SR	0.402	0.12	400	−93	OOK
[63]	2016	40	SR	0.9	1	320	−87	OOK
[60]	2010	180	SR	2.4	2	315	−70	BFSK
[64]	2010	90	IL	0.3	1	120	−34	BPSK
[65]	2011	65	IL	0.75	5	228	−43	BPSK
[56]	2011	180	IL	0.92	5	420	−73	BFSK
[66]	2014	180	IL	0.435	10	1770	−63	BPSK
[67]	2014	130	IL	0.915	3	250	−63	BFSK
[58]	2015	180	IL	0.08	0.312	45	−62	BFSK

receivers, whereas frequency modulation is more suitable for injection-locking-based receivers. This distinguishable preference stems from the characteristics of the oscillators. Section 8.6.1 discusses the fundamentals of superregenerative architectures. Chen et al. [59] and Ayers et al. [60] are the two case studies in this section. Section 8.6.2 shows injection-locking-based receivers.

8.6.1   Low-power super-regenerative receivers

Categorized into noncoherent receivers, superregenerative architecture guarantees low complexity and consequently low-power consumption. Instead of incorporating an low noise amplifier (LNA), a mixer and a set of amplifiers, in superregenerative architectures, an oscillator, controlled by a periodically "quenched" signal, is deployed after a more relaxed LNA to eliminate the need of the mixer and the amplifiers. The first superregenerative receiver was invented by Armstong [68], 95 years ago, in 1922. It was immediately adopted in wireless applications due to the large gain provided by regenerative mechanism. The poor sensitivity, however, is the main reason that it was replaced by the superheterodyne architecture later, which was invented by the same person. Recently, with the advent of emerging ultra-low-power applications, the superregenerative achitecture becomes one of the candidates to reduce cost and power consumption. It is most widely used to demodulate OOK signals [61, 59, 62, 69, 63]. Some recent designs prove that advanced modulation schemes such as FSK and PSK can also be implemented based upon the superregenerative architecture, [60–71].

Figure 8.22 depicts the architecture of a superregenerative receiver, consisting of an isolation low noise amplifier, a VCO, an envelop detector, and a comparator. The main purpose of first stage is to reduce coupling between the oscillator and the antenna. The large signal generated by the oscillator tends to radiate to space through substrate and wire coupling to the antenna. Therefore, there is a need for isolating the antenna from the following stages while amplifying weak RF signals.

The key component of the receiver is the oscillator, best understood by a one-port mode in Figure 8.23 [14]. The LC tank consists of an inductor L, a capacitor C, a resistor representing the loss of the tank, and an active device that provides a negative resistance to compensate for the loss of the tank. We know that the oscillator starts amplifying thermal noise at any node of the circuit and subsequently oscillating once $g_m \geq 1/R$. On the other hand, if $g_m < 1/R$, the active device does not provide sufficient energy to compensate for the loss of tank such that the oscillation dies out. These two conditions are defined as "operation mode" and "quenched mode" in superregenerative receivers. The transconductance of the oscillator in this case becomes time varying, controlled by the quench signal. The response of the oscillator is triggered either by internal thermal noise or by external excitation.

Given the incident signal, $A \sin \omega t$, the output voltage of the LC tank can be found by solving the differential equation:

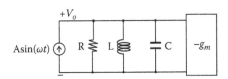

FIGURE 8.22
Superregenerative receiver.

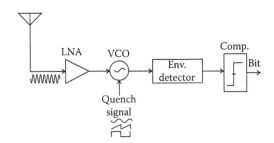

FIGURE 8.23
One-port mode of the oscillator.

$$A\sin(\omega t) = C\frac{dV_o}{dt} + GV_o + \frac{1}{L}\int V_o d\tau, \qquad (8.18)$$

where $G = 1/R - g_m$. Solving for V_o [60],

$$V_o = e^{-\alpha t}(A_1\cos\omega_d t + A_2\sin\omega_d t) + \frac{A\sin\omega t}{\sqrt{G^2 + (\omega t - 1/\omega L)^2}}, \qquad (8.19)$$

where $\alpha = G/2C$ is defined as a damping factor. The damping frequency ω_d is $\sqrt{\omega_o^2 - \alpha^2}$. And $\omega_o = 1/\sqrt{LC}$.

The exponential term in Eq. (8.19) represents the natural response of the oscillator. It is independent of the incident signal, and it describes the transient behavior at w_d. The second term describes the voltage response due to the input signal. If the oscillator is in the quenched mode, Barkhausen criteria does not hold, giving rise to two results: (i) oscillation is not able to be built up and (ii) previous oscillation dies out. Therefore, the first term is negligible and the second term dominates the output voltage of the LC tank. If the oscillator is in the oscillation mode, the exponential term becomes dominant, the output voltage grows exponentially in time. Under this condition, injection signal leads to a different startup time. For instance, if an OOK-modulated signal is injected to the oscillator in the oscillation mode, bit "1" causes the oscillation to start up much faster than that of bit "0". The time difference is then captured by the envelop detector and the comparator. The number of bits "1" at the output of the comparator is used to determine the received signal to be either "1" or "0". After this is done, the quench signal is active to let the oscillator die out. The operation periodically occurs as shown in Figure 8.24, where $t_2 > t_1$. In practice, the quench frequency is either aligned with or above the symbol rate. One may notice that in Figure 8.22, there are different types of quench signals. The shape of the quench signal is tightly associated with the selectivity and maximum data rate. More insights of the circuit behavior can be found in ref. [72].

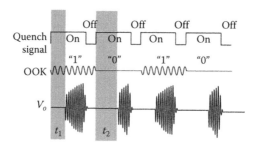

FIGURE 8.24

Response of the oscillator according to the quench signal and OOK signal.

8.6.1.1 Q enhancement superregenerative receiver

Preceding discussion shows that the demodulation of the superregenerative receiver is realized by observing the different time-to-oscillation in the oscillation mode. In other words, the startup time of the oscillator indicates the strength of excited input signal. Ideally, if only the desired signal is received by the antenna, the super-regenerative architecture will be a perfect solution for any wireless data access. Unfortunately, in practice, incoming signal is often accompanied by in-band or out-of-band interferences due to more crowded frequency bands in the lower gigahertz. Even worse, the interferers are several tens of decibels stronger than the desired signal. As a result, the oscillator is more prone to the external "noise" signal, resulting in poor sensitivity and low bit-error-rate (BER).

Chen et al. proposed an Q-enhancement approach to improve the selectivity, as shown in Figure 8.25a [59], One may notice that the second term in Eq. (8.19) is in the form of a 2nd-order band-pass filter with the center frequency at the free-running frequency of the oscillator and the quality factor $Q = R/\sqrt{L/C}$. We also know that the selectivity of the band-pass filter depends on the value of Q. The quality factor of on-chip spiral inductors is usually smaller than 10. Therefore, the overall Q of LC tank is limited by the quality factor of the inductor. To have a high selectivity, Q must be improved. This was achieved using adjustable g_m to compensate the loss of the tank while still keeping the oscillator not to oscillate, i.e. making g_m close to $1/R$, but $g_m < 1/R$ still holds. In the proposed design, the quench cycle was designed

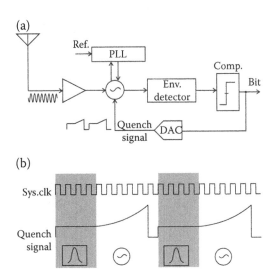

FIGURE 8.25
(a) Simplified architecture of Q enhancement superregenerative receiver and (b) timing diagram of the quench signal.

to have two phases: (i) Q enhancement duration and (ii) superregenerative duration. Each of the quench signal has 10 cycles of the system clock, 5 MHz. The first four clock cycles were reserved for Q enhancement mode. During this duration, the bias current of the cross-coupled oscillator was configured lower than the critical current, I_{crt}, which was the minimum required current to build up the oscillation, such that the oscillator was treated as a 2nd-order band-pass filter with an enhanced Q. The bias current was precisely generated from a digital-to-analog converter (DAC) with an least significant bit (LSB) size of 2 μA. The next five clock cycles were used to set the oscillator in the superregenerative mode in which the oscillator was operating at 2.4 GHz to detect the OOK signal. The oscillator decayed in the last clock cycle. Figure 8.25b shows the timing of the quench signal.

The design was implemented in a CMOS 130 nm process, covering frequencies from 2.35 to 2.53 GHz. The measurements showed that the adjacent interferer rejection was 30 dB at 10 MHz offset and 10 dB at 3 MHz offset, leading to estimated Q of 1400. The minimum selectivity was -90 dBm at 500 kbps while it was deteriorated to -60 dBm at 1 Mbps. The entire receiver drew 2.4 mA current in average with 1.2-V supply voltage.

8.6.1.2   BFSK superregenerative receiver

Due to the nature of self-sustaining autonomous systems, time-to-oscillation is subject to excitation strength, resulting in distinguishable bit "1" and "0" from OOK modulation. OOK modulation combines the low power and low cost, and low data rate as well. To improve receiver performance and energy efficiency, binary-frequency-shift-keying (BFSK) modulation can be considered as one of the underlying solutions. We know that digital signals are modulated to two frequencies in BFSK:

$$y(t) = a \cos \omega_1 t + b \cos \omega_2 t, \tag{8.20}$$

where $a, b = 0,1$ or $1,0$, and $\omega_{1,2} = 2\pi f_{1,2}$ are two carrier frequencies. The difference of ω_1 and ω_2 is dependent on modulation index. To demodulate FSK signals, the receiver in general incorporate a PLL to implement coherent demodulation.

Ayers et al. [60] demonstrated a superregenerative-based BFSK receiver by periodically alternating the center frequency of the oscillator between f_1 and f_2. Unlike Chen's design, the power-hungry building block, PLL, was eliminated as the loose frequency tolerance was adopted. In addition, there was no isolation amplifier to decouple the strong signal from the oscillator to the antenna for the sake of low energy consumed. Since the receiver was directly connected to the antenna, which exhibits a 50 Ω impedance, a matching network was performed in the quenched mode when the oscillation was not built up. In the operation mode, there is no need for the matching network. The bias current of the oscillator was controlled by the output of a DAC, reserving five bits for the quench mode and three bits for the operation mode. The receiver is shown in Figure 8.26a. To demodulate the BFSK signal, once the quench signal was released, the oscillator was entering the oscillation mode where the operation frequency was setting at f_1 first. The received signal was used as the excitation to start up the oscillation. The time-to-oscillation was then recorded by the digital logic after the envelop detector and the comparator. Then, the quench signal forced the oscillation to be decayed, and the second frequency, f_2, was set thereafter. The quench signal was released again, and the received signal was detected in an analogous way to the prior detection, except that the time-to-oscillation shows difference value. The key point here is these two operations were completed within the same bit period. For instance, in Figure 8.26b, the first bit period, bit "0", was associated with f_1. The quench signal turned ON/OFF twice in a single bit period, getting two time-to-oscillation values to be stored. Only the corresponding frequency f_1 resulted in the shorter startup time t_1. Longer time-to-oscillation t_2 indicated unmatched carrier frequency. Similarly, for bit "1", as the carrier frequency was f_2, the faster startup at f_2 frequency could be expected. The demodulation was depending on finding out the smaller value in t_3 and t_4. We can also understand the behavior from a sampling point of view. Each single bit has two sampled signals, and the information was stored in the sampled data for demodulation.

The design was implemented in a 180 nm process with supply voltage 0.6 V. To overcome the narrow tuning range of the varactor due to the supply voltage drop, a couple DC–DC converters were designed to enlarge the control voltage of the varactor. A −70 dBm sensitivity was achieved at the data rate 2 Mbps

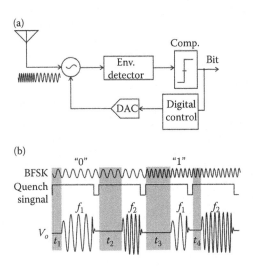

FIGURE 8.26

(a) Simplified architecture of the superregenerative receiver for BFSK modulation and (b) timing diagram of demodulation ($t_1 < t_2$: bit "0", $t_4 < t_3$: bit "1").

while consuming 350 μW. At the data rate 250 kbps, the power consumption
was 215 μW, and the sensitivity was -86 dBm.

8.6.2 Low-power injection-locking receivers

One of the major drawbacks of superregenerative receivers is low data
rate due to the mechanism of demodulation. Recently, FSK and PSK have
received more attention [70,71,60]. OOK modulation scheme is still dominant.
By contrast, injection-locking-based receivers offer a higher data rate while
consuming a comparable amount of power. Further, the complexity of the
superregenerative brings more design efforts than that of injection-locking-
based receivers. In particular, the generation of the quench signal affects the
performance of the receivers. Reference [72] has shed the light on the role of the
quench signal in the detection. It is crucial to generate an optimal waveform
to control the oscillator, thereby achieving better signal detection. In addition,
the circuits that are used to generate the quench signal also consume extra
power. All of aforementioned can be overcome by injection-locking techniques
deployed in the receiver to demodulate signals.

Figure 8.27 illustrates an injection-locked receiver, being composed of four
main building blocks, e.g. the LNA, the oscillator, the envelop detector, and
the 1-bit comparator. It is a typical noncoherent demodulator except the
inserted oscillator, which is the vital component of the architecture. Readers
may have noticed that, in comparison with Figure 8.22, the only difference is
the missing quench signal, turning the oscillator ON/OFF in the superregenerative receivers. With this observation, using the time-to-oscillation to detect
incoming signals is inapplicable to injection-locked receivers as the oscillator is
always running, leading to the loss of the timing information from the startup
of the oscillator. Thus, we have to figure out another way to discriminate the
modulated signals. In ref. [56], a feedback oscillator was analyzed to attain a
close-form expression for the output voltage of the oscillator. Equation (8.21)
shows that output voltage is proportional to the amplitude of the excitation
and the phase difference of oscillation signal and the incident signal.

$$\sqrt{V_{osc}^2 + 2V_{osc}V_{inj}\cos\theta + V_{inj}^2}. \tag{8.21}$$

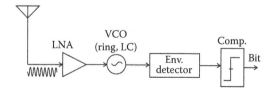

FIGURE 8.27
Injection-locked receiver.

where θ is the phase difference. Given $V_{inj} \ll V_{osc}$, and within the locking range, Eq. (8.21) holds, which is the basis of frequency-to-amplitude conversion. Figure 8.28 depicts how a injection-locked oscillator react if it is locked to the BFSK signals. An example in Figure 8.28b shows a BFSK signal modulated at carrier frequency $\omega_{1,2}$. The output voltage of the injection-locked oscillator varies in response to locked two carrier frequencies, such that the envelop detector generates two level signals.

8.6.2.1 BFSK body sensor injection-locked receiver

Emerging wireless body sensor networks offer seamless connection between various places of patients, enhancing the healthcare services. The sensor types must be wearable, injectable, and implantable. Among design challenges, a limited power budget is crucial, e.g. less than mW for wearable sensors. We have seen many PLL-based solutions in the superregenerative architecture, which apparently are not suitable for these applications. In [56], Bae et al. demonstrated a frequency-synthesizer-less, BFSK injection-locked transceiver.

Figure 8.29a depicts the architecture of the injection-locked transceiver. The key building block is the injection-locked frequency divider, with division of 2. In accordance with the earlier discussion, within the locking range, the amplitude of the output voltage is proportional to the incident frequency, which is the root of BFSK demodulation. According to Eq. (8.20), the modulated signal in 900 MHz was first amplified by the LNA. A simple cascode and interver circuits offer a 30 dB gain, low noise performance, and good reverse isolation. The amplified signal was injected into the gate of the biasing transistor in the same fashion as that in Figure 8.15. As the natural frequency of the LC oscillator was designed at 450 MHz, the divide-by-2 frequency was obtained. To increase the locking range, the shunt-peaking technique [32] was utilized to increase the internal injection power, thereby enlarging the locking range. The output of the oscillator exhibited two DC levels associated with

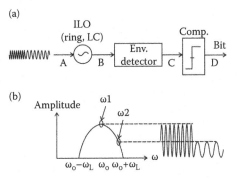

FIGURE 8.28
Frequency-to-amplitude conversion.

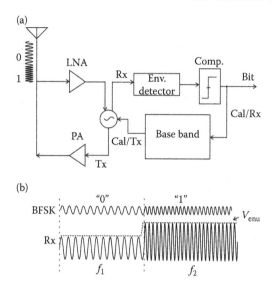

FIGURE 8.29

(a) Diagram of [56]'s transceiver, and (b) envelope of frequency-to-amplitude.

two carrier frequencies. The envelope detector followed by the comparater was therefore performing analog-to-digital conversion. In the transmitter side, the oscillator was used as a digital-controlled oscillator to modulate the digital signal. As the carrier frequency was in 900 MHz range, the common-source node of the cross-coupled transistor in the oscillator was therefore used as the output port. The signal was sending out through a switching power amplifier.

With the mitigated sensitivity, PLL and external crystal oscillator were eliminated. The compensation for frequency drift due to process, voltage and temperature (PVT), however, is still necessitated. Open-loop frequency calibration was implemented by injection locking via the receiving path. The base station sent out a clean carrier signal located at the center of the locking range periodically. The digital-controlled oscillator first swept the code from lower frequency to higher frequency until the status of the oscillator was changed from pulling to locking. The code C_1 was recorded by the digital logic. Then, keep increasing the code until the pulling occurs again. Record the code, C_2, which is the upper bound of the locking range. The average of these two codes was taken as the calibrated result with resolution of 100 kHz/code.

The design was implemented in a CMOS 180 nm process with 0.7 V supply voltage. Operating at 920 MHz, the author claimed -73 dBm sensitivity was achieved at 5 Mbps, only consuming 420 μW. Because of the use of LC-based ILFD as the frequency-to-amplitude, it has a narrow locking range of 5 MHz. The blocker measurements showed that, at 2.5 MHz offset, 10 dB rejection can be obtained. All the inductors were off-chip low-cost inductors.

8.6.2.2 BFSK body sensor injection-locked wake-up receiver

In WSNs or other applications that demand low-power, a promising solution to reduce power consumption is to implement a duty-cycling protocol, where there is a main receiver and a wake-up receiver. In such a sensor node, the main receiver is turned ON/OFF depending upon the signal that is from the wake-up receiver. This special timing scenario requests that the wake-up receiver be always active while consuming a minimum amount of power as only signaling is transferred from the base station [73]. The design challenge again comes to the minimizing energy consumed by the wake-up receiver.

The same group from Korea Advanced Institute of Science and Technology in ref. [58] presented an ultra-low-power wake-up receiver using dual-mode injection locking demodulation. The realization of the wake-up receiver was operating at 80 MHz, consisting of a receiving mode and a calibration mode. The carrier frequencies were 72 and 80 MHz. Instead of detecting radio signal, in this design, BFSK signals were received from electrode. Unlike prior work, injection pulling was adopted to demodulate the BFSK signals so as to decrease hardware complexity. In other words, the frequency difference, $\Delta\omega$ = 82–72 MHz, was out of the locking range. In the design, data "1" and "0" were modulated with 80 and 72 MHz, respectively. A 72 MHz carrier frequency was associated with injection pulling, whereas 80 MHz was within the locking range. The characteristic of the amplitude of the injection pulling signal played a key role in demodulation. In comparison with injection locking signal, which has constant amplitude theoretically, the phase of injection pulling signal exhibited time-varying property and subsequently amplitude variation. As a result, a constant DC voltage at the output of the envelop detector was the condition of the determination of the FSK signal. Injection lock and injection pulling signal are shown in Figure 8.30.

The BFSK demodulation was demonstrated in a CMOS 180 nm process. To reduce area and the power consumption, an injection-locked ring oscillator was designed. With the power consumption of 45 μW, 312 kbps data rate was achieved. Once the receiver entered the calibration mode, information

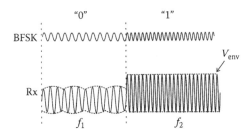

FIGURE 8.30
BFSK demodulation in ref. [58]. Note, f_1 is injection pulling and f_2 is injection locked.

was stopped from the base station. Instead, a "clean" 80 MHz RF signal was entering the receiver to stabilize the frequency drift of the oscillator due to PVT.

8.6.2.3 BPSK injection-locked receivers

So far, we have seen OOK/FSK modulation schemes in superregenerative receivers and injection-locked receivers. For the sake of high spectral efficiency and high data rate, PSK is more preferable. BPSK can be found in numerous wireless systems, for example, Bluetooth, RFID, and Global Positioning System (GPS). The digital bits are denoted by the phase changes, e.g. 180°, of the carrier frequency instead of frequency and amplitude variations. It is, therefore, ideally very suitable for nonlinear power amplifiers in the transceiver. But in practice, the pulse shaping filter is often inserted to limit signal bandwidth, leading to a large change of the amplitude each time the phase makes a transition. As a result, a linear power amplifier is usually inevitable [14]. For receivers, coherent demodulation provided a 3 dB advantage in signal-to-noise ratio (SNR) in comparison with noncoherent demodulation that requires more complex architecture, subsequently high power consumption. The frequency-to-amplitude property of injection-locked oscillators inspires new solutions for low-power BPSK receivers.

The demodulation of injection-locked receiver was analyzed in ref. [74] and demonstrated in ref. [75] using discrete components. References [64] and [65] verified the design in a 90- and 65 nm CMOS process, respectively. Due to the nature of differential pair with a tail current source, two divide-by-2 injection-locked frequency dividers, designed to operate at f_1 and f_2, were

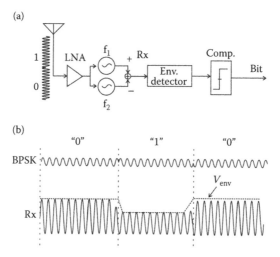

FIGURE 8.31

(a) The simplified BPSK receiver and (b) the output signal after the combiner.

employed. Notice that $f_1 < 1/2 f_{inj}$, and $f_2 > 1/2 f_{inj}$ must be within the lock range. Each time, digital transition induces a phase transition, 180°, which must be compensated by the phase shift from two oscillators. One introduced leading 90° while the other provided lagging 90°, which in total was 180°. The output of the oscillator was combined after the oscillator. The amplitude of the combination exhibited envelop variation from maximum to minimum according to the phase shift of the dividers. A simplified receiver is shown in Figure 8.31.

Short-range, low-rate applications necessitates power/area-conscious design. Reference [64] proposed to use two ring oscillators to demodulate BPSK signals. In addition, extremely low-power, 120 μW, 1 Mbps at 300 MHz frequency band makes this design attractive for WSNs and implantable devices. Low sensitivity, e.g. −34 dBm, was obtained. A similar design [65] demonstrated 5 Mbps at 750 MHz carrier frequency. The sensitivity was improved to −43 dBm with power consumption of 228 μW.

8.7 Summary

A comprehensive treatment of fundamental, circuit topologies, and performance of injection-locking techniques was presented. We showed that a linear-feedback model can be applied to analyze both harmonic and nonharmonic injection-locked oscillations.

The implementation and the performance of injection-locked frequency dividers were investigated. A number of reported designs were examined with an emphasis on their advantages in comparison with conventional frequency dividers. Injection-locked frequency dividers enjoy low-power, high-speed, and flexibility.

Injection-locked oscillators as the timing reference of wireless systems to substitute PLL were investigated. The crystal-less design is very attractive for low-power and low-cost applications.

The use of injection-locking techniques in wireless transceivers was also investigated. The demodulation of OOK, BFSK, and BPSK based upon injection-locked oscillators or injection-locked frequency dividers were presented and their advantages over superregenerative receivers were explored.

Bibliography

[1] H. M. Oliveira and L. V. Melo, "Huygens synchronization of two clocks," *Scientific Reports/online*, vol. 5, no. 11548, pp. 1–12, Jul. 2015.

[2] R.Adler, "A study of locking phenomena in oscillators," *Proceedings of IEEE (reproduced from 1946 version)*, vol. 61, no. 10, pp. 1380–1385, Oct. 1973.

[3] H. Stover, "Theoretical explanation for the output spectra of unlocked driven oscillators," *Proceedings of the IEEE*, vol. 54, pp. 310–311, Feb. 1966.

[4] H. Rategh and T. H. Lee, "Superharmonic injection-locked frequency dividers," *IEEE Journal of Solid-State Circuits*, vol. 34, no. 6, pp. 813–821, Jun. 1999.

[5] S. Verma, H. R. Rategh, and T. H. Lee, "A unified model for injection-locked frequency dividers," *IEEE Journal of Solid-State Circuits*, vol. 38, no. 6, pp. 1015–1028, Jun. 2003.

[6] A. Mazzanti, P. Uggetti, and F. Svelto, "Analysis and design of injection-locked LC dividers for quadrature generation," *IEEE Journal of Solid-State Circuits*, vol. 39, no. 9, pp. 1425–1433, Sept. 2004.

[7] B. Razavi, "A study of injection locking and pulling in oscillators," *IEEE Journal of Solid-State Circuits*, vol. 39, no. 9, pp. 1415–1424, Sept. 2004.

[8] J. Chien and L. Lu, "Analysis and design of wideband injection-locked ring oscillators with multiple-input injection," *IEEE Journal of Solid-State Circuits*, vol. 42, no. 9, pp. 1906–1915, Mar. 2007.

[9] A. Mirzaei, M. Heidari, R. Bagheriand, S. Chehrai, and A. Abidi, "The quadrature LC oscillator: A complete portrait based on injection locking," *IEEE Journal of Solid-State Circuits*, vol. 42, no. 9, pp. 1916–1932, Sept. 2007.

[10] L. N. S. Patnaik and R. Harjani, "Understanding the transient behavior of injection lcoked LC oscillators," in *Proceedings of IEEE Custom Integrated Circuits Conference*, pp. 667–670, San Jose, CA, 16–19 Sept. 2007.

[11] N. Soltani and F. Yuan, "Analysis and design of non-harmonic injection-locked phase-locked loops with applications in remote frequency calibration of passive wireless transponders," *IEEE Trans. Circuits and Systems- I: Redular Papers*, vol. 57, no. 12, pp. 2381–2393, Sept. 2010.

[12] P. Maffezzoni, D. Amore, S. Daneshgar, and M. Kennedy, "Analysis and design of injection locked frequency dividers by means of a phase do- main macromodel," *IEEE Transactions on Circuits and Systems-I: Redular Papers*, vol. 57, no. 11, pp. 2956–2966, Nov. 2010.

[13] B. Razavi, "A study of phase noise in CMOS oscillators," *IEEE Journal of Solid-State Circuits*, vol. 31, no. 3, pp. 331–343, Mar. 1996.

[14] ——, *RF Microelectronics- 2nd edition*. New York: Prentice-Hall, 2012.

[15] Y. Zhou and F. Yuan, "A study of lock range of injection-locked CMOS active-inductor oscillators using a linear control system approach," *IEEE Transactions on Circuits and Systems II - Express Briefs*, vol. 58, no. 10, pp. 627–631, Oct. 2011.

[16] S. Mitra, *Digital Signal Processing: A Computer-Based Approach*. Boston MA: McGraw-Hill, 2006.

[17] M. Schetzen, *The Volterra and Wiener Theory of Nonlinear Systems*. New York: John Wileys and Sons, 1981.

[18] P. Wampacq and W. Sansen, *Distortion Analysis of Analog Integrated Circuits*. Boston MA: Kluwer Academic Publishers, 1998.

[19] F. Yuan and A. Opal, "Distortion analysis of periodically switched nonlin- ear circuits using time-varying volterra series," *IEEE Transactions on Circuits and Systems I - Fundamental Theory and Applications*, vol. 48, no. 6, pp. 726–738, Jun. 2001.

[20] ——, "An efficient transient analysis algorithm for mildly nonlinear cir- cuits," *IEEE Transactions on Computer-Aided Design of Integrated Cir- cuits System*, vol. 21, no. 6, pp. 662–673, Jun. 2002.

[21] Q. Li and F. Yuan, "Time domain response and sensitivity of periodi- cally switched nonlinear circuits," in *IEEE Transactions on Circuits and Systems I - Fundamental Theory and Applications*, vol. 50, no. 11, pp. 1436–1446, Nov. 2003.

[22] F. Yuan and Y. Zhou, "A phasor-domain study of lock range of har- monic oscillators with multiple injections," *IEEE Transactions on Cir- cuits and Systems II - Express Briefs*, vol. 59, no. 8, pp. 466–470, Oct. 2012.

[23] ——, "Frequency-domain study of lock range of non-harmonic oscilla- tors with multiple multi-tone injections," *IEEE Transactions on Circuits and Systems I - Regular Papers*, vol. 60, no. 6, pp. 1395–1406, Jun. 2013.

[24] Y. Zhou and F. Yuan, "A study of injection-locked non-harmonic oscil- lators using Volterra series," *IET Circuits Devices and Systems*, vol. 9, no. 2, pp. 119–130, Mar. 2015.

[25] Q. Huang and R. Rogenmoser, "Speed optimization of edge-triggered CMOS circuits for Gigahertz single-phase clocks," *IEEE Journal of Solid-State Circuits*, vol. 31, no. 3, Mar. 1996.

[26] M. Tiebout, "A 480 μW 2 GHz ultra-low power dual-modulus prescaler in 0.25 μm standard CMOS," in *Proceedings of IEEE Custom Integrated Circuits Conference*, vol. 5, 2000, pp. 741–744, Geneva, Switzerland, 28–31 May 2000.

[27] F. Liu, Z. Wang, Z. Li, Q. Li, G. Yang, and T. Guo, "Low-power 25.4–33.5 GHz programmable multi-modulus frequency divider," *Electronics Letters*, vol. 50, no. 15, pp. 1067–1068, Jul. 2014.

[28] J. Luo, L. Zhang, L. Zhang, Y. Wang, and Z. Yu, "Behavioral analysis and optimization of CMOS CML dividers for millimeter-wave applications," *IEEE Transactions on Circuits and Systems II: Express Briefs*, vol. 62, no. 3, pp. 256–260, Mar. 2015.

[29] R. L. Miller, "Fractional-frequency generators utilizing regenerative modulation," *Proceedings of the IRE*, vol. 27, no. 7, pp. 446–457, Jul. 1939.

[30] R. G. Harrison, "Theory of regenerative frequency dividers using double-balanced mixers," *Proceedings of MTT-S International Microwave Symposium Digest*, pp. 459–462 vol.1, Jun. 1989.

[31] H. Rategh, H. Samavati, and T. H. Lee, "A CMOS frequency synthesizer with an injection-locked frequency divider for a 5 GHz wireless LAN receiver," *IEEE Journal of Solid-State Circuits*, vol. 35, no. 5, pp. 780–787, May 2000.

[32] H. Wu and A. Hajimiri, "A 19 GHz, 0.5mW, 0.35 μm CMOS frequency divider with shunt-peaking locking range enhancement," in *Proceedings of IEEE International Solid-State Circuits Conference.*, pp. 412–413, San Francisco, CA, 4–7 Feb. 2001.

[33] M. Tiebout, "A CMOS direct injection-locked oscillator topology as high-frequency low-power frequency divider," *IEEE Journal of Solid-State Cir- cuits*, vol. 39, no. 7, pp. 1170–1174, Jul. 2004.

[34] K. Yamamoto and M. Fujishima, "A 44 μW 4.3 GHz injection-locked frequency divider with 2.3 GHz locking range," *IEEE Journal of Solid-State Circuits*, vol. 40, no. 3, pp. 671–677, Mar. 2005.

[35] C. Wang, C. Chen, M. Lei, M. Chuang, and H. Wang, "A 66-72 GHz divide-by-3 injection-locked frequency divider in 0.13μm CMOS technology," in *Proceedings of IEEE Asian Solid-State Circuits Conference*, pp. 344–347, Jeju, South Korea, 12–14 Nov. 2007.

[36] S. Cheng, H. Tong, J. Silva-Martinez, and A. l. Karsilayan, "A fully differential low-power divide-by-8 injection-locked frequency divider up to 18 GHz," *IEEE Journal of Solid-State Circuits*, vol. 42, no. 3, pp. 583–591, Mar. 2007.

[37] T. N. Luo and Y. J. E. Chen, "A 0.8 mW 55 GHz dual-injection-locked CMOS frequency divider," *IEEE Transactions on Microwave Theory and Techniques*, vol. 56, no. 3, pp. 620–625, Mar. 2008.

[38] C. C. Chen, H. W. Tsao, and H. Wang, "Design and analysis of CMOS frequency dividers with wide input locking ranges," *IEEE Transactions on Microwave Theory and Techniques*, vol. 57, no. 12, pp. 3060–3069, Dec. 2009.

[39] M. W. Li, H. C. Kuo, T. H. Huang, and H. R. Chuang, "60GHz CMOS divide-by-5 injection-locked frequency divider with an open-stub-loaded floating-source injector," in *Proceedings of IEEE Radio Frequency Integrated Circuits Symposium*, pp. 1–4, Baltimore, MD, 5–7 June 2011.

[40] P. K. Tsai, T. H. Huang, and Y. H. Pang, "CMOS 40 GHz divide-by-5 injection-locked frequency divider," *Electronics Letter*, vol. 46, no. 14, pp. 1003–1004, Jul. 2010.

[41] J. H. Huang, X. P. Yu, S. Y. Xu, J. Jin, and F. X. Yu, "19.1 GHz 18 mW divide-by-3 heterodyne injection locking frequency divider in 0.18 μm CMOS technology," *Electronics Letter*, vol. 52, no. 12, pp. 1076–1078, Jun. 2016.

[42] S. L. Jang, W. C. Lai, S. S. Tzeng, and C. W. Hsue, "A wide-band divide-by-3 injection-locked frequency divider using tunable MOS resistor," in *2015 IEEE Asian Solid-State Circuits Conference (A-SSCC)*, pp. 1–5, Xiamen, China, 9–11 Nov. 2015.

[43] Y. H. Chang and Y.-C. Chiang, "A divide-by-3 injection-locked frequency divider in 0.18 μm CMOS process for K band applications," in *2015 IEEE MTT-S International Microwave Symposium*, pp. 1–3, May 2015.

[44] L. Wu and H. C. Luong, "Analysis and design of a 0.6 v 2.2 mW 58.5- to-72.9 Ghz divide-by-4 injection-locked frequency divider with harmonic boosting," *IEEE Transactions on Circuits and Systems I: Regular Papers*, vol. 60, no. 8, pp. 2001–2008, Aug. 2013.

[45] S. L. Jang, T. C. Kung, and C. W. Hsue, "Wide-locking range divide-by-4 injection-locked frequency divider using linear mixer approach," *IEEE Microwave and Wireless Components Letters*, vol. 27, no. 4, pp. 398–400, Apr. 2017.

[46] Y. H. Kuo, J. H. Tsai, H. Y. Chang, and T. W. Huang, "Design and analysis of a 77.3% locking-range divide-by-4 frequency divider," *IEEE Transactions on Microwave Theory and Techniques*, vol. 59, no. 10, pp. 2477–2485, Oct. 2011.

[47] K. F. Chang and K. K. M. Cheng, "Speed enhancement technique for divide-by-4 injection-locked frequency divider and its application to frac- tional division," in *Proceedings of MTT-S International Microwave Sympo- sium*, pp. 1–4, Baltimore, MD, 5–10 June 2011.

[48] S. L. Jang, W. C. Cheng, C. W. Hsue, and M. H. Juang, "Triple-resonance rlc-tank divide-by-2 injection-locked frequency divider," *Electronics Letters*, vol. 52, no. 8, pp. 624–626, 2016.

[49] J. Yuan and C. Sevensson, "High-speed CMOS circuit technique," *IEEE Journal of Solid-State Circuits*, vol. 24, pp. 62–70, Feb. 1989.

[50] S. L. Jang, W. C. Cheng, and C. W. Hsue, "Wide-locking range divide-by-3 injection-locked frequency divider using sixth-order RLC resonator," *IEEE Transactions on Very Large Scale Integration (VLSI) Systems*, vol. 24, no. 7, pp. 2598–2602, Jul. 2016.

[51] H. Wu and L. Zhang, "A 16 to 18 GHz 0.18 μm EPI-CMOS divid- by-3 injection-locked frequency divider," in *IEEE International Solid-State Circuits Conference, Digest of Technical Papers*, pp. 602–603, San Francisco, CA, 6–9 Feb. 2006.

[52] C. Y. Wu and C. Y. Yu, "Design and analysis of a millimeter-wave direct injection-locked frequency divider with large frequency locking range," *IEEE Transactions on Microwave Theory and Techniques*, vol. 55, no. 8, pp. 1649–1658, Aug. 2007.

[53] K. Yamamoto and M. Fujishima, "70 GHz CMOS harmonic injection-locked divider," in *2006 IEEE International Solid State Circuits Conference - Digest of Technical Papers*, pp. 2472–2481, San Francisco, CA, 6–9 Feb. 2006.

[54] H.-H. Hsieh et. al., "A V-band divide-by-three differential direct injection- locked frequency divider in 65-nm CMOS," in *Proceedings of IEEE Custom Integrated Circuits Conference*, pp. 1–4, Sept. San Jose, CA, 19–22 Sept. 2010.

[55] F. Kocer and M. Flynn, "A new transponder architecture with on-chip ADC for long-range telemetry applications," *IEEE Journal of Solid-State Circuits*, vol. 41, no. 5, pp. 1142–1148, May 2006.

[56] J. Bae, L. Yan, and H. Yoo, "A low energy injection-locked FSK transceiver with frequency-to-amplitude conversion for body sensor appli- cations," *IEEE Journal of Solid-State Circuits*, vol. 46, no. 4, pp. 928–937, Apr. 2011.

[57] J. Bae, K. Song, H. Lee, H. Cho, and H. Yoo, "A low-energy crystal-less double-FSK sensor node transceiver for wireless body-area-netwrok," *IEEE Journal of Solid-State Circuits*, vol. 47, no. 11, pp. 2678–2692, Nov. 2012.

[58] J. Bae and H. Yoo, "A 45 μW injection-locked FSK wake-up receiver with frequency-to-envelope conversion for crystal-less wireless body area network," *IEEE Journal of Solid-State Circuits*, vol. 50, no. 6, pp. 1351–1360, Jun. 2015.

[59] J. Chen, M. Flynn, and J. Hayes, "A fully integrated auto-calibrated super-regenerative receiver in 0.13 μm cmos," *IEEE Journal of Solid-State Circuits*, vol. 42, no. 9, pp. 1976–1985, Sep. 2007.

[60] J. Ayers, K. Mayaram, and T. S. Fiez, "A ultralow-power receiver for wireless sensor networks," *IEEE Journal of Solid-State Circuits*, vol. 45, no. 9, pp. 1759–1769, Sep. 2010.

[61] B. Otis, Y. H. Chee, and J. Rabaey, "A 400 μW-RX, 1.6 mW-TX super-regenerative transceiver for wireless sensor networks," in *Proceedings of IEEE International Solid-State Circuits Conference*, pp. 396–606 Vol. 1, San Francisco, CA, 7–10 Feb. 2005.

[62] J. Bohorquez, A. Chandrakasan, and J. Dawson, "A 350 μW CMOS MSK transmitter and 400 μW OOK super-regenerative receiver for medical impant communications," in *IEEE Journal of Solid-State Circuits*, vol. 44, no. 4, pp. 1248–1259, Apr. 2009.

[63] V. D. Rezaei, S. J. Shellhammer, M. Elkholy, and K. Entesari, "A fully integrated 320 pj/b OOK super-regenerative receiver with -87 dbm sensitivity and self-calibration," in *2016 IEEE Radio Frequency Integrated Cir- cuits Symposium (RFIC)*, pp. 222–225, San Francisco, CA, 22–24 May 2016.

[64] H. Yan, J. G. Macias-Montero, A. Akhnoukh, L. C. N. de Vreede, J. R. Long, J. J. Pekarik, and J. N. Burghartz, "A 120 μW fully-integrated BPSK receiver in 90nm CMOS," in *Proceedings of IEEE Radio Frequency Integrated Circuits Symposium*, pp. 277–280, Anaheim, CA, 23–25 May 2010.

[65] Q. Zhu and Y. Xu, "A 228 μW 750 MHz BPSK demodulator based on injection locking," *IEEE Journal of Solid-State Circuits*, vol. 46, no. 2, pp. 416–423, Feb. 2011.

[66] Y. L. Tsai, J. Y. Chen, B. C. Wang, T. Y. Yeh, and T. H. Lin, "A 400MHz 10Mbps D-BPSK receiver with a reference-less dynamic phase-to-amplitude demodulation technique," in *IEEE Symposium on VLSI Circuits Digest of Technical Papers*, pp. 1–2, Honolulu, HI, 10–13 June 2014.

[67] M. Zgaren and M. Sawan, "Frequency-to-amplitude converter based FSK receiver for ultra-low power transceivers," in *Proceedings of IEEE North-East Workshop on Circuits and System*, Trois-Rivieres, QC, 22–25 June 2014.

[68] E. H. Armstrong, "Some recent developments of regenerative circuits," *Proceedings of IRE*, vol. 10, pp. 244–260, 1922.

[69] P. Thoppay, C. Dehoolain, M. Green, and M. Declercq, "A 0.24-nJ/bit super-regenerative pulsed UWB receiver in 0.18 μm CMOS," *IEEE Journal of Solid-State Circuits*, vol. 46, no. 11, pp. 2623–2634, Nov. 2011.

[70] P. Pala-Schonwalder, F. X. Moncunill-Geniz, J. Bonet-Dalmau, F. del Aguila-Lopez, and R. Giralt-Mas, "A BPSK superregenerative receiver. preliminary results," in *Proceedings of IEEE International Symposium on Circuits and Systems*, pp. 1537–1540, Taipei, Taiwan, 24–27 May 2009.

[71] D. G. Lee and P. P. Mercier, "Noise analysis of phase-demodulating receivers employing super-regenerative amplification," *IEEE Transactions on Microwave Theory and Techniques*, vol. PP, no. 99, pp. 1–13, 2017.

[72] F. O. Fernandez-Rodriguez and E. Sanchez-Sinencio, "Advanced quenching techniques for super-regenerative radio receivers," *IEEE Transactions on Circuits and Systems I: Regular Papers*, vol. 59, no. 7, pp. 1533–1545, Jul. 2012.

[73] I. Demirkol, C. Ersoy, and E. Onur, "Wake-up receivers for wireless sensor networks: benefits and challenges," *IEEE Wireless Communications*, vol. 16, pp. 688–696, 2009.

[74] J. M. López-Villegas and J. J. SieiroCordoba, "BPSK to ASK signal conversion using injection-locked oscillators-part i: theory," *IEEE Transactions on Microwave Theory and Techniques*, vol. 53, no. 12, pp. 3757–3766, Dec. 2005.

[75] J. M. López-Villegas, J. G. Macias-Montero, J. A. Osorio, J. Cabanillas, N. Vidal, and J. Samitier, "BPSK to ASK signal conversion using injection-locked oscillators-part ii: experiment," *IEEE Transactions on Microwave Theory and Techniques*, vol. 54, no. 1, pp. 226–234, Jan. 2006.

9

Low-Power RF Digital PLLs with Direct Carrier Modulation

Salvatore Levantino

Politecnico di Milano, Italy

Carlo Samori

Politecnico di Milano, Italy

CONTENTS

9.1 Introduction

This chapter reviews the key ideas and the main issues related to the implementation of directly modulated phase-locked loops (PLLs) in digital radio-frequency (RF) transmitters in modern complementary metal-oxide semiconductor (CMOS) processes. Unlike conventional Cartesian ones, digital RF transmitters, relying on out-phasing (OP) or polar architectures, avoid mixers and filters, and, therefore, enable low-power implementations in highly scaled CMOS processes. Of course, using this approach, new issues arise: the phase or the frequency of the carrier has to be digitised at high resolution and stringent linearity specifications over a wide modulation bandwidth.

PLLs that are conventionally adopted for frequency synthesis in wireless transceivers can be used to modulate the frequency or the phase of the carrier in RF transmitters. The application of those frequency or phase modulators is not only confined to communications based on frequency or phase modulations of the carrier, such as, for instance, Gaussian minimum-shift keying (GMSK) modulation. Even in the case of non-constant envelope modulations such as quadrature amplitude modulation (QAM), polar or OP transmitter architectures are based on low-noise, linear and wideband phase modulators.

When a closed-loop system, such as PLL, is used to generate the carrier and directly modulate its frequency, good linearity can be achieved. Unfortunately, the PLL being a feedback system, has intrinsically a limited bandwidth. Therefore, PLLs used as frequency modulators have an intrinsic trade-off between modulation linearity and modulation rate. This trade-off can be greatly relaxed by modifying the basic architecture of a PLL and exploiting the power of background digital calibration methods.

This chapter is organised as follows: Section 9.2 describes alternative architectures of a RF transmitter, starting from the conventional Cartesian transmitter to the polar and OP schemes. Section 9.3 deals with the possibile architecture of phase modulators to be used in a polar or OP transmitter, while Section 9.4 introduces digital PLLs (DPLLs). Section 9.5 shows how a DPLL can be used as a direct frequency modulator and compares the two main techniques to achieve wide modulation bandwidth: the two-point injection and the pre-emphasis scheme. Impact of the digitally controlled oscillator (DCO) non-linear characteristic and means to linearise that block are discussed in Section 9.6. Finally, a practical implementation of a low-power high-linearity 3.6 GHz frequency synthesiser, able to generate arbitrary phase and frequency modulations (FMs), is described in Section 9.7.

9.2 RF Transmitters Based on Phase Modulators

The popular Cartesian transmitter whose block diagram is shown in Figure 9.1a requires digital-to-analog conversion of the baseband signals, $I(t)$ and $Q(t)$ and low-pass filtering to properly suppress spectral replica. Moreover, a 0/90 degree phase shifter is needed to generate the quadrature carriers and analog mixers to upconvert the baseband signal. The signal after summation has in general a non-constant envelope. Therefore, it has to be amplified by a linear power amplifier (PA) to limit spectral regrowth and signal-to-noise ratio degradation of transmitted signal. The need for linear PAs greatly limits the achievable power efficiency, because class-A or class-AB PAs with

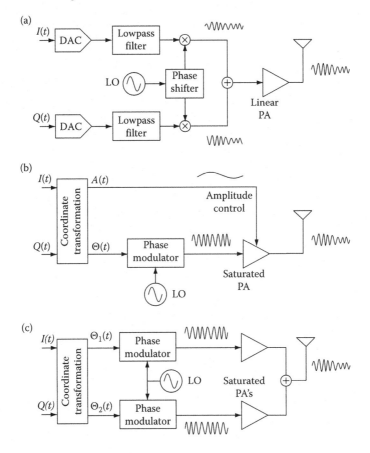

FIGURE 9.1
Simplified diagram of (a) Cartesian, (b) direct polar (DP) and (c) out-phasing (OP) radio transmitter.

sufficient backoff have to be used. Polar and OP transmitter architectures are being investigated and implemented in recent years to outperform the power efficiency of Cartesian transmitters.

In a polar transmitter, as the one illustrated in the block diagram in Figure 9.1b, the carrier generated by a local oscillator (LO) passes through a phase modulator that provides the correct phase modulation to the carrier and then, is fed to a highly efficient saturated PA, whose supply dynamically varies following the signal envelope [1]. Unfortunately, given the non-linear relationship $\Theta(t) = \tan^{-1}[Q(t)/I(t)]$ between the signal phase, $\Theta(t)$ and the Cartesian components of the signal, $I(t)$ and $Q(t)$, the bandwidth of $\Theta(t)$ is much larger than the original signal bandwidth. In typical modulation schemes, the bandwidth scales up by a factor between 5 and 10. This obviously represents a challenge for the phase modulator block. A variant of the polar transmitter concept is the original idea of the envelope-elimination and restoration technique as proposed by Kahn in 1952 [2]. The envelope-elimination and restoration technique consists in extracting the amplitude and phase signals from the modulated carrier and then re-applying them to the carrier in a polar fashion. For this reason, the polar architecture in Figure 9.1b is sometimes referred to as direct polar (DP) [3], since the carrier is modulated directly via its polar components. In recent examples of DP implementations, that can be found in ref. [4–7], a digitised amplitude signal, $A(t)$, and phase signal, $\Theta(t)$, directly modulate the saturated PA and the phase modulator, respectively.

The OP scheme, originally proposed by Chireix in 1935 [8], is illustrated in the block diagram in Figure 9.1c. Instead of a non-constant-envelope modulated carrier, two carriers with constant envelope and differently modulated phases are generated and finally combined at the antenna. Doing so, the two-constant envelope carriers can be amplified by means of two highly efficient saturated PAs. Thus, this scheme is often referred to as linear amplification with non-linear components [9]. As in the DP case, the OP scheme relies on the use of saturated, or almost saturated, PAs that entail good efficiency. Similar to the DP case, the non-linear relationship between the Cartesian components, $I(t)$ and $Q(t)$, and the two phases, $\Theta_1(t)$ and $\Theta_2(t)$, that modulate the two carriers, makes the bandwidth of the phase signals much larger than the bandwidth of the original components. Again, this represents a challenge for the two phase modulators. Recent examples of high-performance OP transmitters can be found in refs. [10], [11].

Among the reasons for the renewed interest in these techniques, the main one is the inherently high efficiency of these architectures and the elimination of mixers and filters typically employed in Cartesian transmitters. This is, however, only a part of the story. Another key point is that these architectures seem naturally suited to the digitally intensive approach to circuit design discussed earlier. Both in DP and OP systems, the phase modulator is typically driven directly by a digital control word, without the need for an explicit voltage- or current-domain digital-to-analog converter (DAC) like in the Cartesian architecture, and acts effectively as digital-to-phase converter. The DP approach can be pushed even to a *more digital* implementation by relying on a digital PA (DPA), where the amplitude of the PA output is controlled by a digital word [12–14]. The net result is a higher efficiency of the overall transmitter over a conventional Cartesian architecture.

9.3   Architectures of Phase Modulators

As discussed in the previous section, both polar and OP transmitters rely on a wideband phase modulator. The two most typical approaches to the design of a phase modulator are illustrated in Figure 9.2. The first one in

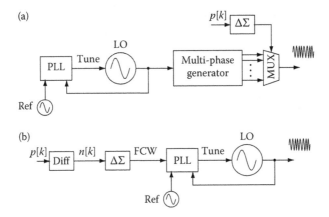

FIGURE 9.2
Phase modulator architecture: (a) direct and (b) indirect phase modulation.

Figure 9.2a can be referred to as *direct phase modulation*, and it is based on the generation of a certain number of phases of the carrier with a constant phase shift among them. The phases may be derived from the output of the frequency synthesiser through frequency division, polyphase filtering or by means of a regulated delay line. However, owing to power constrains, only a limited number of phases (typically no more than few tens) can be practically generated in those ways, while a fine-phase resolution (in the order of few degrees) is often needed. The typical methods to refine phase resolution is either to use an analog phase interpolator [5] or to digitally select one of the phases via a multiplexer (MUX) and dither the phase selection signal by means of a $\Delta\Sigma$ modulator [15], as shown in figure. The quantisation noise introduced by the $\Delta\Sigma$ can be cancelled out by employing more advanced structures as proposed in ref. [16]. Alternatively, the fine phase shift may be achieved by relying on a tuned resonator [17]. More recently, the digitally controlled phase shifter has been implemented by means of a digital-to-time converter (DTC) [18].

The main advantage of the direct phase modulator is the high achievable speed of the phase signal. On the other hand, one of the problems is that the generation of several high-frequency phases of the carrier and the subsequent multiplexing operation cost high power dissipation, especially at high frequency. The linearity of the modulator, which is affected by the mismatches between the time shifts among the phases, is another critical issue [11].

Since the instantaneous frequency of the carrier is given by the first derivative of the carrier phase, the alternative way to implement a phase modulator is shown in the block diagram in Figure 9.2b. The phase signal is differentiated and, then, the result is used to modulate the frequency control word of a PLL. This solution that can be referred to as *indirect phase modulation*, is in principle more power efficient than the previous one, since only one high-frequency signal (i.e., the LO signal) has to be generated at a time. Furthermore, it theoretically entails better linearity, thanks to the linearisation provided by the PLL closed loop. The challenges of this approach are as follows: (i) the limited achievable speed of the modulation and (ii) the wide range of the FM signal.

As regards the first issue, stability issues constrain the bandwidth of a PLL to be lower than about one-tenth of its reference frequency, f_{ref}. Typically, a bandwidth narrower than $f_{ref}/10$ is chosen to properly suppress the reference spur. Thus, to widen the achievable modulation bandwidth, it would be desirable to increase the reference frequency of the PLL.

Concerning dynamic range, the phase signal in polar and OP transmitters is in general unconstrained and can vary up to $\pm\pi$ [rad]. This variation of the phase may be required in just one clock period, being the clock period equal to the period T_{ref} of the reference signal[1]. The step increment of the phase signal can be equivalently produced by a pulse of angular frequency, whose integrated value is equal to $\pm\pi$. The phase waveform as well as the the equivalent frequency waveform are shown in Figure 9.3. If we impose that

$$\phi[kT_{ref}] - \phi[(k-1)T_{ref}] = \pm\pi \,, \tag{9.1}$$

then it follows that $\omega[kT_{ref}] = \pm\pi/T_{ref}$, or equivalently, the frequency pulse should be as wide as as

$$\Delta f_p = \pm\frac{1}{2T_{ref}} = \pm\frac{f_{ref}}{2} \,, \tag{9.2}$$

Hence, the larger the reference frequency, the wider the FM pulse and, thus, the wider the required PLL tracking range. Moreover, any non-linearity introduced by the PLL in the FM would degrade the phase of the carrier and would increase the error-vector magnitude (EVM) of the modulation. Therefore, the PLL needs a controlled oscillator covering the frequency range in (9.2) with margin and tight linearity.

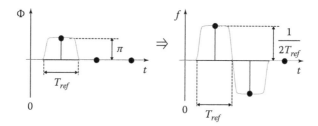

FIGURE 9.3
Waveforms of phase- and frequency-modulation signal.

---

[1]Such a large phase variation in one sample period occurs for instance in an unfiltered quadrature phase shift keying (QPSK)-modulated carrier.

On this respect, it would be preferable to decrease the reference frequency of the PLL, a requirement which is in contrast with the maximum achievable speed of the modulation. We can conclude that in a standard PLL a trade-off exists in the choice of the reference frequency and that the modulation bandwidth of the PLL has to be traded with its linear modulation range.

The remainder of this chapter will be devoted to discuss the implementation of the second approach and, in particular, the main issues related to the design of phase modulators based on digital PLLs (DPLLs).

9.4 Digital PLLs

The most widely adopted architecture for a frequency synthesiser is the analog charge-pump PLL in Figure 9.4a. Though largely employed in industry for their good performance, analog PLLs do not so easily take advantage of today's scaled CMOS processes. The charge pump dissipates non-negligible power which cannot be reduced without impairing noise performance and

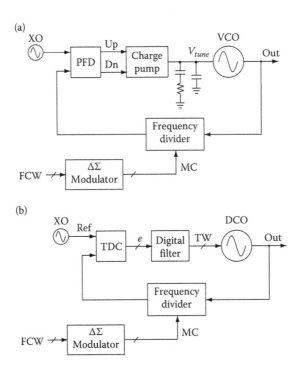

FIGURE 9.4
PLL architectures: (a) analog and (b) digital.

which represents a limit in low-power applications. It also adds non-linearity to the loop, which worsens noise and spur performance. Besides, the analog loop filter may require a relatively large capacitor to be implemented on chip, in the order of nF.

CMOS technology scaling is changing the way electronic circuits are designed. While intrinsic analog performance are degraded because of lower voltage headroom and intrinsic gain of the transistors, digital-signal processing techniques can be applied to boost those performance at insignificant power and area overhead. Furthermore, the digital assistance of analog sub-blocks reduces the overall cost of implementation and the required design time, as it benefits from the design automation of digital sections. The natural way to take advantage of CMOS scaling in the design of a PLL is to replace the charge pump and the analog filter with a digital filter based on standard cells. In this way, not only the filter benefits from scaling to new technology nodes but also the noise sources associated to the charge pump and the analog filter, as well as the non-linearity associated to the charge pump are removed. The resulting system is shown in Figure 9.4b and is often referred to as DPLL or all-DPLL (ADPLL). The relative time difference between the edges of a reference signal and the frequency-divided output of a DCO is detected and digitised through the use of a time-to-digital converter (TDC). The TDC output is fed to a digital loop filter whose output is in turn used as the digital tuning control of the DCO.

9.4.1   TDC-based PLL

As in analog PLLs for frequency synthesis, the modulus control of the frequency divider is dithered by a digital $\Delta\Sigma$ modulator that quantises the frequency control word (FCW) and realises a fractional-N division. Since the phase error induced by $\Delta\Sigma$ quantisation is a deterministic signal, it can be cancelled out by subtracting it (after proper scaling) from the output of the TDC, as shown in the block diagram in Figure 9.5 [9]. This technique, often referred to as *digiphase technique* [20], was first introduced in analog fractional-N PLLs [21], [22] and then applied to DPLLs [19], where the calibration of the scaling factor (the a signal in Figure 9.5) is easily implemented by a digital loop (the shadowed block in Figure 9.5). At steady state, the a gain tends to the value that nulls the product between e and q and, in turn, the correlation between phase error and $\Delta\Sigma$ quantisation error. This loop can be regarded as a simplified implementation of a least mean square (LMS) algorithm [23].

Unfortunately, such an implementation of DPLL requires a TDC with large number of bits, especially at wide PLL bandwidths. In fact, on one hand, it has to accommodate the quantisation error introduced by the $\Delta\Sigma$ and convert it linearly, and, on the other hand, should add no significant intrinsic quantisation noise. The analyses and simulations reported in ref. [24] show that the required number of TDC bits is as high as 10 to guarantee a level of residual fractional spurs below -60 dBc, assuming a first-order $\Delta\Sigma$

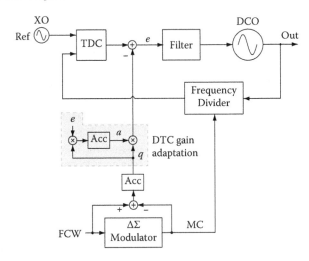

FIGURE 9.5
Block schematic of a DPLL.

modulator dithering the divider modulus control. Furthermore, the integral-non-linearity of the TDC over this wide dynamic range must be as low as one least significant bit (LSB). If a third-order $\Delta\Sigma$ modulator is employed, the dynamic range of the TDC must be as wide as four DCO periods and requires two additional bits. This numerical example leads to 12 equivalent bits required to the TDC. Of course, flash-type TDCs satisfying these specifications would produce excessive power dissipation. For this reason, different types of TDC (such as the oversampling or pipeline) [25–30] have been proposed and investigated, as well as several linearisation techniques [31], [32]. Although considerable effort has been done to improve power efficiency, the TDC still remains one of the main power hungry blocks of the loop.

9.4.2 DTC-based PLL

To solve this issue and improve PLL noise/power trade-off, a new class of DPLLs has been recently introduced [33], [34]. The idea is to relax substantially TDC specifications by adding a DTC in the PLL feedback branch. The DTC allows to subtract the $\Delta\Sigma$ quantisation error and, thus, reduce its amplitude down to DTC resolution. As a result, the required dynamic range of the subsequent TDC is reduced as well. In ref. [33], a TDC and DTC, both with equivalent number of bits equal to four, are employed. In this way, the implementation of the TDC, its linearity requirement and power consumption are greatly reduced. The DTC is implemented as a delay-locked loop with 16 delay elements whose delay is automatically tuned. A more drastic simplification to TDC design and substantial improvement in the PLL noise/power figure of merit was presented in ref. [34]. In that case, the resolution of the

TDC is reduced to only one bit, as shown in the block diagram in Figure 9.6. A single-bit TDC, also known as bang-bang TDC (BB-TDC), detects simply which one of the two input signals leads or lags the other one. In practice, a TDC with coarse mid-rise quantisation (whose characteristic is shown in the upper plot in Figure 9.7) is employed to speed up lock transient. However, even in such a case, only two levels of the TDC characteristic are exploited when the PLL is in lock and, thus, the loop is equivalently controlled by a bang-bang detector. The design of the TDC, which is implemented as a time arbiter, is greatly simplified and power dissipation is substantially reduced. Furthermore, having a single-bit output, the non-linearity issue is removed, like it happens to the comparator of an analog $\Delta\Sigma$ modulator. Of course, it introduces a hard non-linearity in the loop, which potentially gives rise to limit cycles and in turn unwanted spurs in the spectrum.

In this context, the presence of random noise has a positive effect. Under proper conditions, it dithers the time delay detected by the BB-TDC avoiding, limit cycles in the PLL. More specifically, as demonstrated in ref. [35], when the random jitter induced by the thermal noise sources is larger than the deterministic error induced by quantisation, the low-pass-filtered TDC output (i.e. averaged in time $\langle e \rangle$) as a function of the input phase error ϕ is given by the integration of the probability density function $p(\phi)$. This result is schematically shown in Figure 9.7. In the case of Gaussian phase error, the $\langle e \rangle$ curve versus ϕ is given by an error function, whose slope around zero is inversely proportional to the standard deviation σ_ϕ of the phase error. Thus, in practice, the mid-rise quantiser in the presence of dominant random noise

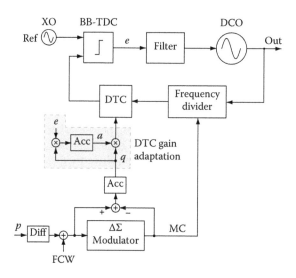

FIGURE 9.6
Block schematic of a DTC-based DPLL.

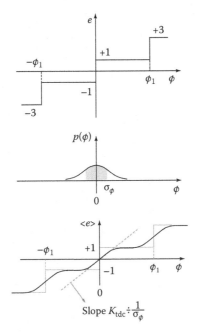

FIGURE 9.7
Input/output characteristic of a mid-rise TDC, probability of input phase difference and average characteristic of the TDC.

at its input can be linearized, and its linear gain is inversely proportional to the root mean square (RMS) value of the input jitter. As discussed earlier, the linearisation of the hard-limiting TDC characteristic holds as long as the random component of jitter dominates over the deterministic one. This mode of operation is referred as to *random-noise regime*, and it is opposed to the *limit-cycle regime* in which quantisation error dominates over random jitter [36]. For an integer-N synthesised channel, the deterministic component is produced by the limited DCO frequency resolution, and in turn by the truncation of the filter output word. Thus, in practice, this condition can be verified by improving DCO resolution [37].

By contrast, for a fractional-N channel, the quantisation is dominated by the $\Delta\Sigma$ modulator dithering the divider modulus control. The resulting deterministic quantisation error that is as wide as a few multiples of T_{dco} is always much larger than the typical random noise jitter at the input of the detector. The latter cannot be increased, since it would raise the output jitter as well. Thus, in the DPLL in Figure 9.6, instead of cancelling the $\Delta\Sigma$ quantisation at TDC output, the cancellation is performed at TDC input via the DTC. The DTC allows us to subtract the phase error induced by the $\Delta\Sigma$ modulator, as in the PLL in Figure 9.5. Similarly, the amplitude of the subtracted signal is automatically estimated by means of an LMS-type feedback loop. However, in contrast to that case, since the quantisation error at TDC input is cancelled out, there is no need for a multi-bit TDC. We can thus rely on a BB-TDC and yet get the detector to work in the random-noise regime.

The main advantages of the DTC-based PLL architecture over the conventional one are still the higher power efficiency and the lower design complexity. The BB-TDC is implemented as a single flip-flop and the DTC as a digital buffer stage with switched capacitor load. Thus, if we compare the cascade of the BB-TDC and the DTC against a multi-bit TDC, we have a single time arbiter instead of many. Furthermore, the DTC-based topology is also favourable to the implementation of automatic pre-distortion algorithms [33], [24]. This allows to adopt a segmented structure with scaled capacitor banks used as load of the DTC, yet reaching very good linearity.

As a final point, it should be noted that the gain of the BB-TDC depends on its input jitter, which in turn depends on the noise transfer function of the PLL, that depends again on the BB-TDC gain. Even if this vicious circle appears as a peculiar issue of the BB-TDC approach, it is only one particular version of a very general and standard problem of PLLs. As a matter of fact, the noise transfer function of any PLL depends on analog parameters (e.g. the TDC time resolution when a multi-bit TDC is employed in a DPLL) that are difficult to control with the required accuracy. This issue is solved by adopting the automatic loop-gain control scheme disclosed in ref. [38]. This automatic calibration normalises the loop gain so that the whole frequency response and the bandwidth of the PLL are repeatable regardless of the spread of the analog parameters, including TDC gain. This solution makes the phase noise profile and the integrated phase noise (or jitter) independent of PLL parameters such as TDC gain, but dependent on digital coefficients. This property will also be useful in the following, when the PLL is used as direct-FM modulator with pre-emphasis filter.

9.5   Digital PLLs for Wideband FM

9.5.1   Direct FM

To derive the transfer function of the DPLL in Figure 9.6, we may refer to the linear phase model in the z-domain shown in Figure 9.8. This model is a discrete-time model at reference rate. The phase-modulation signal p is differentiated, and the result n is added to the FCW and fed to the $\Delta\Sigma$ modulator. For the sake of simplicity, we considered a first-order $\Delta\Sigma$ modulator, although in practice a second- or higher-order $\Delta\Sigma$ is employed. In the model of the $\Delta\Sigma$, the quantiser is linearised, and a proper quantisation error $-q(z)$ is added. The output of the $\Delta\Sigma$ drives the modulus control of the divider. The latter is modelled as a discrete-time integrator, whose gain is $2\pi/N$. This follows from

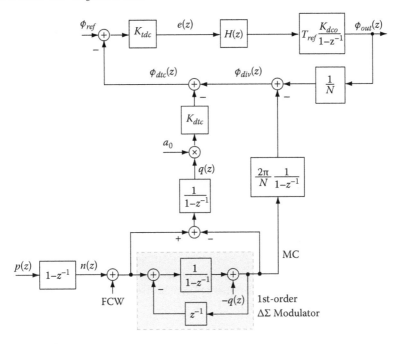

FIGURE 9.8
Equivalent model of the DTC-based DPLL in Figure 9.6.

the fact that if modulus control (MC) is incremented by one, a time delay equal to the output period T_{dco} is added to the *div* signal and in turn a shift of $2\pi/N$ is added to the phase. The DTC is simply modelled as a gain K_{dtc} [rad/bit] and an adder that adds/subtracts a certain phase shift in the feedback branch. The TDC is modelled as a detector of the phase error with gain K_{tdc} [bit/rad] and the DCO as an integrator with gain K_{dco} [rad/sec/bit]. This means that, as the DCO tuning word is incremented by one, the excess phase increases by $T_{ref} \cdot K_{dco}$ [rad] (after T_{ref} [sec]). It is easy to verify that after proper choice of the gain:

$$a_0 \cdot K_{dtc} = 2\pi/N, \qquad (9.3)$$

this scheme allows to cancel out the effect of the $\Delta\Sigma$ quantisation $-q$. The value of a_0 is set automatically in the background by the LMS loop (not shown in this model).

On the basis of the phase model, let us derive the transfer function from the phase-modulation signal $p(z)$ to the output phase $\phi_{out}(z)$, which is given by

$$F(z) = \frac{\phi_{out}(z)}{p(z)} = 2\pi \cdot \frac{G_{loop}(z)}{1 + G_{loop}(z)}, \qquad (9.4)$$

being G_{loop} the PLL loop gain:

$$G_{loop}(z) = \frac{K \cdot H(z)}{1 - z^{-1}},$$
(9.5)

and

$$K = \frac{K_{tdc} K_{dco} T_{ref}}{N}.$$
(9.6)

At low frequency, the transfer function $F(z)$ is the desired one, that is a flat transfer function, equal to 2π. It means that increasing MC by one produces a time shift at the output equal to one output period T_{dco} (i.e., 2π shift in the phase). However, the bandwidth of $F(z)$ is the bandwidth of PLL, which is constrained to be much lower than f_{ref}. This is a severe limitation, since in both polar and OP transmitters, the bandwidth of the phase-modulation signal is typically larger than the signal bandwidth at RF. Moreover, PLL bandwidth is typically chosen to optimise the integrated phase noise of a PLL and to filter out the reference spur. Therefore, conventional values of bandwidths of PLL-based frequency synthesisers range between 100 kHz and 1 MHz.

9.5.2   Two-point injection scheme

The modulation bandwidth can be substantially incremented by exploiting the two-point injection scheme, originally proposed in ref. [39], and recently adopted in ref. [40–42]. In this topology, the phase-modulation signal, after being differentiated, is injected both into the divider (in the feedback branch) and the DCO (in the forward branch). Applying this technique to the DTC-based DPLL so far discussed, the resulting block scheme is the one sketched in Figure 9.9.

Neglecting the DTC, the *DTC gain adaption* block and the *DCO gain adaptation* block, that play no role in the system response, the phase model of the system in the z-domain is shown in Figure 9.10. We will assume that the gain g has reached a constant value g_0. On the basis of this model, we can calculate the new transfer function from $p(z)$ to the output phase $\phi_{out}(z)$, which is

$$T(z) = \frac{\phi_{out}(z)}{p(z)} = 2\pi \cdot \frac{G_{loop}(z)}{1 + G_{loop}(z)} + g_0 K_{dco} T_{ref} \cdot \frac{1}{1 + G_{loop}(z)},$$
(9.7)

and where G_{loop} has the same expression (9.5) as in the previous system.

While the first term in (9.7) has a low-pass shape in frequency, the second term, that comes from the DCO injection path, is high-pass. If we manipulate (9.7) and impose that it is equal to 2π for all z:

$$T(z) = 2\pi \cdot \frac{\frac{g_0 K_{dco} T_{ref}}{2\pi} + G_{loop}(z)}{1 + G_{loop}(z)} = 2\pi,$$
(9.8)

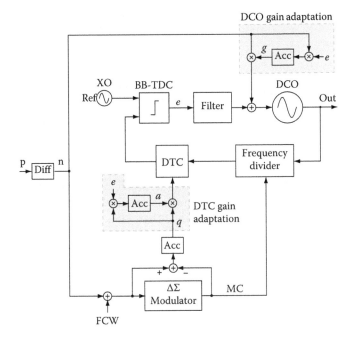

FIGURE 9.9
DPLL with the two-point injection scheme for direct FM.

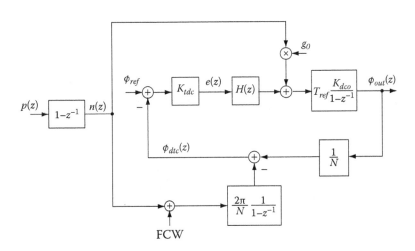

FIGURE 9.10
Equivalent model of two-point injection scheme in Figure 9.9.

we obtain that, as long as the following equality holds

$$\frac{g_0 K_{dco} T_{ref}}{2\pi} = 1, \tag{9.9}$$

$T(z)$ becomes an all-pass transfer function, which allows the direct modulation of wideband signals. The modulation rate is simply limited by Nyquist theorem, since the system is sampled at f_{ref}. Hence, the maximum signal bandwidth (at baseband) is limited to $f_{ref}/2$.

The validity of the previous equality (9.9) can be guaranteed by a proper choice of the gain g_0. Unfortunately, the DCO gain, K_{dco}, is hardly controllable, as it varies over process and temperature, and over the synthesised channel frequency. Any mismatch in the equality would result is a non-ideal zero-pole cancellation in $T(z)$ and, therefore, in a linear distortion of the signal.

To solve this impairment and guarantee the ideal all-pass shape of the system response over parameter variations, the automatic regulation of g shown in Figure 9.9 has been conceived. Its working principle can be understood by relying on the model in Figure 9.10. Applying the superposition principle, we can calculate the effect of $n(z)$ on the error $e(z)$:

$$e(z) = n(z) \cdot (2\pi f_{ref} - g_0 K_{dco}) \cdot \frac{G_{loop}(z)}{1 + G_{loop}(z)} \cdot \frac{1}{K_{dco} H(z)}. \tag{9.10}$$

From the previous expression, we therefore realise that the error induced by the injection of $n(z)$ is proportional to the gain imbalance $(2\pi - g_0 K_{dco} T_{ref})$ between the two injection paths. In fact, if (9.9) holds, this difference is null.

Equation (9.10) also implies that the correlation between the error signal $e(z)$ and $n(z)$ provides a measure of the gain imbalance. Thus, if we force this correlation to be null, we will get perfect gain balance and all-pass-shaped system response. The DCO gain adaptation in Figure 9.9 implements this concept: the gain g will tend to the g_0 value that nulls the correlation between e and n (i.e., the product between e and n). This automatic gain adaptation can be easily implemented in the digital domain with insignificant resources or design effort.

9.5.3 Pre-emphasis scheme

A second architecture allowing wideband direct FM is the PLL with pre-emphasis filter [43], [44]. The basic idea is borrowed from a classical concept exploited in communications systems, and the block scheme is sketched in Figure 9.11, where a multi-bit TDC is employed. The phase-modulation signal $p(z)$ after differentiation is fed to a digital pre-emphasis filter (DPF), which amplifies the high-frequency components to compensate for the roll off of the PLL transfer function. So, the total transfer function from $p(z)$ to $\phi_{out}(z)$ is ideally an all-pass one.

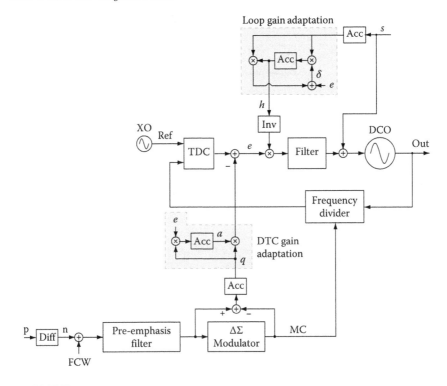

FIGURE 9.11
DPLL with pre-emphasis scheme for direct FM.

The loop gain adaptation circuit normalises the loop gain by estimating a proper coefficient, h, and multiplying the digital signal at the TDC output by $1/h$. Neglecting, for a moment, that block, that play no role in the system response, the phase model of the system in the z-domain is shown in Figure 9.12. We will assume that the gain h has reached a constant value h_0. So, a multiplication by $1/h_0$ is accounted for in the model in Figure 9.12. According to the model, the transfer function from the modulating signal to the output phase is

$$T_{pe}(z) = H_{pe}(z) \cdot F(z), \tag{9.11}$$

where $H_{pe}(z)$ is the transfer function of the pre-emphasis filter DPF. The transfer function $F(z)$ has the same formal expression achieved in (9.4) for the PLL, and reported here for convenience

$$F(z) = \frac{\phi_{out}(z)}{p(z)} = 2\pi \cdot \frac{G_{loop}(z)}{1 + G_{loop}(z)}. \tag{9.12}$$

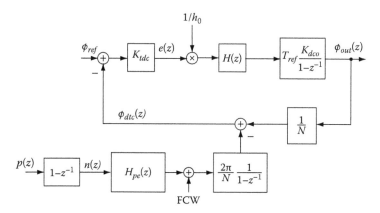

FIGURE 9.12
Equivalent model of the DPLL with pre-emphasis scheme in Figure 9.11.

Unlike the previous case, the loop gain of the system in Figure 9.12 includes the coefficient $1/h_0$:

$$G_{loop}(z) = \frac{K}{h_0} \cdot \frac{H(z)}{1 - z^{-1}}. \tag{9.13}$$

where $H(z)$ is the transfer function of the digital loop filter.

To derive the transfer function of the pre-emphasis filter, we first derive a normalised expression of the loop gain. In practice, if we assume that the gain h_0 in (9.13) is chosen to be equal exactly to $K = (K_{tdc}K_{dco}T_{ref})/N$, the loop gain can be written as follows

$$G_{loop,n}(z) = \frac{H(z)}{1 - z^{-1}}, \tag{9.14}$$

which only depends on digital parameters and is not affected by process, voltage and temperature (PVT) variations or, in general, by spreads of analog parameters. If we use a pre-emphasis filter with the following transfer function

$$H_{pe}(z) = \frac{1 + G_{loop,n}(z)}{G_{loop,n}(z)} = \frac{1 + \frac{H(z)}{1-z^{-1}}}{\frac{H(z)}{1-z^{-1}}} = \frac{1 - z^{-1}}{H(z)} + 1, \tag{9.15}$$

then the PLL transfer function in (9.11) is flat and equal to 2π. The most interesting feature of this approach is that (9.15) is not affected by the typical analog issues, such as mismatch or PVT spreads, because $H(z)$ is a digital filter.

Of course, flat response of the PLL is achieved only when the value of the coefficient h_0 equals K. This is the role of the *loop gain adaptation* circuit in Figure 9.11 [38]. Its operating principle can be illustrated as follows. A random signal s with high-pass-shaped power spectral density (PSD) is added to the digital tuning word at the output of the digital filter. This sequence s is essentially used as a training sequence to estimate the loop gain. Having $s[k]$ a high-pass shaped PSD, we can assume that most of the power of the sequence is concentrated at high frequencies, where the transfer function of the loop filter $H(z)$ is practically zero. In practice, this means we can consider the loop as it was open and we can calculate the expression of the error e relying on the equivalent model in Figure 9.13. Using this open-loop approximation, the error $e[k]$ at the output of the TDC induced by $s[k]$ can be calculated just by integration of $s[k]$ (because of the DCO) and multiplication by the gains of the DCO, divider and TDC:

$$e(z) \approx -K \cdot \frac{1}{1 - z^{-1}} \cdot s(z). \tag{9.16}$$

where $K = (K_{tdc} K_{dco} T_{ref})/N$.

In the presence of the earlier calculated component of the error $e[k]$ correlated with $s[k]$, the loop gain adaptation circuit comes into play. The sequence $s[k]$ is first fed to an accumulator to match the DCO intrinsic integration. Then, it is multiplied by h_0 and added to $e[k]$. Therefore, the resulting signal, δ, has a z-transform given by the following expression

$$\delta(z) = (h_0 - K) \cdot \frac{1}{1 - z^{-1}} \cdot s(z), \tag{9.17}$$

which is proportional to the difference between h_0 and K.

The rest of the loop gain adaptation circuit in Figure 9.11 is a standard LMS loop. If the output of the LMS accumulator is exactly equal to K, i.e. $h_0 = K$ in (9.17), then the input of the LMS accumulator is null. That is, the typical consistence condition of an LMS loop. The gain h_0 is then inverted, to multiply the output of the TDC, as discussed earlier, and make the loop gain insensitive to the analog parameters.

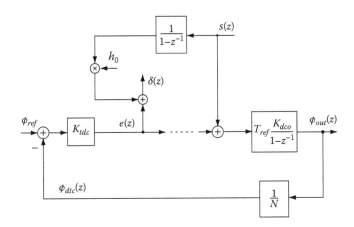

FIGURE 9.13
Model for the description of the loop gain adaptation block.

Two remarks are in order: First, this loop gain adaptation circuit can be realised both for DPLL embedding a BB-TDC, as demonstrated in [38], and for a DPLL employing a multi-bit TDC. Second, it is not necessary to introduce a training sequence s. Actually, in most DPLLs, the output of the digital loop filter is quantised by a digital $\Delta\Sigma$ modulator, whose quantisation noise is injected in the loop and has an intrinsically high-pass shaped PSD [38]. Therefore, the $\Delta\Sigma$-DCO quantisation error can serve as a training sequence for the loop gain adaptation circuit.

9.5.4   Two-point vs. pre-emphasis: A brief comparison

It is of interest to briefly compare the two techniques presented, to point out the strengths and weaknesses of the two-point injection scheme and the pre-emphasis approach.

A first issue concerns the transfer function of the DPF, $H_{pe}(z)$, derived in the ideal case in (9.15). Typically, $H(z)$ is the transfer function of proportional-integral filter. Hence, the resulting pre-emphasis filter cannot be realised in practice because it would not be causal. Therefore, to obtain a physically realisable filter, an additional high-frequency pole has to be added to the filter [43]. Obviously, the additional pole will make the total transfer function $T_{pe}(z)$ depart from the ideal flat one at high frequency, and the maximum achievable modulation bandwidth will be limited.

A more important aspect concerns the resilience of the two systems to the error relative to the estimated calibration gains. In both cases, in fact, to achieve wide modulation bandwidth, it is necessary to match two different gains. In the two-point injection scheme, the gains of the low-pass and high-pass paths have to be matched via the coefficient g_0, see (9.7). Similarly, in the pre-emphasis case, the pre-emphasis filter has to compensate for the roll off of the PLL transfer function via the coefficient h_0. The problem is that in both cases these coefficients essentially estimate an analog gain that can vary over a wide range. This is in particular true for the DCO gain K_{dco}, which enters both in the expression of g_0 and h_0. The LMS algorithm in its basic version estimates an average gain. Therefore, if the modulation signal produces a large variation of the output frequency, the K_{dco} can change substantially because of DCO non-linear characteristic, and the estimated average gains, g_0 and h_0, cause an error in the output modulation.

It is possible to show that the impact of this error is the same in the two techniques. Starting with the two-point injection scheme, we have that in the ideal case it should be $2\pi = g_0 K_{dco} T_{ref}$. Now, let us assume that K_{dco} varies by a relative error ϵ_K. It is useful to rewrite (9.7), using $K_{dco} \cdot (1 + \epsilon_K)$ in place of K_{dco}. This gives the following expression of the PLL transfer function

$$T(z) = \frac{\phi_{out}(z)}{p(z)} = 2\pi \cdot \left(\frac{G_{loop}(z)}{1 + G_{loop}(z)} + (1 + \epsilon_K) \cdot \frac{1}{1 + G_{loop}(z)} \right), \quad (9.18)$$

The frequency responses relative to the two terms in the previous equation are schematically shown in Figure 9.14a, where $|G(f)|$ is the magnitude of the second term in (9.18). At low frequency, the impact of the error is zero, since $G_{loop}(z)$ goes to infinite and the second term including the error goes to zero, giving $T(z) = 2\pi$. On the other hand, at high frequency, $G_{loop}(z)$ goes to zero, and it is $T(z) = 2\pi(1 + \epsilon_K)$. In practice, this means that the error affects the modulation spectrum only out of the PLL bandwidth.

Concerning the pre-emphasis scheme, we combine (9.11), (9.12) and (9.14), and obtain the following new expression for the $T_{pe}(z)$

$$T_{pe}(z) = \frac{\phi_{out}(z)}{p(z)} = 2\pi \cdot \frac{\left(1 + \frac{H(z)}{1-z^{-1}}\right) \cdot \frac{K}{h_0}}{1 + \frac{H(z)}{1-z^{-1}} \cdot \frac{K}{h_0}}. \tag{9.19}$$

To obtain a flat frequency response of the pre-emphasis scheme, i.e. $T_{pe}(z) = 2\pi$, we need to impose that $h_0 = K$. If we assume that the gain K departs from its bias value by ϵ_K (with respect to K), rewriting the earlier expression with $K \cdot (1 + \epsilon_K)$ in place of K, we have

$$T_{pe}(z) = \frac{\phi_{out}(z)}{p(z)} = 2\pi \cdot \frac{\left(1 + \frac{H(z)}{1-z^{-1}}\right) \cdot (1 + \epsilon_K)}{1 + \frac{H(z)}{1-z^{-1}} \cdot (1 + \epsilon_K)}, \tag{9.20}$$

The frequency response relative to the pre-emphasis case is schematically shown in Figure 9.14b. In this case too, the impact of ϵ_K at low frequency

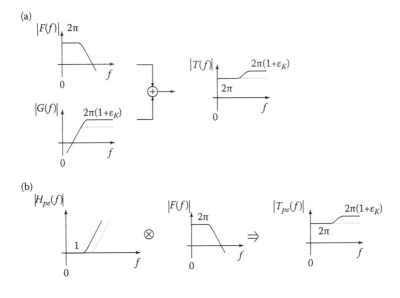

FIGURE 9.14
Frequency responses for (a) two-point injection and (b) pre-emphasis schemes.

is null, because the term $H(z)/(1 - z^{-1})$ in (9.20) tends to infinity and T_{pe} tends to 2π. At high frequency, $H(z)/(1 - z^{-1})$ tends to zero and we get $T_{pe}(z) = 2\pi \cdot (1 + \epsilon_K)$. The impact of the error is therefore identical in the two schemes. In both cases, it entails no error at low frequency, and it is proportional to the relative variation of the analog parameter at high frequency.

Finally, it is interesting to compare the range required to the TDC block in the two systems. We are assuming for the moment that a standard multi-bit TDC range is employed in both cases. To this purpose, we can use the transfer function linking the differentiated phase signal $n(z)$ to the TDC output $e(z)$ for the two-point modulator in Figure 9.9 achieved in (9.10) and reported here for convenience

$$e(z) = n(z) \cdot (2\pi f_{ref} - g_0 K_{dco}) \cdot \frac{G_{loop}(z)}{1 + G_{loop}(z)} \cdot \frac{1}{K_{dco}H(z)}. \qquad (9.21)$$

When g_0 is ideally calibrated, it is $g_0 = 2\pi f_{ref}/K_{dco}$ and $e(z) = 0$ at every frequency (the conclusion does not change substantially if a small variation affects K_{dco}). The TDC, in this condition, essentially detects only thermal noise and DCO quantisation noise. This result makes it possible to implement a single-bit TDC, i.e. the BB-TDC, even when the two-point injection scheme is adopted in DPLL.

In the pre-emphasis topology, instead, the range of the error the TDC has to convert is substantially larger. Similar to the case of the two-point injection scheme, it is possible to compute the TDC output induced by the differentiated phase-modulation signal, $n(z)$ in Figure 9.12, obtaining:

$$e(z) = n(z) \cdot H_{pe}(z) \cdot \frac{G_{loop}(z)}{1 + G_{loop}(z)} \cdot \frac{1}{K_{dco}H(z)}. \qquad (9.22)$$

Assuming that the pre-emphasis filter perfectly cancels the factor $G_{loop}(z)/(1 + G_{loop}(z))$ in (9.22), the TDC output is then given by

$$e(z) = \frac{n(z)}{K_{dco}H(z)}. \qquad (9.23)$$

In this case, the error $e(z)$ is not zero for every frequency, but follows a filtered version of the modulating sequence $n[k]$. This property was already pointed out in ref. [43], for an analog charge-pump-based PLL. Therefore, in the DPLL with pre-emphasis scheme, the TDC needs to account for an additional range to accomodate this signal. This makes the TDC design more challenging and increases its power consumption.

Intuitively, the loop filter transfer function $H(z)$ can be approximated at low frequency with that of a pure integrator. From (9.23), it follows that the signal $e[k]$ in the same frequency range is proportional to the first difference of $n[k]$. Considering, for instance, a typical FM signal featuring a positive and negative swing around a central frequency, the first difference of

$n[k]$ would go from positive to negative values as well, substantially exploiting a wide TDC dynamic range. Note that the DTC-based PLL scheme in Figure 9.6 does not solve this problem, since the role of the DTC is to cancel out the quantisation noise introduced by the $\Delta\Sigma$ modulator, while the error signal component evaluated in (9.23) comes from the pre-emphasised modulation signal.

9.6 Automatic DCO Predistortion Scheme

The comparison between the two architectures of phase modulators presented in the previous section has evidenced that the two-point injection scheme is potentially more power-efficient with respect to the pre-emphasis one, since the former allows using a BB-TDC, indisputably a low-power element when compared with a high-resolution TDC. We will therefore focus on the two-point approach for the rest of this chapter.

As we have already mentioned, one potential problem in a DPLL-based wideband phase modulator that may compromise modulation accuracy and degrade the EVM of the constellation is DCO non-linearity. To analyse this issue, we have to take into account that an additional constraint in DPLLs is the so-called frequency granularity induced by the finite number of bits of the DCO [45], which introduces quantisation noise in the loop. For this reason, the LSB of the DCO has to be sufficiently small to add insignificant truncation error. In practical systems, DCO resolution should be in the order of 10 kHz/bit, which can be achieved in modern CMOS processes. So, this is typically not a serious issue when the DPLL is employed as a standard frequency synthesiser. However, when the DPLL is used as a wideband modulator, the largest variation of the output frequency has to reach $\pm f_{ref}/2$, to produce a phase variation up to $\pm\pi$ in one reference clock. Therefore, the DCO full-scale range should exceed a 12-bit plus margin (if we assume, for instance, f_{ref} equal to 40 MHz). Of course, it is not easy to guarantee a linear tuning characteristic of the DCO over such a wide dynamic range.

The main source of non-linearity in the DCO characteristic is the mismatch among the switched capacitors in the resonator. Hence, a thermometric weighting of the capacitors should be preferred since it guarantees the monotonicity of the characteristic or, equivalently, a low differential non-linearity (DNL). Unfortunately, in this coding scheme, the number of connections increases exponentially with the number of bits, resulting impractical in a DCO with 12 or more bits. Note that this problem is analogous to the one encountered in the design of a DAC. A standard solution that guarantees a good compromise between DNL and number of connections is to use several thermometrically weighted banks of capacitors in the LC resonator of the DCO.

In Figure 9.15, we consider, as an example, the use of two banks of switched capacitors that digitally tune the DCO. The fine-tuning bank is implemented employing a unit capacitance C driven by the digital word tw_0, while the coarse one has unit capacitance $M \cdot C$ (with $M > 1$) driven by tw_1. This design reduces the number of capacitors and thus the area occupation with respect to a full thermometric coding approach. As the number of interconnections goes down, the parasitic capacitance decreases and the tuning range of the DCO increases as well. In addition to that, different types of capacitors can be chosen for different banks and, in this way, the resonator quality factor be optimised. As pointed out, this solution is somehow similar to the segmentation employed in the design of DACs, although in that case the LSBs are usually binary weighted.

Unfortunately, this segmentation does not assure the DCO characteristic to be monotonic. In fact, as shown in Figure 9.15, also in the ideal case of a perfect matching within each single capacitor bank, the gains of the two banks, indicated as K_{dco0} and K_{dco1}, may be different. In other words, it is not guaranteed that the LSB of the coarse bank is perfectly identical to the full-scale range of the fine bank. This is induced by a mismatch between the fine and the coarse banks of capacitors. A means to mitigate the impact of this analog impairment is to design the coarse bank deliberately smaller than Mx and to perform an automatic digital calibration that essentially predistort the DCO tuning curve in the background. The idea is to modify the two-point injection architecture, as shown in Figure 9.16. In the new scheme, two gains g_0 and g_1 are used to compensate for the mismatch between K_{dco0} and K_{dco1}, and make the gains of the two DCO paths equal. This operation can be seen

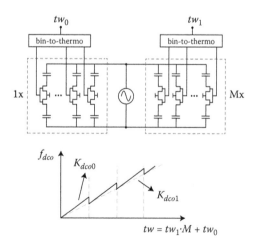

FIGURE 9.15
Segmented DCO topology and resulting non-linear tuning characteristic.

as a pre-distortion of the DCO tuning characteristic. We will refer to this calibration block as *DCO multi-gain adaptation* circuit.

Neglecting for the moment the role of the LMS blocks in Figure 9.16, we see that the FM signal n is scaled by the gain g_1 to generate the digital coarse tuning tw_1. The $\Delta\Sigma$ simply quantises this signal since the number of bits in the product $n \cdot g_1$ is clearly much larger than the number of bits of the switched capacitor bank. The $\Delta\Sigma$ quantisation noise r is multiplied by the gain g_0 and the resulting signal is added to the output of the digital filter to produce the fine-tuning word tw_0. The operating principle of this scheme can be explained with the help of the linear model in Figure 9.17. According to this scheme, the error at TDC output (in the z-domain), $e(z)$, can be written as a function of $n(z)$ and $r(z)$:

$$
e(z) = \left[n(z) \cdot (2\pi f_{ref} - g_1 K_{dco1}) + r(z) \cdot (g_0 K_{dco0} - K_{dco1}) \right] \cdot
$$
$$
\times \frac{G_{loop}(z)}{1 + G_{loop}(z)} \cdot \frac{1}{H(z) K_{dco0}}. \tag{9.24}
$$

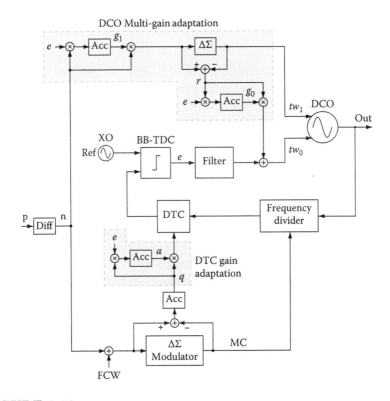

FIGURE 9.16
DPLL with two-point injection scheme and multi-gain DCO predistortion.

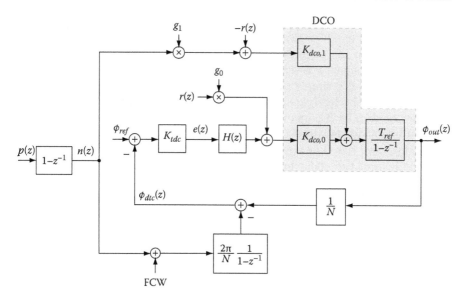

FIGURE 9.17
Model of DPLL with two-point injection scheme and multi-gain DCO predistortion.

As we found out in the plain implementation of the two-point injection scheme, the error $e(z)$ induced by the signal injection has to be null, to achieve an all-pass frequency response. The same thing should be fulfilled in this scheme. So, if we impose that $e(z) = 0$ in (9.24), we get

$$K_{dco1} = g_0 K_{dco0} \tag{9.25}$$

$$f_{ref} = g_1 K_{dco1}. \tag{9.26}$$

Intuitively, the first condition in (9.25) implies matching between the overall gains of the coarse, K_{dco1}, and the fine, $g_0 K_{dco0}$, banks of the DCO. Instead, the second condition, in (9.26), implies matching of the overall gains of the DCO path, $g_1 K_{dco1}$, and the divider path, f_{ref}.

Having said that, it is clear that an automatic calibration circuit is necessary to regulate both g_0 and g_1 automatically and in the background. This function is performed by the two LMS loops in Figure 9.16, in the same fashion already described for the other calibration loops. From (9.24), we have that $e(z)$ is a linear combination of both $n(z)$ and $r(z)$. When the correlation between $e(z)$ and $n(z)$ and the correlation between $e(z)$ and $r(z)$ is null, the outputs of the two accumulators, g_0 and g_1, are constant, implying that both gains have reached the desired value. Of course, in a practical implementation, this multi-gain scheme can embed more than two capacitor banks to realise more than two levels of segmentations. Correspondingly, for each additional level of segmentation, one additional LMS calibration should be added and an additional gain estimated. One example of practical implementation is described in the following section.

9.7 Example of Practical Implementation

The block diagram of a practical implementation of phase modulator is sketched in Figure 9.18 [42]. It is based on a DPLL with two-point injection and the DCO predistortion schemes discussed earlier. The DPLL employs a BB-TDC(or single bit), realised as a D-type flip-flop, and a DTC in the feedback path (as a consequence of the presence of the BB-TDC). The DCO is an oscillator relying on an LC resonator segmented in capacitor banks.

The complete modulator, which is designed to synthesise carriers with frequency tunable between 2.9 and 4.0 GHz from a 40-MHz reference oscillator, embeds three additional blocks (not shown in Figure 9.18): a coarse frequency loop, an automatic bandwidth control (already mentioned earlier) and another digital circuit running in background that corrects for the delay mismatches between the two injection paths [42].

In this implementation, the multi-gain adaptation block that compensates for DCO non-linearity features three levels. The input FM signal $n(z)$ is first multiplied by a proper factor g_2, then it is quantised via a digital $\Delta\Sigma$ modulator and fed to the coarse capacitor bank. The quantisation error of the first $\Delta\Sigma$ quantiser is then multiplied by a factor g_1 and fed to the fine capacitor bank. Finally, the residual quantisation error is multiplied by another factor g_0 and fed to the input of the voltage-mode DAC driving the bias voltage of a metal-oxide-semiconductor (MOS) varactor. Let us assume that the gains relative to the three DCO banks from the coarsest to the finest are K_{dco2}, K_{dco1} and K_{dco0}, respectively. Following the previous discussion, it is easy to demonstrate that an all-pass-shaped response of the whole modulator systems is assured by the following equalities:

$$2\pi = g_2 K_{dco2} T_{ref} \tag{9.27}$$

$$g_1 \cdot K_{dco1} = K_{dco2} \tag{9.28}$$

$$g_0 \cdot K_{dco0} = K_{dco1}. \tag{9.29}$$

Also in this case, while (9.27) expresses the balance between the gain of the two injection paths (divider and DCO), the other two equalities, (9.28) and (9.29), get the finer banks to have the same equivalent gain as the coarsest one. The three conditions are satisfied by proper choice of the three gains, g_2, g_1 and g_0, that are automatically regulated in the background by the three LMS loop shown in Figure 9.18.

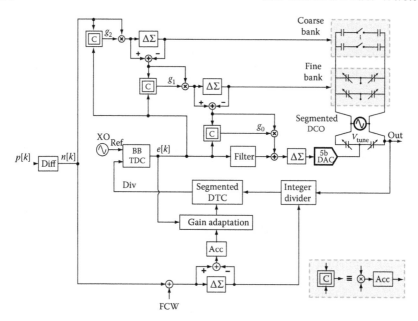

FIGURE 9.18
Block schematic of practical DPLL with two-point injection scheme and automatic DCO predistortion.

9.7.1 DTC design

The DTC is in principle a buffer stage, whose input-output delay is controlled by varying its capacitive load. In practice, the required number of bits of the DTC is equal to the required number of TDC bits in the conventional DPLL architecture in Figure 9.5. As we have already observed, this number may vary between 10 and 13 in high-performance wireless applications. Thus, as for the DCO, a thermometric coding of the load capacitors of the DTC would be impractical. A power-efficient implementation of the DTC at circuit level is illustrated in Figure 9.19. It adopts a segmented architecture. The output of the integer divider (labelled as *div output*) is resampled by a cascade of three latches, which are clocked by the two outputs of the differential DCO. Those outputs *dco* and $\overline{dco}$ are 180 degrees out of phase. In this fashion, the time shift between the signals $P0$ and $P1$ is equal to $T_{dco}/2$, i.e., half the DCO period. This value represents the coarser delay generated by the circuit or, in other words, the delay corresponding to the most significant bit, $a_0[k]$. The analog MUX controlled by $a_0[k]$ selects one of the $P0/P1$ signals and drives the capacitive load. The latter is divided into two banks of thermometrically weighted switched MOS capacitors: a fine and a coarse bank of 64 capacitors each.

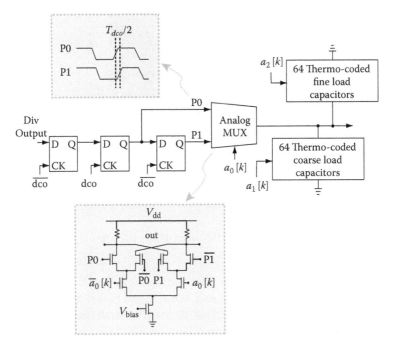

FIGURE 9.19
Circuit schematic of the DTC block.

The two banks are controlled by $a_1[k]$ and $a_2[k]$, respectively. To guarantee the overlap between fine and coarse characteristics in the presence of PVT spreads, the delay range of the finest bank covers with margin the resolution of the coarse bank. Similarly, the total delay range of the coarse bank exceeds the delay generated by the most significant bit (i.e. the $T_{dco}/2$ delay). The digital signals $a_0[k]$, $a_1[k]$ and $a_2[k]$ are automatically regulated in the background by three LMS loops [34], whose scheme is essentially the one just described earlier for the segmentation of the DCO banks. In essence, the DTC control word is further quantised by means of a digital $\Delta\Sigma$ modulator, whose quantisation noise is exploited to regulate the following segmentation level.

The differential topology of the MUX circuit provides better immunity to supply bounces and also features a reduced injection of disturbances in the supply rail itself, if compared to a CMOS implementation. Clearly, with respect to a simple CMOS inverter, the static power contribution of the differential topology leads to a higher consumption.

9.7.2 Experimental verification

The modulator described in Figure 9.18 has been fabricated in a 65 nm CMOS technology. The die photograph is shown in Figure 9.20. The area of the core

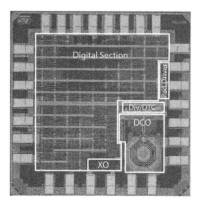

FIGURE 9.20
Die photo of the phase modulator fabricated in a 65 nm CMOS process.

circuits (excluding pads used for testing purposes) is slightly larger than 0.5 mm^2 and includes the pad drivers and the reference oscillator driven by an external crystal (XO). It is interesting to observe that most of the area is occupied by digital circuits, realised with standard cells. In addition to the described digital blocks, this section includes two different baseband signal generators, for GMSK and QPSK modulation, respectively. The overall power consumption (excluding XO reference oscillator and pad drivers) is 5 mW from the 1.2-V voltage supply. The divider DTC and BB-TDC contributes for 2.2 mW, while the DCO burns 0.7 mW and the following buffer dissipates 1 mW.

The first test consists in the measurement of the output spectrum of the DPLL when a fractional channel is synthesised but no modulation signal is applied. This is instrumental to verify if the regulated DTC is really capable to cancel the quantisation noise induced by the dithering of the division factor and enforce the random-noise regime described earlier. Figure 9.21 shows the measured output spectrum for a near-integer-N channel. In this case, the fractional-N spur, that is visible in the spectrum, is unfiltered by the PLL, because it is inside the PLL bandwidth. So, it reaches the highest value of -52 dBc. The total RMS jitter, including random noise and fractional spurs integrated from 3 kHz to 30 MHz, is 503 fs for this worst-case channel. Equivalently, the integrated phase error is equal to -39 dB. The jitter perfomance is at par with the state-of-the-art of analog PLLs, a clear signature that the designed DPLL entails no degradation of noise arising from quantisation noise or limit cycles. In other words, as desired, the BB-TDC behaves as a linear block. The achieved jitter-power product, a common figure of merit for PLLs, is particularly low, evidencing the benefits in terms of power efficiency of the BB-TDC approach.

Then, the performance of the DPLL as a phase modulator is also tested. To this purpose, two digital generators of modulation signals (QPSK and GMSK, respectively) were also integrated on the same die. Figure 9.22a shows the

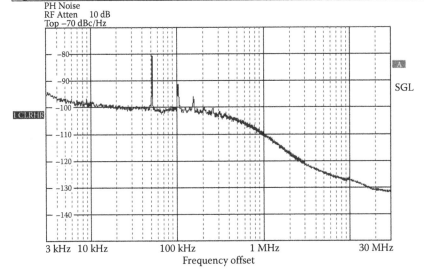

		Phase noise		
	Settings	Residual noise		Spot noise(T1)
Signal freq: 3.600902 GHz		Evaluation from 3 kHz to 30 MHz	1 kHz	**Not valid**
Signal level: −5.52 dBm		Residual PM 0.652°	10 kHz	−99.19 dBc/Hz
Signal freq Δ: −38.41 Hz		Residual FM 40.594 kHz	100 kHz	−99.48 dBc/Hz
Signal level Δ: −0.48 dBm		RMS jitter 0.5028 ps	1 MHz	−110.00 dBc/Hz

FIGURE 9.21

Measured spectrum of the implemented DPLL for a near-integer-N synthesised channel.

measured performance of the phase modulator (signal constellation, phase trellis diagram, spectrum), when the 3.6 GHz carrier is modulated with a 20 Mb/s QPSK signal. Probably, the most important parameter to verify in this case is the EVM of the modulation. In fact, the EVM accounts not only for the phase noise but also for the distortion of the modulator. The measured EVM is below 1.6% or −36 dB, again a very low value, considering the low level of dissipated power. The efficiency of the modulator, that measures the energy required for transmitting one bit, is equal to 0.25 nJ/bit. From the value of the integrated phase noise (−39 dB) and the value of the measured EVM (−36 dB), we can infer that there is a second source of EVM degradation in the order of −39 dB or, equivalently, 1.1%. The latter can be ascribed to a residual distortion of the phase modulator.

Similar EVM values are measured when the carrier is modulated with a continuous-phase signal, such as a 10 Mb/s GMSK modulation. The measured performance are shown in Figure 9.22b. The total RMS value of the EVM is still −36 dB and the energy per bit is 0.5 is nJ/bit. For both modulations, the efficiency and the EVM are much better than the figures obtained in phase modulators based on analog PLLs or phase-switching technique.

(a)

(b)

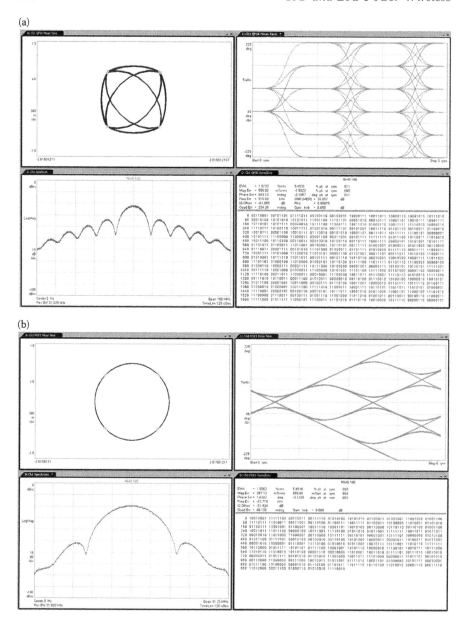

FIGURE 9.22

Measured performance of the phase modulator: (a) 20 Mb/s QPSK modulation and (b) 10 Mb/s GMSK.

9.8 Conclusions

In this chapter, we have shown how a DPLL can be designed to realise a low-power, linear, wideband phase modulator suitable for wireless transmitters. Both conventional TDC-based and recently proposed DTC-based DPLLs have discussed and compared, showing that the adoption of the DTC together with a BB-TDC allows lower power consumption at same jitter performance. Moreover, the two main techniques to widen the modulation bandwidth of a PLL, such as the two-point injection and the pre-emphasis schemes have been analysed and compared. The two-point injection scheme while offering the same degree of modulation accuracy enables the adoption of the BB-TDC even when the DPLL is used as a phase modulator. A fabricated CMOS DPLL employing both the BB-TDC with DTC and the two-point injection scheme has been shown to achieve −36 dB EVM at 0.25 nJ/bit of efficiency. The very good EVM and energy-per-bit achieved demonstrate that the digitally intensive design approach for analog circuits is becoming more than just a curiosity, at least for some important RF building blocks.

Bibliography

[1] J. Groe. Polar transmitters for wireless communications. *IEEE Commun. Mag.*, 45(9):58–63, Sept. 2007.

[2] L. R. Kahn. Single sideband transmission by envelope elimination and restoration. *Proc. IRE*, 40(7):803–806, Jul. 1952.

[3] E. McCune. Envelope tracking or polar—which is it? *IEEE Microw. Mag.*, pages 34–56, Jun. 2012.

[4] Z. Boos, A. Menkhoff, F. Kuttner, M. Schimper, J. Moreira, H. Geltinger, T. Gossmann, P. Pfann, A. Belitzer, and T. Bauernfeind. A fully digital multimode polar transmitter employing 17b RF DAC in 3G mode. In *IEEE ISSCC Digest of Technical Papers*, Feb. 20–24 2011, San Francisco, CA, pages 376–377.

[5] Lu Ye, Jiashu Chen, Lingkai Kong, P. Cathelin, E. Alon, and A. Niknejad. A digitally modulated 2.4GHz WLAN transmitter with integrated phase path and dynamic load modulation in 65nm CMOS. In *IEEE ISSCC Digest of Technical Papers*, Feb. 17–21 2013, San Francisco, CA, pages 330–331.

[6] J. Chen, L. Rong, F. Jonsson, G. Yang, and L.-R. Zheng. The design of all-digital polar transmitter based on ADPLL and phase synchronized $\Delta\Sigma$ modulator. *IEEE J. Solid-State Circuits*, 47(5):1154–1164, May 2012.

[7] Y.-H. Liu, X. Huang, M. Vidojkovic, K. Imamura, P. Harpe, G. Dolmans, and H. De Groot. A 2.7nJ/b multi-standard 2.3/2.4GHz polar transmitter for wireless sensor networks. In *IEEE ISSCC Digest of Technical Papers*, Feb. 19–23 2012, San Francisco, CA, pages 448–449.

[8] H. Chireix. High power outphasing modulation. *Proc. IRE*, 23(11): 1370–1392, Nov. 1935.

[9] D.C. Cox. Linear amplification with nonlinear components. *IEEE Trans. Commun.*, 22(12):1942–1945, Dec. 1974.

[10] M. E. Heidari, M. Lee, and A. A. Abidi. All-digital outphasing modulator for a software defined transmitter. *IEEE J. Solid-State Circuits*, 44(4):1260–1271, Apr. 2009.

[11] A. Ravi, P. Madoglio, H. Xu, K. Chandrashekar, M. Verhelst, S. Pellerano, L. Cuellar, M. Aguirre-Hernandez, M. Sajadieh, J. Zarate-Roldan, O. Bochobza-Degani, H. Lakdawala, and Y. Palaskas. A 2.4-GHz 20–40-MHz channel WLAN digital outphasing transmitter utilizing a delay-based wideband phase modulator in 32-nm CMOS. *IEEE J. Solid-State Circuits*, 47(12):3184–3196, Dec. 2012.

[12] S.-M. Yoo, J. S. Walling, E. C. Woo, B. Jann, and D. J. Allstot. A switched-capacitor RF power amplifier. *IEEE J. Solid-State Circuits*, 46(12):2977–2987, Dec. 2011.

[13] D. Chowdhury, L. Ye, E. Alon, and A. M. Niknejad. An efficient mixed-signal 2.4-GHz polar power amplifier in 65-nm CMOS technology. *IEEE J. Solid-State Circuits*, 46(8):1796–1809, Aug. 2011.

[14] T. Nakatani, J. Rode, D. F. Kimball, L. E. Larson, and P. M. Ashbeck. Digitally controlled polar transmitter using a watt-class current-mode class-D CMOS power amplifier and Guanella reverse balun for handset applications. *IEEE J. Solid-State Circuits*, 47(5):1104–1112, May 2012.

[15] Y.-H. Liu, C.-L. Li, and T.-H. Lin. A 200-pJ/b MUX-based RF transmitter for implantable multichannel neural recording. *IEEE Trans. Microwave Theory Tech.*, 57(10):2533–2541, Oct. 2009.

[16] P. Su and S. Pamarti. A 2.4GHz wideband open-loop GFSK transmitter with phase quantization noise cancellation. *IEEE J. Solid-State Circuits*, 46(3):615–626, Mar. 2011.

[17] G. Yahalom and J. L. Dawson. A low-Q resonant tank phase modulator for outphasing transmitters. In *IEEE Symposium on Radio Frequency Integrated Circuits Symposium (RFIC)*, Jun. 2–4 2013, Seattle, WA, pages 221–224.

[18] P. Madoglio, H. Xu, K. Chandrashekar, L. Cuellar, M. Faisal, W. Yee Li, H. Seok Kim, K. M. Nguyen, Y. Tan, B. Carlton, V. Vaidya, Y. Wang, T. Tetzlaff, S. Suzuki, A. Fahim, P. Seddighrad, J. Xie, Z. Zhang, D. S. Vemparala, A. Ravi, S. Pellerano, and Y. Palaskas. A 2.4GHz WLAN digital polar transmitter with synthesized digital-to-time converter in 14 nm trigate/FinFET technology for IoT and wearable applications. In *IEEE ISSCC Digest of Technical Papers*, Feb. 5–9 2017, San Francisco, CA, pages 226–227.

[19] C.-M. Hsu, M. Z. Straayer, and M. H. Perrott. A low-noise wide-BW 3.6-GHz digital $\Delta\Sigma$ fractional-N frequency synthesizer with a noise-shaping time-to-digital converter and quantization noise cancellation. *IEEE J. Solid-State Circuits*, 43(12):2776–2786, Dec. 2008.

[20] A. L. Lacaita, S. Levantino, and C. Samori. *Integrated frequency synthesizers for wireless systems*. Cambridge University Press, Cambridge, 2007.

[21] S. Pamarti, L. Jansson, and I. Galton. A wideband 2.4-GHz delta-sigma fractional-N PLL with 1-Mb/s in-loop modulation. *IEEE J. Solid-State Circuits*, 39(1):49–62, Jan. 2004.

[22] M. Gupta and Bang-Sup Song. A 1.8-GHz spur-cancelled fractional-N frequency synthesizer with LMS-based DAC gain calibration. *IEEE J. Solid-State Circuits*, 41(12):2842–2851, Dec. 2006.

[23] A. H. Sayed. *Adaptive filters*. Wiley-IEEE Press, Piscataway Township, NJ, 2008.

[24] S. Levantino, G. Marzin, and C. Samori. An adaptive pre-distortion technique to mitigate the DTC non-linearity in digital PLLs. *IEEE J. of Solid-State Circuits*, 7, Aug. 2014.

[25] M. Z. Straayer and M. H. Perrott. A multi-path gated ring oscillator TDC with first- order noise shaping. *IEEE J. Solid-State Circuits*, 44(4): 1089–1098, Apr. 2009.

[26] L. Vercesi, A. Liscidini, and R. Castello. Two dimensions vernier time-to-digital converter. *IEEE J. Solid-State Circuits*, 45(8):1504–1512, Aug. 2010.

[27] Y. Cao, P. Leroux, W. De Cock, and M. Steyaert. A 1.7mW 11b 1-1-1 MASH $\Delta\Sigma$ Time-to-Digital Converter. In *IEEE ISSCC Digest of Technical Papers*, Feb. 20–24 2011, San Francisco, CA, pages 480–482.

[28] A. Elshazly, S. Rao, B. Young, and P.K. Hanumolu. A 13b 315fsrms 2mW 500MS/s 1MHz bandwidth highly digital time-to-digital converter using switched ring oscillators. In *IEEE ISSCC Digest of Technical Papers*, Feb. 19–23 2012, San Francisco, CA, pages 464–466.

[29] J.-P. Hong, S.-J. Kim, J. Liu, N. Xing, T.-K. Jang, J. Park, J. Kim, T. Kim, and H. Park. A $0.004mm^2 250\mu W$ $\Delta\Sigma$ TDC with time-difference accumulator and a $0.012mm^2$ 2.5mW Bang-Bang digital PLL using PRNG for low-power SoC applications. In *IEEE ISSCC Digest of Technical Papers*, Feb. 19–23 2012, San Francisco, CA, pages 240–242.

[30] P. Lu, A. Liscidini, and P. Andreani. A 3.6 mW, 90 nm CMOS gated-Vernier time-to-digital converter with an equivalent resolution of 3.2 ps. *IEEE J. Solid-State Circuits*, 47(7):1626–1635, Jul. 2012.

[31] E. Temporiti, C. Weltin-Wu, D. Baldi, R. Tonietto, and F. Svelto. A 3 GHz fractional all-digital PLL with a 1.8 MHz bandwidth implementing spur reduction techniques. *IEEE J. Solid-State Circuits*, 44(3):824–834, Mar. 2009.

[32] E. Temporiti, C. Welti-Wu, D. Baldi, M. Cusmai, and F. Svelto. A 3.5 GHz wideband ADPLL with fractional spur suppression through TDC dithering and feedforward compensation. *IEEE J. Solid-State Circuits*, 45(12):2723–2736, Dec. 2010.

[33] M. Zanuso, S. Levantino, C. Samori, and A. L. Lacaita. A wideband 3.6 GHz digital $\Delta\Sigma$ fractional-N PLL with phase interpolation divider and digital spur cancellation. *IEEE J. Solid-State Circuits*, 46(3):627–638, Mar. 2011.

[34] D. Tasca, M. Zanuso, G. Marzin, S. Levantino, C. Samori, and A. L. Lacaita. A 2.9-to-4.0GHz fractional-N digital PLL with Bang-Bang phase detector and 560fsrms integrated jitter at 4.5mW power. *IEEE J. Solid-State Circuits*, 46(12):2745–2758, Dec. 2011.

[35] N. Da Dalt. Linearized analysis of a digital Bang-Bang PLL and its validity limits applied to jitter transfer and jitter generation. *IEEE Trans. Circuits Syst. I*, 55(11):3663–3675, Nov. 2008.

[36] M. Zanuso, D. Tasca, S. Levantino, A. Donadel, C. Samori, and A. L. Lacaita. Noise analysis and minimization in Bang-Bang digital PLLs. *IEEE Trans. Circuits Syst. II*, 56(11):835–839, Nov. 2009.

[37] G. Marucci, S. Levantino, P. Maffezzoni, and C. Samori. Analysis and design of low-jitter digital Bang-Bang phase-locked loops. *IEEE Trans. Circuits Syst. I*, 61(1):26–36, Jan. 2014.

[38] G. Marzin, S. Levantino, C. Samori, and A. Lacaita. A background calibration technique to control bandwidth in digital PLLs. In *IEEE ISSCC Digest of Technical Papers*, Feb. 9–13 2014, San Francisco, CA, pages 54–55.

[39] M. J. Underhill and R. I. H. Scott. Wideband frequency modulation of frequency synthesisers. *IEEE Electron. Lett.*, 15(13):393–394, Jun. 1979.

[40] S.-A. Yu and P. Kinget. A 0.65-V 2.5-GHz fractional-N synthesizer with two-point 2- Mb/s GFSK data modulation. *IEEE J. Solid-State Circuits*, 44(9):2411–2425, Sept. 2009.

[41] M. Youssef, A. Zolfaghari, B. Mohammadi, H. Darabi, and A.A. Abidi. A low-power GSM/EDGE/WCDMA polar transmitter in 65-nm CMOS. *IEEE J. Solid-State Circuits*, 46(12):3061–3074, Dec. 2011.

[42] G. Marzin, S. Levantino, C. Samori, and A. Lacaita. A 20Mb/s phase modulator based on a 3.6GHz digital PLL with -36dB EVM at 5mW power. *IEEE J. Solid-State Circuits*, 47(12):2974–2988, Dec. 2012.

[43] M. H. Perrott, T. L. Tewksbury, and C. G. Sodini. A 27-mW CMOS fractional-N synthesizer using digital compensation for 2.5-Mb/s GFSK modulation. *IEEE J. Solid-State Circuits*, 32(12):2048–2060, Dec. 1997.

[44] D. McMahill and C. G. Sodini. A 2.5-Mb/s GFSK 5.0-Mb/s 4-FSK automatically calibrated $\Delta\Sigma$ frequency synthesizer. *IEEE J. Solid-State Circuits*, 37(1):18–26, Jan. 2002.

[45] F. M. Gardner. Frequency granularity in digital phase-lock loops. *IEEE Trans. Commun.*, 44(4):749–758, Jun. 1996.

10

Frequency Synthesis Technique for 60 GHz Multi-Gbps Wireless

Teerachot Siriburanon

University College Dublin

Hanli Liu, Kenichi Okada, and Akira Matsuzawa

Tokyo Institute of Technology

Wei Deng

Apple Inc.

Satoshi Kondo and Makihiko Katsuragi

Toshiba Corporation

Kento Kimura

Fujitsu Limited

CONTENTS

10.1 Introduction

Today, wireless communication has changed the way people lived and turned a mobile device into a necessary equipment in our daily lives. Due to a continuous increase of demands for higher data rate, it challenges the design of low-cost complementary metal oxide semiconductor (CMOS) process that would satisfy stringent requirements in a small, low-cost prototype with low power consumption. This has prompted research into new radio frequency integrated circuits system architectures and design approaches. Increasing demand for unlimitedly higher data rate in portable devices has driven the research community to seek for new wireless spectrums for a support of larger capacity. Due to a congestion of frequency below 3 GHz, utilization of higher frequency will enable higher data rates by larger bandwidth. Ultra-wide band also allows the transmission of information over a wide bandwidth that is larger than 500 MHz. This application has been aimed for high data rate and short-range communication while sharing bandwidth with other standards while providing minimal interference. This results to a strict requirement in output power that limits its communication range. Therefore, the trend for future high-speed wireless communication is toward the mm-wave region. In this chapter, we will focus on the 60 GHz frequency band that offers wider channel bandwidth as shown in Figure 10.1 [1].

Unlicensed 9 GHz bandwidth at 60 GHz has been standardized for future wireless local area network known as WiGig and IEEE802.11ad to cooperate with the present IEEE802.11ac for multi-Gbps short-range communications [2]. For the integration of portable devices of the next-generation wireless communication, 60 GHz transceivers in a direct conversion architecture is more preferable due to low power consumption and small area [3–8]. More importantly, to support high modulation scheme in Cartesian

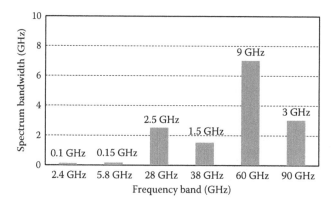

FIGURE 10.1
Bandwidth allocation for different spectrum bands [1].

transceivers, it requires low I/Q mismatches, arisen from layout parasitics and process, voltage and temperature (PVT) variations introduced by radio frequency (RF) and baseband building blocks, in cooperation with low phase noise from a local oscillator. Conventionally, I/Q gain and phase mismatches can be calibrated by variable gain amplifier (VGA) and additional RF-path phase shifter, respectively. However, VGA and RF-path phase shifter could introduce undesired gain or phase shift when operating over wide bandwidth at mm-wave frequency, which further require additional calibration [9]. Digital I/Q calibration can be used for low-frequency synthesizers. However, it is difficult to achieve fine resolution for mm-wave transceivers due to limited resolution of digital-to-analog converter (DAC). An alternative approach is to use a 60 GHz synthesizer for phase calibration, which has proved its effectiveness by achieving as high as 20 Gb/s data rate as shown in Figure 10.2 [10]. Therefore, a 60 GHz frequency synthesizer should not only exhibit low phase noise but also add the ability for I/Q calibration that relaxes requirements for other building blocks. More importantly, solutions for low-power designs are required to enable high-speed data transfer in mobile devices.

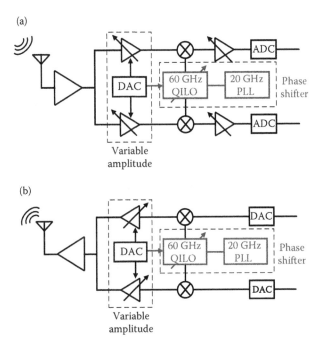

FIGURE 10.2
Simplified block diagram of (a) 60 GHz receiver and (b) 60 GHz transmitter with amplitude and phase calibration using 20 GHz PLL and 60 GHz QILO as phase shifter.

Phase noise performance of previously reported mm-wave and near-mm-wave synthesizers [11–18] is usually in tradeoff with large power consumption. Due to a degradation of quality factor of LC tank at mm-wave frequency, it results in a degradation of out-of-band phase noise and large power [12–15]. Moreover, it requires a chain of high-speed dividers, which further increases power consumption and requires independent tuning [13, 14]. An alternative method is to use a subharmonic injection technique based on a 20 GHz phase-locked loop (PLL) and a 60 GHz quadrature injection-locked oscillator (QILO), which can achieve good out-of-band phase noise due to better overall quality factor of tank. The QILO also adds an ability to adjust phase of quadrature outputs for any I/Q mismatch that arises from other blocks when integrated in the transceiver [10]. However, common constraints that limit the feasibility of low-power mm-wave synthesizers for high data rates are high in-band phase noise, due to high division ratio in PLL feedback path (N_{PLL}), and large power consumption to maintain performance of high-speed building blocks [11, 18].

In this chapter, low-power techniques in a high-performance mm-wave subsampling frequency synthesizer are investigated. In Section 10.2, the architecture of 60 GHz synthesizer is discussed. Then, it is followed by considerations of I/Q mismatch and optimum tracking bandwidth to determine phase noise requirements of IEEE802.11ac/ad. Section 10.3 presents an implementation of a 20 GHz subsampling PLL (SS-PLL), which helps suppress the in-band phase noise [19, 20]. The design and theoretical locking range of the proposed dual-step-mixing injection-locked frequency divider (ILFD) [21] are analyzed and compared with conventional ILFD. Moreover, design and analysis of g_m-enhanced QILO using tail-coupling technique are discussed. In Section 10.4, a 32 mW 60 GHz subsampling frequency synthesizer, which achieves a figure-of-merit (FoM) of −236 dB, is demonstrated as a proof of concept. Finally, conclusions are summarized in Section 10.4.

10.2 Requirements and Architecture Considerations for Multi-Gbps Wireless Communications

10.2.1 Frequency synthesizer jitter and phase noise requirements

The 60 GHz synthesizers require quadrature phases and frequency that could cover four main channels, as well as those in between for channel bonding capability. It should also interoperate with both 36 and 40 MHz reference clock to cope with different symbol rates specified in various standards [2, 22]. More importantly, to support higher data rate, the system should satisfy the phase noise and signal-to-noise ratio (SNR) requirement for higher modulation

TABLE 10.1

Required TX EVM for Different Modulation Schemes in IEEE802.11ad [2]

MCS	Modulation		Data Rate (Mb/s)	TX EVM (dB) (spec)
9	$\pi/2$-QPSK	Single Carrier	2,502.5	−15
12	$\pi/2$-16QAM	Single Carrier	4,620	−21
17	QPSK	OFDM	2,079.00	−13
21	16QAM	OFDM	4,504.50	−20
24	64QAM	OFDM	6,756.75	−26

scheme. The transmitter error vector magnitude (TX EVM) performance should be below −20 dB to support complex quadrature amplitude modulation scheme, i.e. 16QAM as shown in Table 10.1. The TX EVM can be expressed as follows [24, 25].

$$\text{TXEVM} = \sqrt{\frac{1}{\text{SNR}^2} + \varphi^2_{\text{RMS}}} \qquad (10.1)$$

where SNR is the signal-to-noise ratio and φ^2_{RMS} is a double-sideband integrated phase noise of a carrier that can be computed as follows [23, 24]:

$$\varphi^2_{\text{RMS}} = 2 \int_0^{B/2} L(f)\, df \qquad (10.2)$$

where B is the modulation bandwidth. Therefore, to maintain low EVM, SNR should be kept high, whereas root mean square (RMS) jitter should be kept as low as possible. Typical mm-wave PLLs suffer from poor integrated phase noise due to high division ratio in the feedback path and degradation of quality factor of tank that cannot meet 16QAM requirements [11, 18].

Figure 10.3 shows the simplified diagram of the typical TX and RX in direct-conversion architecture. A local oscillator is required to down/up convert the data to be received/transmitted wirelessly. For mm-wave or near-mm-wave applications, phase noise is usually one of the most crucial limiting factor to achieve low EVM performance. Phase noise adds an ambiguity into the data causing a circular distortion of EVM. In the receiver (RX) side, small timing or phase offset of received signal can be recovered in the digital signal processing. Thus, low-offset phase noise can be canceled and EVM performance can be improved. However, the amount of noise to be canceled, i.e., faster tracking bandwidth, requires larger hardware cost and high power consumption. Moreover, maximum tracking bandwidth is restricted by subcarrier spacing in an orthogonal frequency-division multiplexing (OFDM) system. This will be explained in detail later.

To relax phase noise requirement, a decision-directed PLL can be used for symbol-timing recovery in the baseband circuitry to cancel low-offset phase noise. Unlike a single carrier case, subcarrier spacing in an OFDM system

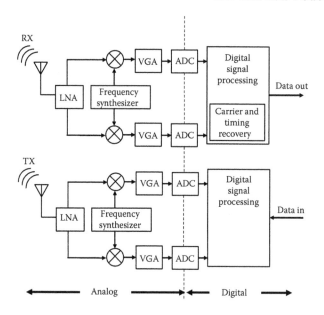

FIGURE 10.3
Simplified diagram of TX and RX with analog and digital baseband.

restricts the use of excessively wide tracking bandwidth (f_{track}) that could cause an intersymbol interference [23, 24]. Assume that the frequency response of carrier tracking $S_{\text{track}}(f)$ loop exhibits a characteristic of a second-order PLL in a form as follows [23, 24]:

$$S_{\text{track}}(f) = \frac{1}{\left[1 + \left(\frac{f}{f_{\text{track}}}\right)^4\right]} \tag{10.3}$$

Therefore, an optimum f_{track} of carrier recover loop can be computed as follows:

$$f_{\text{track}} = \frac{f_{\text{sub}}}{\left(\frac{1}{S_{\text{track}}(f_{\text{sub}})} - 1\right)^{\frac{1}{4}}} \tag{10.4}$$

where f_{sub} is the frequency at the edge of the first occupied subchannel, which is estimated to be half of the subcarrier spacing. If 30 dB suppression is required at that frequency, $S_{\text{track}}(f_{\text{sub}})$ is given to be 10^{-3}. For IEEE802.11ac, the optimum tracking bandwidth should be less than 27.8 kHz due to a subcarrier spacing of 312.5 kHz [23]. In the case of IEEE802.11ad, the subcarrier spacing is 5.15625 MHz, which results in the optimum tracking bandwidth of 458.6 kHz. The baseband output after the demodulator is the difference between that of the receiver input, which exhibits a phase noise from 60 GHz

PLL and the tracking loop output. From Eqs. (10.2) and (10.4), the effective integrated phase noise at the demodulator output can be computed by

$$\varphi^2_{\text{RMS,eff}} = 2 \int\limits_{0}^{B/2} L(f) \left(1 - \frac{1}{1 + \left(\frac{f}{f_{\text{track}}}\right)^4} \right) df \tag{10.5}$$

It can be observed that the low-offset phase noise from the received signal can be canceled up to the tracking bandwidth as shown in Figure 10.4. In case of IEEE802.11ac, in-band phase noise is important even with an assistance

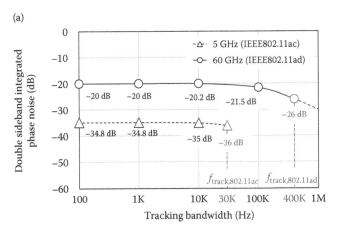

(a)

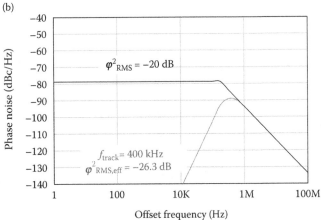

(b)

FIGURE 10.4
(a) Optimum tracking bandwidth and integrated phase noise for IEEE802.11ac and IEEE802.11ad and (b) target phase noise performance for mm-wave PLL to satisfy 16QAM and 64QAM without and with carrier recovery circuit, respectively.

from carrier recovery circuit with optimum bandwidth of 30 kHz as shown in Figure 10.4a. On the other hand, for IEEE802.11ad, integrated noise is -20 dB without carrier recovery. If a tracking bandwidth of 400 kHz can be used, integrated phase noise can be lower. However, in the situation where low-power operation is preferred or wide tracking bandwidth is not available due to a limitation of baseband circuitry, in-band phase noise is also important. Figure 10.4b shows a target performance for IEEE802.11ad where an in-band phase noise should be as low as -78 dBc/Hz and out-of-band phase noise at 10 MHz should be -115 dB/Hz to achieve an EVM lower than -20 and -26 dB without and with carrier recover loop using Eqs. (10.2) and (10.5), respectively.

Another important parameter is the system SNR. To achieve required SNR, received signal should be more than noise floor and image signals. The difference between downconverted received signal and image is called image rejection ratio (IMRR), which should be as high as possible to maintain system SNR. I/Q amplitude and phase mismatch result in a degradation of IMRR. In this work, the IMRR is targeted below -40 dB for 64QAM, which refers to less than 0.15 dB amplitude and less than $1.5°$ phase mismatches.

10.2.2 Phase-locked loop architectures for 60 GHz applications

Conventionally, direct 60 GHz PLL usually requires higher power consumption for main oscillator and high-speed prescaler [12–15]. This is because mm-wave oscillators suffer from inferior quality factor of tank when compared with those at lower frequency. Thus, it requires high power to achieve reasonable phase noise. Moreover, inductorless prescaler divider consumes high power when directly operating at mm-wave frequency. An alternative solution is to adopt the use of frequency multipliers for mm-wave PLLs to maintain high quality factor of main oscillators [11,16–20, 26]. Unlike push–push technique [16–17, 26], an approach using a 20 GHz PLL and 60 GHz QILO is more preferable due to capability of fine phase adjustment. The QILO in Ref. [27] uses a polyphase filter for 20 GHz quadrature injection but does not allow phase calibration. Alternatively, a single-sided injection enables the control of I/Q phase offset with fine resolution by adjusting the free-running frequency of QILO [10]. Figure 10.5a shows an equivalent diagram of QILO composed of a cross-coupled oscillator with a G_m stage to represent an injection transistor. Impedances of tank I and Q are denoted as Z_i and Z_q, respectively. Figure 10.5b shows current and voltage vectors of coupled oscillators for a single-ended equivalent circuit. By injecting a current into one of the coupled oscillator, output frequency of QILO will be locked to the third harmonic of injected signal shown in Figure 10.6. However, if there are any differences in free-running frequency of QILO and an injected signal is in locked condition, phase difference between $\overline{I_i}$ and $\overline{V_i}$ will be generated as shown in Figure 10.5b. This can be represented as follows [37].

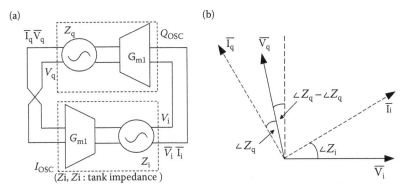

FIGURE 10.5
(a) Simplified block diagram of quadrature injection-locked oscillator and
(b) its phasor diagram.

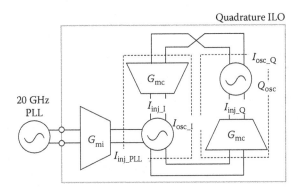

FIGURE 10.6
Simplified block diagram of the 20 GHz PLL and 60 GHz QILO with single-
sided injection.

$$\phi_i = \angle Z_i = \sin^{-1}\left(2Q\frac{\omega_{0I} - \omega_{inj\_PLL(20\ GHz)}}{\omega_{0I}}\frac{I_{osc\_I}}{I_{inj\_I}}\right) \quad (10.6)$$

where Q_I is quality factor of tank I, ω_{0I} is free-running frequency of tank
I of QILOs, $\omega_{inj\_PLL(20\,GHz)}$ is the injection frequency near 20 GHz. From
Eq. (10.6) and Figure 10.6, phase difference between I/Q oscillators can be
computed from an angle between impedance of tank I/Q as follows.

$$\Delta\phi = \angle Z_i - \angle Z_q$$
$$= \sin^{-1}\left(2Q_I\frac{\omega_{0I} - \omega_{inj\_PLL(20\ GHz)}}{\omega_{0I}}\frac{I_{osc\_I}}{I_{inj\_I(60\,GHz\_Q)} + \eta I_{inj\_PLL(20\,GHz\_Q)}}\right)$$
$$- \sin^{-1}\left(2Q_Q\frac{\omega_{0Q} - \omega_{inj\_PLL(20\ GHz)}}{\omega_{0Q}}\frac{I_{osc\_Q}}{I_{inj\_Q(60\,GHz\_I)}}\right) \quad (10.7)$$

where subscript Q denotes the parameter of tank Q and η is an injection efficiency. It can be seen that a QILO can be used as a fine phase shifter by tuning ω_{0I} and ω_{0Q} through varactor of a QILO. When cooperating with VGA for amplitude correction, it can satisfy the requirement for high modulation scheme [10]. To further ensure locked operation, QILO calibration scheme in Ref. [18] can be implemented and operate before fine I/Q phase calibration is in action. Small frequency drift for I/Q phase calibration can cause degradation of injection efficiency, but it does not result in significant phase noise degradation due to large bandwidth of injection locking scheme [35]. Even though a subharmonic injection locking technique is capable of achieving low out-of-band phase noise and offers a solution for I/Q calibration of 60 GHz transceivers [11] and [18] still suffer from poor in-band phase noise and high power consumption. In the next section, the design of 60 GHz subsampling synthesizer and techniques to achieve lower power consumption will be discussed.

10.3   Proposed 60 GHz Subsampling Synthesizer Architecture

10.3.1   20 GHz subsampling PLL

By sampling the slope of oscillator waveform, high gain in phase detection can be achieved. Thus, it can significantly suppress oscillator phase noise with wide bandwidth. Moreover, noise in the phase detection path does not amplify by the divide ratio in the feedback path. Unlike conventional approach in Figure 10.7a that has no problem for an architecture using direct oscillator at low frequency, directly sampling voltage from voltage-controlled oscillator (VCO) outputs with a sampling switch [28, 29] would directly load the tank and result in a reduction of VCO output amplitude [15], which decreases an injection efficiency and degrades locking range of 60 GHz QILO and 20 GHz–5 GHz ILFD. This could result in a failure to lock both oscillators, as shown in Figure 10.7b. In this work, Figures 10.7c and 10.8 show the simplified diagram and the detailed diagram of the proposed 60 GHz PLL with subsampling operation, which samples the output of the 20 GHz ILFD, respectively. The proposed synthesizer includes a 20 GHz SS PLL and a 60 GHz QILO. The 20 GHz PLL can perform both conventional phase detection, which uses phase frequency detector and charge pump (PFD/CP), and proposed subsampling phase detection. For a PFD/CP mode, E_n is set to 0. The subsampling loop is disabled as CP_1 is off and PFD in lower loop works without any dead zones. A divide-by-2 divider is placed after reference clock to support channel bonding. Selecting code (SEL_1) of the multiplexer (MUX_1) can control division ratio to support both 36 and 40 MHz reference frequency as shown in Figure 10.8.

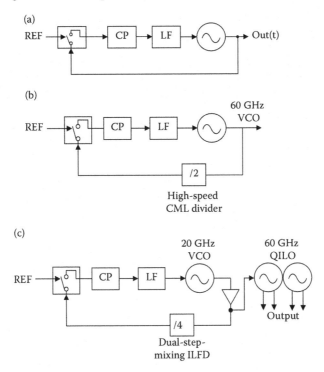

FIGURE 10.7
Simplified architecture of (a) SS-PLL, (b) mm-wave direct SS-PLL, and (c) mm-wave SS-PLL in subharmonic injection architecture.

In subsampling mode, E_n is set to 1, a PFD dead zone is created so that it does not interfere with subsampling loop. The subsampling phase detector (SSPD) samples the pseudodifferential divider outputs from an ILFD with the selected reference clock from MUX_2 and converts phase error into voltage variation. The slope from an output from ILFD is shaped by RC circuits to maintain sinusoidal characteristics and relax differential mismatches. The feedback gain of SSPD is controlled by a pulsewidth controller [28] as follows:

$$\beta_{SSPD} = \frac{K_{SS,d}}{N_{SS-PLL}} = \frac{2A_{VCO}}{\tau_{RC}} \frac{T_{DIV}}{2\pi} \frac{2I_{CP_1}}{V_{gs,eff}} \frac{\tau_{pul}}{T_{REF}} \frac{1}{N_{SS-PLL}} \qquad (10.8)$$

where A_{VCO} is an amplitude of VCO, τ_{RC} is an RC time constant of wave shaper, T_{DIV} is the period of output of ILFD, I_{CP_1} is charge-pump current in SSPD, $V_{gs,eff}$ is effective gate bias of n-channel MOSFET (NMOS) transistor, T_{REF} is a period of reference clock, τ_{pul} and is a controlled pulse width. The in-band phase noise of SSPD can be computed by $L_{in\text{-}band,SSPD} = S_{i,CP_1}/2\beta_{SSPD}^2$. Since a division ratio is reduced from 1200 to 4 in the subsampling feedback path, this helps to alleviate SSPD noise that can be brought below noise of

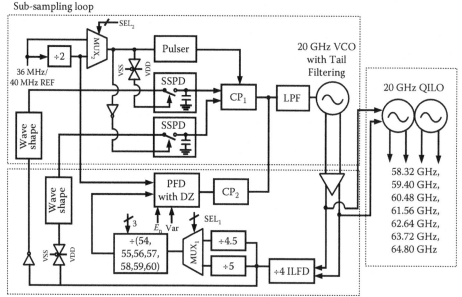

FIGURE 10.8
Detailed block diagram of the proposed 60 GHz subsampling frequency synthesizer.

reference clocks. In this work, loop bandwidth is set to approximately 400 kHz to optimize both RMS jitter and out-of-band phase noise performance. To further lower power consumption, a divide-by-4 dual-step-mixing ILFD is used in the 20 GHz SS-PLL, and a 60 GHz g_m-enhanced QILO is implemented for subharmonic injection locking. In the next section, the design of key building blocks will be discussed in detail.

10.3.2 Dual-step-mixing ILFD using second-harmonic direct injection

Instead of using two cascading current mode logic (CML) dividers [18], only a single-stage divide-by-4 ILFD is used to downconvert 20–5 GHz. In a case of using traditional single-step mixing ILFD shown in Figure 10.9a, an injection signal is applied through a switch that is placed directly across the fundamental output nodes of the differential pair [31–34]. Unfortunately, this approach has limited locking range for higher than 2 division ratio, because there are two disturbing injections shown as gray dots in every cycle as shown in Figure 10.10a. It is clear that if disturbing injections are eliminated, locking range can be extended [34]. This can be done by applying an injection

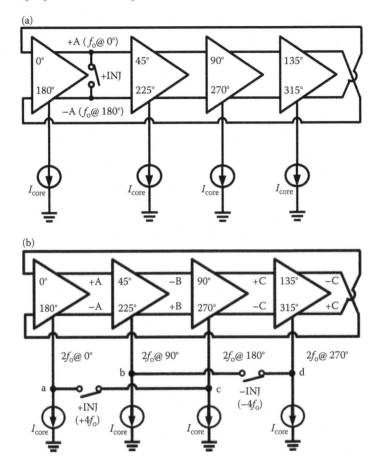

FIGURE 10.9
(a) Simplified diagram of conventional direct injection inductorless ILFD and
(b) Simplified diagram of even-harmonic-enhanced direct injection inductor-
less ILFD.

across the second harmonic nodes at the common nodes of each delay cells
as shown in Figure 10.9b called "dual-step mixing" [20, 21]. This gave more
headroom and reliability over the method using additional cascoded tail transistor [36] to perform a dual-step mixing operation. The phases of the second
harmonic signals at the common nodes a, b, c, d are progressively coupled
from the differential outputs of each delay cell. A switch is placed across common nodes of the first and third delay cells with a phase difference of 180°. As
shown in Figure 10.10b, the first direct injection step happens at every positive peak of the input signal (+INJ) near 4th harmonic of fundamental signal
that is injected into the zero crossing point of the second harmonic signal at
common nodes of the first and third delay without any harmful injection [35–
37]. Once signals at the second harmonic nodes of a and c are synchronized,

(a)

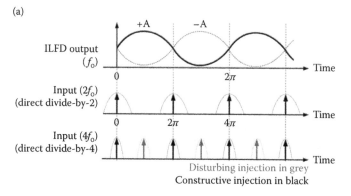

Disturbing injection in grey
Constructive injection in black

(b)

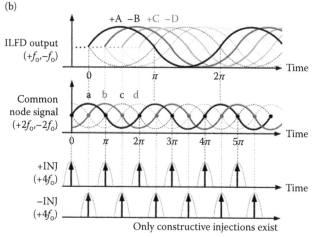

Only constructive injections exist

FIGURE 10.10
Timing injection of (a) conventional ILFD for divide-by-2 and divide-by-4 operation and (b) dual-step-mixing ILFD for a divide-by-4 operation with differential injections.

they further inject every positive peak to zero crossing points of fundamental outputs as a second step. Similarly, an injection with opposite phase (−INJ) synchronizes nodes b and d at common nodes and further to the fundamental outputs. Thus, a differential injection from primary mixer creates quadrature synchronizing signals at secondary mixer which improve the locking range. The proposed ILFD works as two cascoded divide-by-2 frequency dividers, resulting in lower power and larger locking range. This architecture can also work as a cascoded divide-by-2 and divide-by-3 operation results to comparable locking range for divide-by-6 operation when compared with conventional divide-by-3 ILFD as described in Ref. [21].

The locking range of the proposed ILFD is investigated as follows. A typical ring oscillator oscillates at a frequency (f_o) with a phase shift of $-3\pi/4$ at each stage. With an injection applied to either tail transistor of differential pair or injection switch across output nodes, the phase contributed by each stage of an oscillator deviates by θ, which results in a change in the oscillation frequency, to compensate for a phase shift introduced by each mixer ($\phi(\alpha)$). The single-sided locking range of a four-stage ring ILFD can be approximated by [27].

$$\frac{\Delta f}{f_o} = \frac{1}{N}\left(\frac{1+\tan^2\frac{\pi}{N}}{\tan\frac{\pi}{N}}\right)\phi_T \tag{10.9}$$

where N is a number of delay cell and ϕ_T is the total phase shift from injections. It can be observed that phase shift by a mixer is averaged over a number of delay cell. Multiple injection in an ILFD has proved effective to enhance the locking range of an ILFD [31, 34]. Unlike a conventional ILFD that utilizes weak 3rd and 5th harmonics for a divide-by-4 operation, locking range of the proposed dual-step-mixing ILFD is analyzed. An equivalent model of the conventional direct-mixing ILFD and proposed dual-step-mixing ILFD with an assumption of a single injection point is shown in Figures 10.11a and b, respectively. The divide-by-4 operation has been achieved by an injection switch as a primary mixer and a single balanced mixer formed by a differential pair as a secondary mixer shown in Figure 10.9b. The injection signal near 4th harmonic is applied with the current in an oscillator (I_{OSC}) at the primary

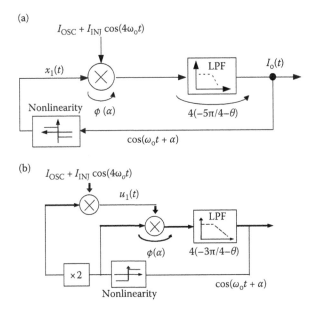

FIGURE 10.11
Equivalent circuit model for (a) conventional direct-mixing ILFD and (b) proposed dual-step-mixing ILFD with an assumption of a single injection point.

mixer, which first mixed with even harmonics of the 2nd and 6th components in the first delay cell results to $u_1(t)$, which can be derived as

$$u_1(t) = I_{OSC}b_2 \cos(2w_ot - 2\alpha) \tag{10.10}$$
$$+ I_{INJ}\frac{\gamma}{2}\left[b_2 \cos(2w_ot - 2\alpha) + b_6 \cos(2w_ot + 6\alpha)\right]$$

where I_{INJ} is the amplitude of injection current, γ is the gain of even harmonic terms in differential pair that is typically equal to 2 [36], w_o is the free-running frequency, α is the relative phase difference of the fundamental component and the injection signal, and b_n is the coefficient of nth harmonic at output. $u_1(t)$ is further mixed with 1st and 3rd harmonics at gates of each differential pair. The current at the output of secondary mixer can be derived as

$$y_1(t) = I_{INJ}e^{jw_ot}e^{j\alpha} \tag{10.11}$$

$$\times \left(\frac{I_{OSC}}{I_{INJ}}(b_1b_2 + b_2b_3) + \frac{\gamma}{4}\left((b_1b_2 + b_3b_6)e^{-j4\alpha} + (b_1b_6 + b_2b_3 + b_1b_4)e^{j4\alpha}\right)\right)$$

By rearranging Eq. (10.11) in the form of $|I_o|\cos(w_ot + \alpha + \phi(\alpha))$, phase shifted by each mixer can be derived in Table 10.2. By letting $\alpha_1 = \alpha$, the mixer output phase of the first stage can be derived as

$$\gamma_1(\alpha) = \alpha + \phi_1(\alpha) = \alpha - \tan^{-1}\left(\frac{k_1 \sin(4\alpha)}{\left(\frac{k_2}{\eta}\right) - \cos(4\alpha)}\right) \tag{10.12}$$

where η is an injection efficiency (I_{INJ}/I_{OSC}), k_1 is $\frac{b_1 - b_3}{b_1 + b_3}$ and k_2 is $\frac{2b_2}{b_2 + b_6}$. Signal $y_1(t)$ passes low-pass filter (LPF) and acts as an input to the secondary mixer. It further mixes with the mixed product from another phase of injection with even harmonics at the second stage as shown in Figure 10.13. Since $\phi_1(\alpha)$ is small, $\alpha_2 \approx \alpha_1 - (3\pi/4)$ [31]. The secondary mixer at the second stage contributes to an additional phase shift of $\phi_2(\alpha)$. The process continues and

TABLE 10.2
Comparison of Theoretical Locking Range of a Divide-by-4 ILFD in a Four-Stage Ring Topology

Type	Phase Introduced by Each Mixer ($\phi(\alpha)$)	Single-Sided Locking Range	Norm. Lock. Range
Direct mixing	$-\tan^{-1}\left(\frac{\frac{(b_5 - b_3)}{(b_5 + b_3)}\sin(4\alpha)}{\frac{2b_1}{(b_5 + b_3)\eta} - \cos(4\alpha)}\right)$	$2\tan^{-1}\left(\frac{0.56\eta}{\sqrt{14 - \eta^2}}\right)$	1
Dual-step mixing	$-\tan^{-1}\left(\frac{\frac{(b_1 - b_3)}{(b_1 + b_3)}\sin(4\alpha)}{\frac{2b_2}{(b_2 + b_6)\eta} - \cos(4\alpha)}\right)$	$2\tan^{-1}\left(\frac{0.39\eta}{\sqrt{2.85 - \eta^2}}\right)$	~ 2.5

results to phase shift across the mixers at the 3rd and 4th stages. The total phase shifts can be expressed as

$$\phi_T = -\sum_{i=1}^{4} \tan^{-1} \left(\frac{k_1 \sin\left(4\alpha - 3\left(i-1\right)\pi - \left(i-1\right)\pi\right)}{\left(\frac{k_2}{\eta}\right) - \cos\left(4\alpha - 3\left(i-1\right)\pi - \left(i-1\right)\pi\right)} \right) \qquad (10.13)$$

where i is each stage of delay in an oscillator, $3\left(i-1\right)\pi$ is phase shifted across each stage, $\left(i-1\right)\pi$ and is the phases of injection at each primary mixer. Interestingly, this results to an optimum phase to enhance locking range [21, 31, 36]. By differentiating Eq. (10.13) with respect to α, the maximum phase shift of the mixers can be derived. Thus, theoretical single-sided locking range of the proposed ILFD is compared with the locking range of the conventional single-step ILFD derived from the same procedure [31–34], as shown in Table 10.2. Note that optimum phase shift using differential injection is taken into account for both cases for a fair comparison as discussed in Ref. [31]. The single-sided locking range is proportional to $\tan^{-1}\left(k_1\eta/\sqrt{k_2^2 - \eta^2}\right)$. Thus, wider locking range of the dual-step mixing is derived from smaller attenuation on the injection efficiency (k_2). By substituting b_n obtained from simulation, it shows approximately 2.5 times larger locking range compared with a conventional divide-by-4 ILFD. The detailed schematic of the proposed ILFD is shown in Figure 10.13. A simulation result has been performed for the proposed ILFD with a comparison to a direct mixing ILFD for a divide-by-4 operation by assuming the same injection current with differential injection for both cases. From Figure 10.14, the single-step-mixing divide-by-4 ILFD can provide 5% single-sided locking range with η of 0.25, while the proposed ILFD required only 0.1 to achieve the same locking range. In actual implementation,

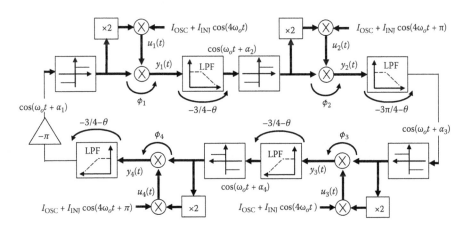

FIGURE 10.12
Equivalent circuit model for the proposed dual-step mixing ILFD with differential injection.

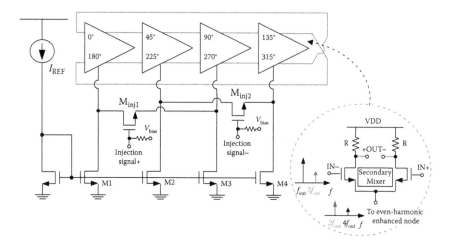

FIGURE 10.13
Detailed schematic of the dual-step-mixing ILFD.

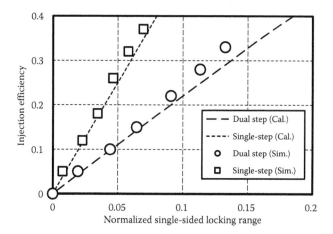

FIGURE 10.14
Theoretical and simulated locking range of conventional single-step mixing and proposed dual-step mixing divide-by-4 ILFD using differential injection.

the differential phases can be directly obtained from the 20 GHz buffer after the VCO. For a divide-by-6 operation in the proposed ILFD, the models in Figures 10.11 and 10.12 can be used with a modified $u_1(t)$, which are derived from the injection signal that is mixed with 4th and 8th harmonics. Then, the same analysis can be applied to obtain modified closed form of locking range. Table 10.3 shows a comparison of the high-speed divider chain in the state-of-the-art mm-wave PLLs. This work achieves low power consumption

TABLE 10.3
Comparison of High-Speed Divider Chain in mm-Wave Frequency Synthesizers

Ref.	Topology	Components	Power Consumption	Robustness	Area
[15]	60 GHz Direct	(/2) LC-CML + (/3) Ring-CML	28 mW	Good	Large
[13]	60 GHz Direct	(/2) LC-ILFD + (/2) Ring-CML + (/4) Ring-CML	23 mW	Good	Large
[14]	60 GHz Direct	(/4) Ring-DCML + (/4) Ring-ILFD	12 mW	Normal	Small
[11]	20 GHz PLL + 60 GHz QILO	(/2) Ring-CML + (/2) Ring CML	18 mW	Good	Small
This work	20 GHz PLL + 60 GHz QILO	(/4) Ring-ILFD	2.2 mW (buffer included)	Good	Small

and small area by using only a single ILFD in the high-speed divider chain of the mm-wave PLL. Moreover, the proposed dual-step mixing helps achieving acceptably wide locking range for a robust operation in a mm-wave PLL.

10.3.3 The 20 GHz class-C VCO and 20 GHz cross-coupled buffer

In this work, 20 GHz LC-VCO with tail filter is utilized as shown in Figure 10.15. This allows high oscillation swing while preserving current efficiency that is superior to other topologies [38]. The frequency can be tuned by a varactor and a 5-bit capacitor array. Tail filter is composed of inductor and an array of switched capacitor for second harmonic peaking to preserve the tank from loading at second harmonics [39]. Capacitive cross-coupled 20 GHz buffer is used to cancel parasitic capacitances [6–8]. This results in high swing with low power.

To further reduce the power consumption, an alternative approach is to adopt Class-C VCO [40]. Recent trend in the design of oscillator focuses on the improvement of flicker noise corner in mm-wave frequency, which is usually more than 1 MHz offset frequency. To deal with this problem, one technique is

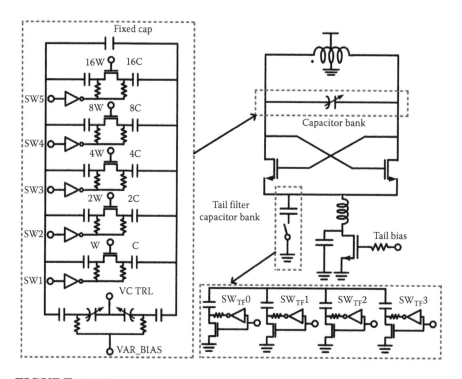

FIGURE 10.15
Detailed schematic of 20 GHz class-B VCO with tunable tail filtering.

careful consideration of the amplitude and phases of harmonics of oscillation frequency [41].

10.3.4 The 60 GHz g_{m}-enhanced quadrature injection-locked oscillator

To satisfy the oscillation conditions at minimum power consumption, larger parallel tank resistance R_{p} is required. However, to maintain large tuning range, a technique to boost negative transconductance of mm-wave oscillators is required. Conventional mm-wave oscillators shown in Figure 10.16a suffer from degradation of $-g_{\mathrm{m}}$ as a result of parasitic capacitances C_{gs}, C_{gd} and the gate resistance R_{g} of the cross-coupled pair as shown in Figure 10.16b. The negative resistance at mm-wave frequency can be approximated by Eq. (10.14) [42].

$$g_{\mathrm{m\_eq1}} \approx -\frac{g_{\mathrm{m,c}}}{2} + \frac{\omega^2 C_{\mathrm{gs}}^2 R_{\mathrm{g}}}{2} \tag{10.14}$$

where $g_{\mathrm{m\_eq1}}$ represents the transconductance of conventional mm-wave NMOS VCOs and $g_{\mathrm{m,c}}$ represents the transconductance of NMOS cross-coupled pair, respectively. Figure 10.17a and b shows the simplified schematic and an equivalent small-signal circuits of the proposed tail-cross-coupled QILO, where an additional $-g_{\mathrm{m}}$ of cross-coupled tail transistors $(-g_{\mathrm{m,tail}})$ are formed at the bottom of the main cross-coupled pair. Due to parasitic capacitance and gate resistance of the cross-coupled pair, the equivalent $-g_{\mathrm{m}}$ of the

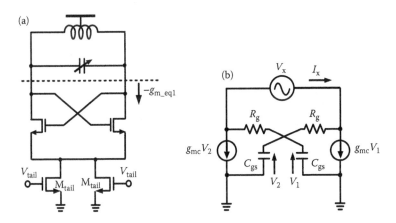

FIGURE 10.16
Simplified circuit schematic of (a) conventional oscillator with conventional cross-coupled pair and (b) its small-signal circuit.

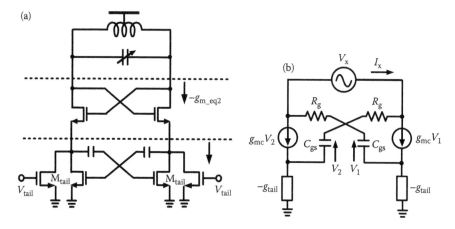

FIGURE 10.17
Simplified circuit schematic of (a) proposed tail-cross-coupling oscillator for g_m-enhancement and (b) its small-signal circuit.

proposed QILO using cross-coupled tail transistors can be ($g_\mathrm{m\_eq2}$) derived as Eq. (10.15).

$$g_{m_{eq2}} \approx -\frac{g_{m,c}}{2\left(1 - g_{m,c}g_{m,tail}^{-1}\right)} - \frac{\omega^2 C_{gs}C_{gd}\left(R_g - 4g_{m,tail}^{-1}\right)}{2\left(1 - g_{m,c}g_{m,tail}^{-1}\right)^2} \tag{10.15}$$

According to Eq. (10.15), it can be observed that the $-g_\mathrm{m}$ degradation effect from gate resistance can be relieved by $-g_\mathrm{m,tail}$. Simulation shows a comparison between simulated $-g_\mathrm{m}$ of conventional QILO and that of the proposed QILO, where the transistor sizes of the cross-coupled pair (M_c) and tail transistors (M_tail) are 24 $\mu m/60$ nm for both cases. For a conventional QILO, $-g_\mathrm{m}$ drops to 5.3 mS at 60 GHz. On the other hand, for the proposed QILO, $-g_\mathrm{m}$ almost reaches 6.9 mS, which is approximately 30% improvement in $-g_\mathrm{m}$ as shown in Figure 10.18a. The parasitic capacitance seen from the tank also reduces by 20% as shown in Figure 10.18b. This allows the proposed QILO to have smaller cross-coupled transistors with less C_PAR. Thus, higher tank inductance can be utilized, which leads to larger parallel tank resistance (R_p), as it can be derived as

$$R_p = \left(\frac{1}{\omega L_p Q_L} + \frac{\omega C_p}{Q_C}\right)^{-1} \approx \left(\frac{1}{Q_L} + \frac{1}{Q_C}\right)^{-1} \omega L_p \tag{10.16}$$

where ω is resonant frequency, C_p and L_p are parallel capacitance and inductance of tank, respectively, Q_C and Q_L are the quality factors of capacitor and inductor, respectively. It is clear that to achieve large R_p, L_p and Q_C should be maximized. Higher L_p, however, gives less room for total capacitance for

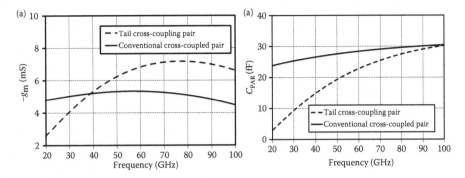

FIGURE 10.18
Comparisons of its (a) negative transconductance of an oscillator and (b) parasitic capacitance (C_{PAR}) seen from the tank using conventional cross-coupled pair and tail cross-coupling pair.

a desired frequency. Moreover, since a 9 GHz tuning range is required for 60 GHz standards, it requires larger on/off ratio of switched capacitor (α_{sw}). Unfortunately, larger α_{sw} degrades quality factor of parallel capacitance (Q_C), as shown in Figure 10.19.

$$L = \frac{\alpha_{SW} - \left(\frac{f_{max}}{f_{min}}\right)^2}{(2\pi f_{max})^2 (\alpha_{SW} - 1) C_{fixed}} \tag{10.17}$$

where C_{fixed} is a capacitance seen into the cross-coupled pair (C_{PAR}) and varactor. For a particular α_{sw}, L can be calculated as Eq. (10.17). From Figure 10.17a, by sweeping α_{sw}, an optimum inductance can be found. Figure 10.19b

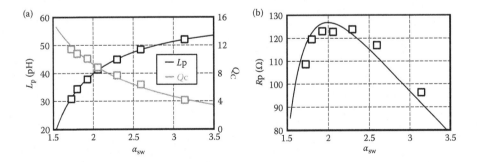

FIGURE 10.19
Tank optimization of (a) inductance and (b) tank resistance (R_p) versus switched ratio (α_{SW}) for power reduction of the 60 GHz QILO.

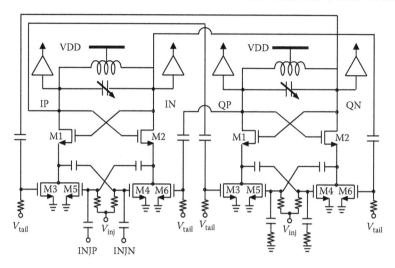

FIGURE 10.20
Detailed schematic of the proposed QILO.

shows the result of R_p, which shows highest peak when α_{sw} is equal to 2. Simulation results are plotted with square symbols and matched well with calculated results plotted in solid line of Figure 10.19. As a result, the oscillators can operate at smaller current by optimizing a tank resistance. Moreover, according to Adler's equation [43], less current in oscillator also helps contributing to larger locking range. The detailed schematic of the proposed 60 GHz g_m-enhanced tail-cross-coupled QILO is depicted in Figure 10.20, composed of two LC tanks and two pairs of cross-coupled NMOS pairs (M1,M2), tail transistors (M3,M4), and cross-coupled tail transistors pairs (M5,M6), where the transistor sizes of M1–M6 are 24 μm/60 nm. An injection from 20 GHz PLL is applied to one side of a QILO for injection, and it also allows I/Q phase calibration as discussed in Section 10.2.2. With 10-bit digital-to-analog converter and tuning voltage of QILO (V_{TUNE}), fine-phase calibration with a resolution of 0.15 degree/code can be achieved.

10.4   Experimental Results

The proposed PLL is designed and implemented using a 65-nm CMOS technology. Figure 10.21a and b shows the chip micrographs of the 20 GHz SS-PLL and the g_m-enhanced 60 GHz QILO, respectively. The QILO occupies a core area of only 60 × 90 μm^2 while the 20 GHz SS-PLL occupies 1.08 mm^2 including decoupling and pads. The measurement is performed on two probe stations with cables and an external buffer to compensate for cable loss

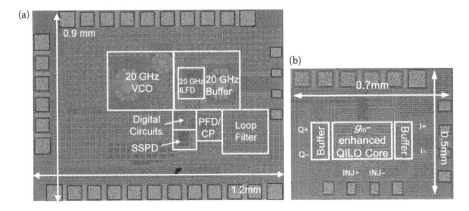

FIGURE 10.21
Chip microphotographs of (a) 20 GHz SS-PLL and (b) g_m-enhanced QILO.

that only adds noise floor to the system and is insignificant. Phase noise performance is measured using Agilent E5052B signal source analyzer, Agilent E5053A microwave down-converter, and 50–75 GHz external mixer. Frequency spectrum is evaluated using Agilent E4448A PSA spectrum analyzer.

The measured locking range of the proposed dual-step-mixing ILFD is shown in Figure 10.22. It achieves 5 GHz locking range while consuming only 2.2 mW, including its buffer to restore output swing for the next-stage digital divider. On the other hand, the proposed g_m-enhanced QILO with R_p optimization achieves about five times larger locking range as current consumption reduces from 14 to 7.8 mW from 1.0 V supply. Through tuning capacitor bank, the QILO can cover for all required channels as shown in Figure 10.23. The proposed synthesizer can support all frequency channels using 36 or 40 MHz

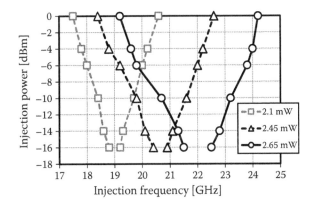

FIGURE 10.22
Measured locking range of the 20 GHz dual-step mixing ILFD.

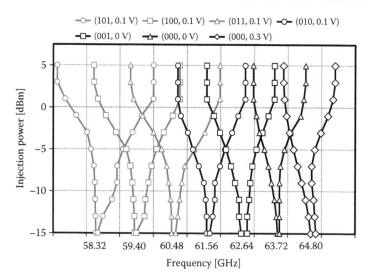

FIGURE 10.23

Measured locking range of the proposed 60 GHz QILO over the frequency range with 3-bit switch and tuning voltage.

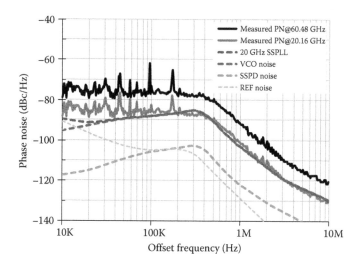

FIGURE 10.24

Simulation and measured phase noise of the 20 GHz subsampling PLL and 60 GHz QILO at a carrier of 20.16 GHz and 60.48 GHz, respectively.

reference. Figure 10.24 shows a comparison of the simulated noise contribution from VCO, SSPD, reference clock, and measured phase noise characteristics at 20.16 and 60.48 GHz from 20 GHz SS-PLL and 60 GHz QILO when locked to 20 GHz SS-PLL, respectively. In the subsampling mode, the out-of-band

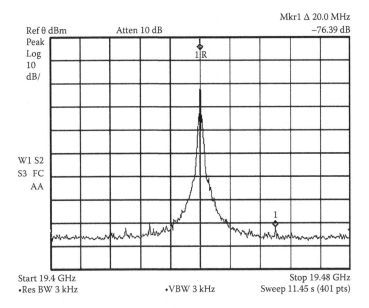

FIGURE 10.25
Measured spectrum of 20 GHz SS-PLL at a carrier frequency of 19.44 GHz.

phase noise of 60 GHz QILO locked to 20 GHz SS-PLL is maintained at −122 dBc/Hz at 10-MHz offset, where in-band phase noise reduces to −78.5 dBc/Hz at 100 kHz offset. To maintain low out-of-band phase noise with relatively lower in-band phase noise comparatively [11–18], the maximum loop bandwidth of this PLL is designed to be 400 kHz. Figure 10.25 shows measured spectrum at 19.44 GHz of 20 GHz SS-PLL. The reference spur is −76.3 dBc without other significant harmonics. The calibrated output power of 20 GHz PLL is approximately −4 dBm, which is large enough to ensure locking to the 60 GHz QILO when integrated together with the QILO. The calibrated output power from the 60 GHz QILO is −10 dBm. From 1 V supply, the power consumption of the SS-PLL is 24.2 mW, where VCO and its buffer, ILFD, and digital circuits consume 18.9, 2.2, and 3.1 mW, respectively. This PLL consumes 2.5 times lower power when compared to Ref. [11, 18]. The proposed ILFD and QILO contribute to 20 mW of power reduction [11, 18]. Due to the absence of on-chip I/Q mixers, an I/Q phase imbalance of 60 GHz signals cannot be accurately measured since any connectors and cables can result in significant delays.

Table 10.4 summarizes the comparison of the proposed work with the state-of-the-art 60 GHz PLLs. Subharmonic injection method shows the lowest out-of-band phase noise comparatively [13–15, 44, 45]. However, the work in Refs. [11, 18] and PFD/CP mode of this work achieves higher in-band phase noise [13–15]. This SS-PLL reduces the division ratio and suppresses in-band phase

TABLE 10.4
Performance Comparison with the State-of-the-Art 60 GHz Frequency Synthesizers

Ref.	Feature	Tech. (nm)	REF (MHz)	Freq. (GHz)	Spur (dBc)	Norm PN[a]	PN[b] (ps)	σRMS (ps)	σRMS, eff (ps)	P_{DC} (mW)	FoM (dB)
[44]	40 GHz PFD/CP PLL	65	36	40.16	–	–	–110	–	–	80	–
[13]	60 GHz AD-PLL	65	100	56.0–62.0	–74	–69	–109	0.50	0.28	48	–229
[14]	60 GHz PFD/CP PLL	65	135	57.9–68.3	–55	–77	–109	0.24	0.23	25	–238
[15]	60 GHz SS-PLL	40	40	53.8–63.3	–40	–90	–108	0.22	0.21	42	–236
[45]	60 GHz PFD/CP PLL	90	60	61.0–6.03	–	–	–110	–	–	78	–
[11,18]	20 GHz PLL + 60GHz QILO	65	36/40	58.1–65.0	–52	–60	–117	9.00	0.12	86	–202
This work	**20 GHz SS-PLL 60 GHz + gm-enhanced QILO**	65	36/40	55.6–65.2	–73	–78.5	**–122**	0.29	**0.15**	32	–236

[a] Normalized to 40 MHz reference clock: PN@100kHz (dBc/Hz) + 20log(f_{REF}/40 MHz).

[b] PN@1MHz (dBc/Hz).

σ_{RMS} is integrated from [10 kHz:40 MHz].

$\sigma_{RMS,eff}$ is an effective integrated phase noise Eq. (10.5) with 400 kHz tracking bandwidth.

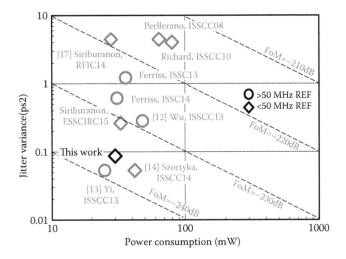

FIGURE 10.26
Performance comparison with state-of-the-art mm-wave frequency synthesizers.

noise to -78.5 dBc at 100 kHz offset. This is equivalent to 0.29 ps, as it is integrated from 10 kHz to 40 MHz offset frequency without any assistance from digital carrier recovery circuits.

$$\text{FoM} = 20 \log \left(\frac{\sigma_\text{t}}{1s} \right) + 10 \log \left(\frac{P_\text{DC}}{1\text{mW}} \right) \qquad (10.18)$$

where σ_t is the RMS jitter of the PLL and P_DC is the power consumption. Eq. (10.18) is used to compare this work with the state-of-the-art mm-wave PLLs. The proposed synthesizer achieves an FoM of -236 dB while consuming only 32 mW. Figure 10.26 shows performance comparison with other mm-wave frequency synthesizers. This work achieves the lowest out-of-band phase noise with comparable integrated jitter while consuming lower power comparatively [15] without using high reference clock [13, 14]. In the case of 400 kHz tracking bandwidth of baseband carrier recovery, RMS jitter is significantly reduced to 0.15 ps, which can support 16QAM.

10.5 Conclusions

This chapter has presented detailed phase noise requirements in 60 GHz applications, especially for IEE802.11ad and the architecture considerations of mm-wave PLLs. The proposed 60 GHz frequency synthesizer using 20 GHz SS-PLL and 60 GHz QILO was presented with design optimization of high-speed building blocks using dual-step mixing ILFD and g_m-enhanced QILO with tank optimization to achieve low power consumption. It achieved a suppression of in-band phase noise in subsampling mode and also achieved low out-of-band phase noise. It could support various 60 GHz standards with complex modulation scheme.

Bibliography

[1] Federal Communications Commission (2008, May 6), Title 47: Telecommunication Part 2 – Frequency Allocations and Radio Treaty Matters; General Rules and Regulations. Available: https://www.gpo.gov/fdsys/pkg/ CFR-2010-title47-vol1/pdf/CFR-2010-title47-vol1-part2.pdf.

[2] IEEE802.11ad, IEEE Std. Available: http://standards.ieee.org/develop/project/802.11ad.html.

[3] K. Okada et al., "A 60 GHz 16QAM/8PSK/QPSK/BPSK Direct-Conversion Transceiver for IEEE 802.15.3c," *IEEE International Solid-State Circuits Conference (ISSCC), Digest of Technical Papers*, pp. 160–161, 2011.

[4] K. Okada et al., "A 60-GHz 16QAM/8PSK/QPSK/BPSK Direct-Conversion Transceiver for IEEE802.15.3c," *IEEE Journal of Solid-State Circuits*, vol. 46, no. 12, pp. 2988–3004, Dec. 2011.

[5] K. Okada et al., "A Full 4-Channel 6.3 Gb/s 60 GHz Direct-Conversion Transceiver with Low-Power Analog and Digital Baseband Circuitry," *IEEE International Solid-State Circuits Conference (ISSCC), Digest of Technical Papers*, pp. 218–219, 2012.

[6] K. Okada et al., "Full Four-Channel 6.3 Gb/s 60 GHz CMOS Transceiver with Low-Power Analog and Digital Baseband Circuitry," *IEEE Journal of Solid-State Circuits*, vol. 48, no. 1, pp. 46–65, Jan. 2013.

[7] K. Okada et al., "A 64-QAM 60 GHz CMOS Transceiver with 4-Channel Bonding," *IEEE International Solid-State Circuits Conference (ISSCC), Digest of Technical Papers*, pp. 346–347, 2014.

[8] R. Wu et al., "A HCI-Healing 60 GHz CMOS Transceiver," *IEEE International Solid-State Circuits Conference (ISSCC), Digest of Technical Papers*, pp. 350–351, 2015.

[9] W.-T. Li et al., "60 GHz 5-bit Phase Shifter with Integrated VGA Phase-Error Compensation," *IEEE Transactions on Microwave Theory and Techniques*, vol. 61, no. 3, pp. 1224–1235, Mar. 2013.

[10] S. Kawai et al., "A Digitally-Calibrated 20-Gb/s 60-GHz Direct-Conversion Transceiver in 65-nm CMOS," *IEEE Radio Frequency Integrated Circuits Symposium (RFIC)*, pp. 137–140, Jun. 2013.

[11] A. Musa, R. Murakami, T. Sato, W. Chaivipas, K. Okada, and A. Matsuzawa, "A Low Phase Noise Quadrature Injection Locked Frequency Synthesizer for MM-Wave Applications," *IEEE Journal of Solid-State Circuits*, vol. 46, no. 11, pp. 2635–2649, Nov. 2011.

[12] K. Scheir, G. Vandersteen, Y. Rolain, and P. Wambacq, "A 57-to-66 GHz Quadrature PLL in 45 nm Digital CMOS," *IEEE International Solid-State Circuits Conference (ISSCC), Digest of Technical Papers*, pp. 494–495, Feb. 2009.

[13] W. Wu, R. B. Staszewski, and J. R. Long, "A 56.4–63.4 GHz Multi-Rate All-Digital Fractional-N PLL for FMCW Radar Applications in 65 nm CMOS," *IEEE Journal of Solid-State Circuits*, vol. 49, no. 5, pp. 1081–1096, May 2014.

[14] X. Yi, C. C. Boon, H. Liu, J. F. Lin, and W. M. Lim, "A 57.9-to-68.3 GHz 24.6 mW Frequency Synthesizer with In-Phase Injection-Coupled QVCO in 65 nm CMOS Technology," *IEEE Journal of Solid-State Circuits*, vol. 49, no. 2, pp. 347–359, Feb. 2014.

[15] V. Szortyka, Q. Shi, K. Raczkowski, B. Parvais, M. Kujik, and P. Wambacq, "A 42 mW 230fs-Jitter Sub-Sampling 60 GHz PLL in 40 nm CMOS," *IEEE International Solid-State Circuits Conference (ISSCC), Digest of Technical Papers*, pp. 366–367, Feb. 2014.

[16] C. Marcu et al., "A 90 nm CMOS Low-Power 60 GHz Transceiver with Integrated Baseband Circuitry," *IEEE Journal of Solid-State Circuits*, vol. 44, no. 12, pp. 3434–3447, Dec. 2009.

[17] T. Siriburanon, H. Liu, K. Nakata, W. Deng, J. H. Son, D. Y. Lee, K. Okada, and A. Matsuzawa, "A 28-GHz Fractional-N Frequency Synthesizer with Reference and Frequency Doublers for 5G Cellular," *IEEE European Solid-State Circuits Conference (ESSCIRC)*, pp. 76–79, 2015.

[18] W. Deng, T. Siriburanon, A. Musa, K. Okada, and A. Matsuzawa, "A Sub-Harmonic Injection-Locked Quadrature Frequency Synthesizer with Frequency Calibration Scheme for Millimeter-Wave TDD Transceivers," *IEEE Journal of Solid-State Circuits*, vol. 48, no. 7, pp. 1710–1720, Jun. 2013.

[19] T. Siriburanon, T. Ueno, K. Kimura, S. Kondo, W. Deng, K. Okada, and A. Matsuzawa, "A 60-GHz Sub-Sampling Frequency Synthesizer Using Sub-Harmonic Injection-Locked Quadrature Oscillators," *IEEE Radio Frequency Integrated Circuits Symposium (RFIC)*, pp. 105–108, Jun. 2014.

[20] T. Siriburanon, S. Kondo, M. Katsuragi, H. Liu, K. Kimura, W. Deng, K. Okada, and A. Matsuzawa, "A Low-Power Low-Noise mm-Wave Sub-Sampling PLL Using Dual-Step-Mixing ILFD and Tail-Coupling Quadrature Injection-Locked Oscillator for IEEE802.11ad," *IEEE Journal of Solid-State Circuits*, vol. 51, no. 5, pp. 1246–1260, May 2016.

[21] T. Siriburanon, W. Deng, A. Musa, K. Okada, and A. Matsuzawa, "A 13.2% Locking-Range Divide-by-6, 3.1 mW, ILFD Using Even-Harmonic-Enhanced Direct Injection Technique for Millimeter-Wave PLLs," *IEEE European Solid-State Circuits Conference (ESSCIRC)*, pp. 403–406, 2013.

[22] IEEE Std., 802.15.3c-2009, Oct. 2009. Available: http://standards.ieee.org /getieee802/download/802.15.3c-2009.pdf.

[23] J. R. Pelliccio et al., "Phase Noise Effect on OFDM Wireless LAN Performance," *Applied Microwave & Wireless*, vol. 13, no. 7, pp. 68–78, 2011.

[24] K. Okada et al., "60 GHz WiGig Frequency Synthesizer Using Injection Locked Oscillator," *IEEE Radio Frequency Integrated Circuits Symposium (RFIC)*, pp. 109–134, Jun. 2014.

[25] A. Georgiadis, "Gain, Phase Imbalance, and Phase Noise Effects on Error Vector Magnitude," *IEEE Transactions on Vehicular Technology*, vol. 53, no. 2, pp. 443–449, 2004.

[26] S. Emami et al., "A 60 GHz CMOS Phased-Array Transceiver Pair for Multi-Gb/s Wireless Communications," *IEEE International Solid-State Circuits Conference (ISSCC), Digest of Technical Papers*, pp. 164–165, Feb. 2011.

[27] V. Vidojkovic et al., "A Low-Power 57-to-66 GHz Transceiver in 40 nm LP CMOS with −17 dB EVM at 7 Gb/s," *IEEE International Solid-State Circuits Conference (ISSCC), Digest of Technical Papers*, pp. 268–269, Feb. 2012.

[28] X. Gao, E. Klumperink, M. Bohsali, and B. Nauta, "A Low Noise Sub-Sampling PLL in Which Divider Noise is Eliminated and PD/CP Noise is Not Multiplied by N^2," *IEEE Journal of Solid-State Circuits*, vol. 46, no. 11, pp. 2635–2649, Nov. 2009.

[29] T. Siriburanon, S. Kondo, K. Kimura, T. Ueno, S. Kawashima, T. Kaneko, W. Deng, M. Miyahara, K. Okada, and A. Matsuzawa, "A 2.2 GHz − 242 dB-FoM 4.2 mW ADC-PLL Uing Digital Sub-Sampling," *IEEE International Solid-State Circuits Conference (ISSCC), Digest of Technical Papers*, pp. 440–441, Feb. 2015.

[30] T. Siriburanon, S. Kondo, K. Kimura, T. Ueno, S. Kawashima, T. Kaneko, W. Deng, M. Miyahara, K. Okada, and A. Matsuzawa, "A

2.2 GHz -242 dB-FoM 4.2 mW ADC-PLL Using Digital Sub-Sampling," *IEEE Journal of Solid-State Circuits*, vol. 51, no. 6, pp. 1385–1397, Jun. 2016.

[31] J-C. Chien and L-H. Lu, "Analysis and Design of Wideband Injection-Locked Ring Oscillators with Multiple-Input Injection," *IEEE Journal of Solid-State Circuits*, vol. 42, no. 9, pp. 1906–1915, Sep. 2007.

[32] A. Mazzanti, P. Uggetti, and F. Svelto, "Analysis and Design of Injection-Locked LC Dividers for Quadrature Generation," *IEEE Journal of Solid-State Circuits*, vol. 39, no. 9, pp. 1425–1433, Sep. 2004.

[33] K. Yamamoto and M. Fujishima, "A 44-W 4.3-GHz Injection-Locked Frequency Divider with 2.3-GHz Locking Range," *IEEE Journal of Solid-State Circuits*, vol. 40, no. 3, pp. 671–677, Mar. 2005.

[34] A. Mirzaei, M. Heidari, R. Bagheri, and A. Abidi, "Multi-Phase Injection Widens Lock Range of Ring-Oscillator-Based Frequency Dividers," *IEEE Journal of Solid-State Circuits*, vol. 43, no. 3, pp. 656–671, Mar. 2008.

[35] W. Deng, S. Hara, A. Musa, K. Okada, and A. Matsuzawa, "A Compact and Low-Power Fractionally Injection-Locked Quadrature Frequency Synthesizer Using a Self-Synchronized Gating Injection Technique for Software-Defined Radios," *IEEE Journal of Solid-State Circuits*, vol. 49, no. 9, pp. 1984–1994, Sep. 2014.

[36] A. Musa, K. Okada, and A. Matsuzawa, "Progressive Mixing Technique to Widen the Locking Range of High Division Ratio Injection Locked Frequency Dividers," *IEEE Transactions on Microwave Theory and Techniques*, vol. 61, no. 3, pp. 656–671, Mar. 2013.

[37] B. Razavi, "A Study of Injection Locking and Pulling in Oscillators," *IEEE Journal of Solid-State Circuits*, vol. 39, no. 9, pp. 1415–1424, Sep. 2004.

[38] M. Garampazzi et al., "An Intuitive Analysis of Phase Noise Fundamental Limits Suitable for Benchmarking LC Oscillators," *IEEE Journal of Solid-State Circuits*, vol. 49, no. 3, pp. 635–645, Mar. 2014.

[39] E. Hegazi, H. Sjoland, and A. Abidi, "A Filtering Technique to Lower LC Oscillator Phase Noise," *IEEE Journal of Solid-State Circuits*, vol. 36, no. 12, pp. 1921–1930, Dec. 2001.

[40] T. Siriburanon, W. Deng, K. Okada, and A. Matsuzawa, "A Current-Reuse Class-C LC-VCO with an Adaptive Bias Scheme," *IEEE Radio Frequency Integrated Circuits Symposium (RFIC)*, pp. 35–38, Jun. 2013.

[41] Y. Hu, T. Siriburanon, and R. B. Staszewski, " A 30-GHz Class-F23 Oscillator in 28 nm CMOS Using Harmonic Extraction and Achieving

120 kHz 1/f3 Cornerm," *IEEE European Solid-State Circuits Conference (ESSCIRC)*, pp. 87–90, 2017.

[42] B. Razavi, "A 300-GHz Fundamental Oscillator in 65-nm CMOS Technology," *IEEE Journal of Solid-State Circuits*, vol. 46, no. 4, pp. 894–903, Apr. 2011.

[43] R. Adler, "A Study of Locking Phenomena in Oscillators," *Proceedings of the IEEE*, vol. 61, pp. 1380–1385, Oct. 1973.

[44] A. Siligaris et al., "A 65-nm CMOS Fully Integrated Transceiver Module for 60-GHz Wireless HD Applications," *IEEE Journal of Solid-State Circuits*, vol. 46, no. 12, pp. 3005–3017, Dec. 2011.

[45] H. Hoshino, R. Tachibana, T. Mitomo, N. Ono, Y. Yoshihara, and R. Fujimoto, "A 60 GHz Phase-Locked Loop with Inductor-Less Prescaler in 90-nm CMOS," *IEEE European Solid-State Circuits Conference (ESSCIRC)*, pp. 472–475, 2007.

11

60 GHz Multiuser Gigabit/s Wireless Systems Based on IEEE 802.11ad/WiGig

Koji Takinami, Naganori Shirakata, Masashi Kobayashi, Tomoya Urushihara, Hiroshi Takahashi, Hiroyuki Motozuka, Masataka Irie, and Kazuaki Takahashi

Panasonic Corporation

CONTENTS

11.1 Introduction

Due to the shortage of available spectrum in conventional cellular bands below 6 GHz, the spectrum at high-frequency bands from 24.25 to 86 GHz, also known as millimeter-wave (mmWave) spectrum, is becoming a subject of active research toward fifth-generation (5G) mobile communications [1, 2].

By utilizing mmWave spectrum, 1 GHz or more of continuous spectrum can be allocated, which enables the support of multi-Gb/s data rates.

While other candidates for mmWave radio access exist, the 60 GHz short-range wireless local area network (LAN) technology known as IEEE 802.11ad/WiGig [3, 4] is one of the most promising solutions [5–7]. At the 60 GHz band, the multi-GHz unlicensed bandwidth has been allocated, which enables extremely high data rates when compared with those offered by 2.4 and 5 GHz wireless LAN standards. Despite of such superior capabilities, traditionally, the mmWave systems have mainly been deployed for military applications. However, recent technology advancements have enabled a 60 GHz transceiver to be realized by means of the cost-effective complementary metal oxide semiconductor (CMOS) process [8–10], offering opportunities for IEEE 802.11ad/WiGig to be deployed in a variety of applications including the laptop PC, tablet, smartphone, etc.

This chapter aims to give the most up-to-date status of the 60 GHz technology development, with an emphasis on hardware design, network integration, and system evaluation. The rest of the chapter is organized as follows. Section 11.2 gives a brief overview of the target use cases and IEEE 802.11ad/WiGig standards. Section 11.3 discusses CMOS transceiver design. Section 11.4 presents design and experimental evaluation of the prototype system. Section 11.5 gives future perspectives, followed by concluding remarks in Section 11.6.

11.2   A 60 GHz Wireless System

11.2.1   Use case examples

The use of high frequencies comes with both advantages and disadvantages. Large path loss, as well as high attenuation due to obstacles such as human body, limits 60 GHz applications to those suitable for short-range wireless communication. On the other hand, higher frequencies lead to smaller sizes of RF components including antennas, enabling compact realization of an array structure which offers improved antenna gain with high directivity [11, 12]. It is worth mentioning that these properties unique to 60 GHz help reduce interference, offering opportunities to maintain multi-Gb/s throughput even in high density environments.

The first commercially available consumer products using 60 GHz appeared in the early 2010s and were based on the WirelessHD standard, which supports uncompressed high definition (HD) video links between fixed devices such as a set-top box and a TV. The chipset employed sophisticated beamforming technology to extend communication distance [13], but this resulted in high power consumption. Even though substantial progress has been made since then, employing a 60 GHz system in mobile terminals

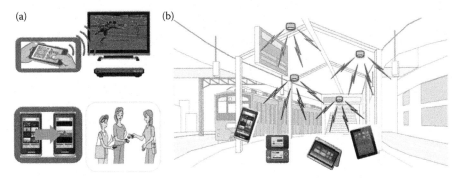

FIGURE 11.1
60 GHz mobile use case examples. (a) Peer-to-peer connection and (b) multiuser access in a dense environment.

such as smart phones and tablets remains a difficult challenge since it requires extensive reduction in power consumption as well as small form factor.

Figure 11.1 shows use case examples. As illustrated in Figure 11.1a, the 60 GHz solution offers a high-speed peer-to-peer connection that can be an alternative to existing wired standards such as high definition multimedia interface (HDMI) or universal serial bus (USB) 3.0. By using 60 GHz, it is possible to transmit the HD video wirelessly without compression, providing a natural responsiveness between the mobile device and the large-screen TV. Another example application is fast file transfer, which enables to share HD video contents in a short period of time. Even though the typical use case in the initial products will be a high-speed peer-to-peer wireless connection, it is also expected to be deployed for multiuser access in a dense environment as shown in Figure 11.1b. According to a recent research, the use of the 60 GHz band for traffic offloading is a key to achieve an efficient network with high capacity required in future 5G networks [6].

11.2.2 Frequency allocation

Figure 11.2 illustrates the global frequency allocation and IEEE 802.11ad/ WiGig channel allocation. Frequency bands around 60 GHz are available worldwide. The International Telecommunication Union Radiocommunication Sector (ITU-R) recommended channelization comprises four channels, each 2.16 GHz wide, centered on 58.32, 60.48, 62.64, and 64.80 GHz, respectively. The frequency allocations in each region do not match exactly, but there is substantial overlap of at least 3.5 GHz of contiguous spectrum in all regions. This allows a single device to operate worldwide without hardware modifications just like the traditional wireless LAN based on the IEEE 802.11 at 2.4/5 GHz. Furthermore, the Federal Communications Commission (FCC) announced a

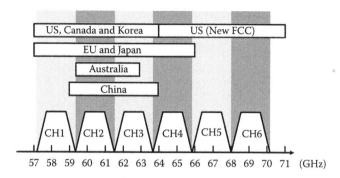

FIGURE 11.2
Frequency allocations by region and channel in IEEE 802.11ad/WiGig [14].

new unlicensed band allocation of 64–71 GHz on July 2016, which allows up to six channels for 60 GHz operation in the United States.

11.2.3   IEEE 802.11ad/WiGig standard overview

There are many different standards for 60 GHz wireless communication. Within the IEEE 802.15 working group, Task Group 3 (TG3) had led the standardization, and the IEEE 802.15.3c was issued in 2009 by Task Group 3c (TG3c). Soon after, the Wireless Gigabit (WiGig) Alliance was formed to develop an industrial standard for 60 GHz. In addition to the physical (PHY) and medium access control (MAC) layers, the WiGig Alliance defined protocol adaptation layers (PALs) for effective audio, video, and data transmissions. The PHY and MAC specification defined by the WiGig Alliance was later proposed to the IEEE 802.11 working group, Task Group 'ad' (TGad), which was the group tasked with defining modifications to the 802.11 MAC and PHY specifications to enable operation in the 60 GHz frequency band capable of a maximum throughput of at least 1 Gb/s. The standard also specifies fast session transfer (FST), which enables wireless devices to seamlessly transit between the 60 GHz frequency band and the legacy 2.4 and 5 GHz bands, to complement optimal performance with wider communication range.

The final IEEE 802.11ad specification was officially approved by the IEEE in December 2012 [3]. In September 2013, the Wi-Fi Alliance, a global non-profit industry association that provides certification programs, announced the integration of the WiGig Alliance, followed by successful launch of the WiGig certification program in October 2016. This placed IEEE 802.11ad/WiGig in the lead position to become the de facto standard for wireless communication at 60 GHz.

In IEEE 802.11ad/WiGig, several technologies have been introduced to improve performance and robustness as listed in Table 11.1 [3, 4, 14]. Considering the need to support various use-case scenarios, both single carrier

TABLE 11.1
Features of the IEEE 802.11ad/WiGig

Layer	Feature	Technology
PHY	Low power operation	SC modulation $\pi/2$-BPSK
	High-efficiency data transmission	LDPC coding
		Hybrid of CSMA/TDMA
MAC	Robust against mobility and blockage	Beamforming

(SC) modulation and orthogonal frequency division multiplexing (OFDM) modulation have been adopted. In general, the SC modulation is suitable for operations with reduced power consumption due to its low peak-to-average power ratio (PAPR), whereas the OFDM modulation offers better multipath tolerance. Table 11.2 shows examples of the modulation and coding schemes (MCSs) where MCS 0 to MCS 4 are mandatory. MCS 0 is primarily used for control channel messages for achieving better robustness. Low density parity check (LDPC) code is employed for efficient error correction and a hybrid of carrier sense multiple access (CSMA) and time division multiple access (TDMA) allows flexible choice of access methods. One of the unique features of IEEE 802.11ad/WiGig is the use of beamforming technology. To compensate for large path loss, the phased array antenna is commonly used in

TABLE 11.2
MCS Examples (Not All Listed) in the IEEE 802.11ad/WiGig.

MCS index	Modulation	Code rate	PHY payload rate (Mb/s)
0	DBPSK (SC)	1/2	27.5
1	$\pi/2$-BPSK (SC)	1/2	385
4	$\pi/2$-BPSK (SC)	3/4	1155
5	$\pi/2$-BPSK (SC)	13/16	1251.25
9	$\pi/2$-QPSK (SC)	13/16	2502.5
12	$\pi/2$-16QAM (SC)	3/4	4620
12.2[a]	$\pi/2$-16QAM (SC)	7/8	5390
12.6[a]	$\pi/2$-64QAM (SC)	7/8	8085
21	16QAM (OFDM)	13/16	4504.5
24	64QAM (OFDM)	13/16	6756.75

MCS 0 and MCS 1 employ spreading factor of 32 with $\pi/2$ rotation and spreading factor of 2, respectively

[a]Extended SC MCSs are added in the latest release [14], offering enhanced throughput in the SC modulation mode.

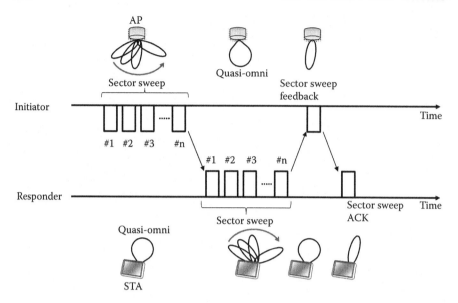

FIGURE 11.3
Operation of beamforming protocol.

mmWave systems. This necessitates that the transmit (Tx) antenna beam and the receive (Rx) antenna beam need to be aligned in appropriate directions.

Figure 11.3 illustrates one of the beamforming operations examples known as sector-level sweep (SLS), which is the most basic type of beam training procedure. In this specific example, the access point (AP) takes a role of an initiator and the mobile station (STA) operates as a responder. At first, the initiator sends sector sweep frames as training signals while the responder measures the quality of the received frames using a quasi-omni beam pattern based on received signal strength indicator (RSSI) or signal-to-noise ratio (SNR). The responder informs the best sector to the initiator while it performs its own training sector sweep. This is followed by best sector feedback from the initiator and an acknowledgement to finish the exchange. A similar process can also be applied to align the Rx beam directions. By executing the SLS procedure periodically, the Tx and Rx beam patterns can be continually aligned in the right directions.

11.3 CMOS Transceiver

This section describes a design example of the 60 GHz CMOS transceiver based on IEEE 802.11ad/WiGig. The transceiver prototype features beamforming technology, while achieving low power consumption that is applicable to mobile usage.

11.3.1 Radio architecture

Two of the most popular choices of radio architecture are super heterodyne and direct conversion [15]. In the super heterodyne architecture, the frequency conversion is performed in two steps. The use of an intermediate frequency (IF) that is lower than the original carrier frequency makes radio design easier, but it requires two phase-locked loop (PLL) frequency synthesizers. In addition to occupying large chip area, oscillators fabricated on the same chip suffer from unwanted coupling. To avoid these issues, a sliding IF architecture is commonly used in modern integrated transceivers [13, 16, 17]. As shown in Figure 11.4a, two local (LO) frequencies are generated from one frequency synthesizer by using frequency multipliers or dividers. The IF is often chosen relatively high so that it provides large channel separation between the carrier frequency and its image frequency, relaxing the requirement for image rejection and eliminating external IF filtering.

Direct conversion, on the other hand, performs the frequency conversion in one step, making the radio architecture very simple as illustrated in Figure 11.4b. However, since the LO frequency is same as the carrier frequency, special care must be taken to avoid frequency pulling from the transmitter (Tx) output to the PLL. Also, the LO leakage in the receiver (Rx) path makes the LO self-mixing, resulting in a dynamic DC offset. In spite of these technical difficulties, recent works [8, 9, 18–23] have realized a 60-GHz transceiver with direct conversion architecture. Compared to the super heterodyne or the sliding IF, the direct conversion architecture typically achieves smaller die area and lower power consumption.

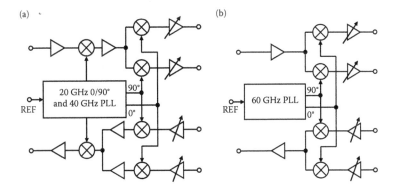

FIGURE 11.4
(a) Sliding IF and (b) direct conversion architectures.

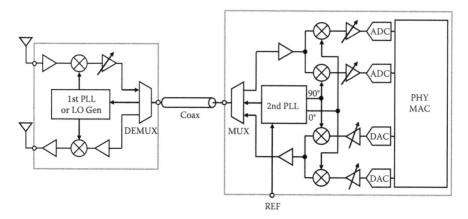

FIGURE 11.5
Heterodyne architecture using coaxial cable.

In some 60 GHz applications, it is preferable to place a radio frequency
(RF) module and a baseband integrated circuit (BBIC) in separate locations.
Figure 11.5 is an example where the RF module and the BBIC are interfaced
with a single coaxial cable [10, 24, 25]. This can be viewed as the super heterodyne architecture split across two chips. The IF and the LO as well as some
control signals are fed through the coaxial cable, which is suitable for PC platforms. For example, in the laptop PC, the RF module should be mounted on
the display frame to ensure better signal radiation from the antenna, whereas
the BBIC must be placed close to the processors to minimize signal routings

11.3.2   Low-power beamforming transceiver

At mmWave frequencies, it is possible to implement beamforming capability
in a small form factor by using a phased array antenna. This enables expansion
of the transmission and reception angles as well as increase of communication
distance by improved antenna gain. Figure 11.6 shows the block diagram of
the transceiver chipset, which consists of the radio frequency integrated circuit (RFIC) and BBIC [26]. The RFIC employs direct conversion architecture
with 4Tx/4Rx RF signal paths. The RF phase shifter is inserted in each RF
path, which introduces the required phase shift. The amount of phase shift
required is determined by a code book, which is a look-up table that stores
the required phase shifts for individual beam directions. The output power is
adjusted by the automatic power control loop that detects the output power
at the power amplifier (PA) output by the envelope detector that is digitized
by an 8-bit 1 MHz analog to digital converter (ADC) and fed back to the
digital signal processor. This feedback loop is also utilized to compensate for
other performance impairments such as the carrier and the image leakages by

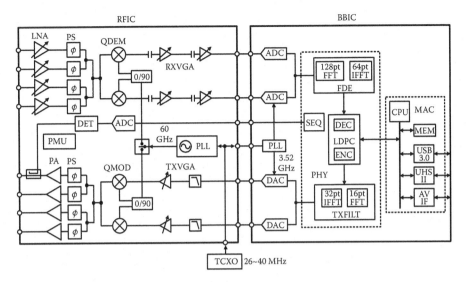

FIGURE 11.6
Block diagram of a transceiver chipset. (From [29], copyright © 2017 IEICE.)

the calibration schemes presented in ref. [27]. The power management unit (PMU) integrates low dropout regulators that provide regulated 1.25/1.4 V DC voltages from a 1.8 V power supply to improve external noise tolerance. The average total power consumption of the chip set is less than 1 W.

The phase shifter utilizes a quadrature hybrid with a digitally controlled vector combiner. As shown in Figure 11.7a, the transformer-based hybrid [28] splits the RF input to 0°/90° signals, which is converted to differential signal and fed to vector combiners. Each vector combiner is composed of a 5-bit current-controlled RF amplifier, providing 32 × 32 outputs for one quadrant. As illustrated in Figure 11.7b, by flipping the polarity of differential signals, it covers four quadrants to achieve a 360° variable range.

Figure 11.8a shows RF signal distribution in the Tx path. Wilkinson divider is used for power splitter. As illustrated in Figure 11.8b, the quarter length transmission line is replaced with lumped elements to shrink the chip size. Similar approach is applied for the receiver as well.

Figure 11.9a shows the die photo of the RFIC. The Tx is on the left and the Rx is on the right side. As explained earlier, they consists of four RF paths to steer the antenna beam. The chip area is 3.2 × 5.1 mm^2, which was fabricated in standard 40-nm digital CMOS without the ultrathick metal option. The RFIC and the BBIC are integrated in the miniaturized antenna module shown in Figure 11.9b, which employs 4-element patch antennas with circular polarization.

Figure 11.10 shows the measured phase shifter performance. Figure 11.10a plots all the outputs available from vector summation, which exhibits distorted

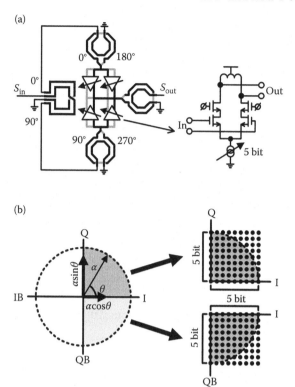

FIGURE 11.7
Phase shifter circuit. (a) Schematic and (b) variable output. (From [26], copyright © 2016 IEICE.)

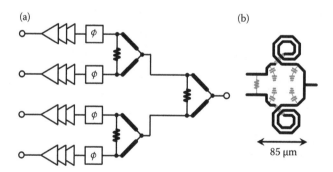

FIGURE 11.8
(a) RF signal distribution and (b) Wilkinson divider. (From [26], copyright © 2016 IEICE.)

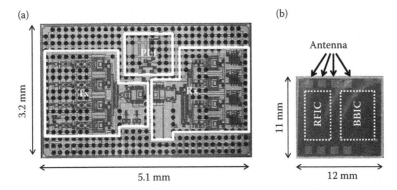

FIGURE 11.9
(a) Die photo of RFIC and (b) miniaturized antenna module. (From [29], copyright © 2017 IEICE.)

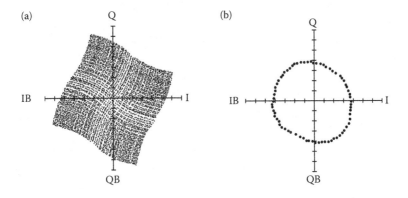

FIGURE 11.10
Measured phase shifter performance. (a) All measured points and (b) unit circle for phase shift. (From [26], copyright © 2016 IEICE.)

output distribution due to nonideality in the vector summation. However, as shown in Figure 11.10b, the extracted unit circle achieves excellent 5° phase resolution over 360° variable range. As shown in Figure 11.11, the simple codebook-based beamforming provides 7-step beam direction, achieving about a 120° steering range.

Figure 11.12 shows the measured area coverage of the miniaturized antenna module. The beamforming capability significantly improves the area coverage, achieving 1.5 Gb/s MAC throughput over 2.5 m and 0.9 Gb/s MAC throughput over 4 m while covering a wide angle of 120° (+45° to −75°). The maximum measured MAC throughput reaches 1.7 Gb/s.

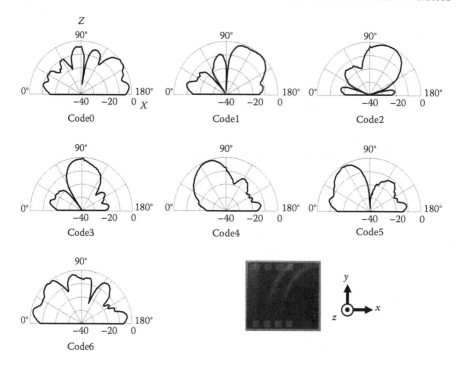

FIGURE 11.11
Measured analog beamforming performance and picture of antenna module.
(From [26], copyright © 2016 IEICE.)

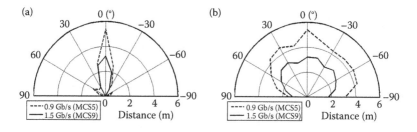

FIGURE 11.12
Comparison of area coverage. (a) Without beamforming and (b) with beamforming. (From [29], copyright © 2017 IEICE.)

11.4   Hardware and Network Design

This section presents design and experimental evaluation of a 60 GHz wireless
system, targeting for multiuser access in dense environment [29]. Both the

STA and AP are developed using the CMOS transceiver chip set explained in the previous section. The AP prototype combines three RF modules with beamforming technology to provide 360° area coverage. The developed 60-GHz wireless system utilizes multiple APs with handover capability to expand area coverage.

11.4.1   System architecture

Figure 11.13 shows an example of the system architectures. Since the communication distance of the 60 GHz AP is limited, it is necessary to install multiple APs to achieve sufficient area coverage. The legacy 2.4/5 GHz AP is used to maintain wireless connections while STAs are outside of the 60 GHz AP coverage. The AP controller (APC) manages handover between multiple APs. Although it is outside the scope of this chapter, STAs are assumed to be connected to the long term evolution (LTE) or the 5G new radio (5G NR) network if they are outside of the 2.4/5 GHz AP coverage in the future 5G network.

11.4.2   Mobile station

For STAs, a USB dongle prototype has been developed. As shown in Figure 11.14, the antenna module is mounted on a printed circuit board that is housed in the 39 × 95 × 15 mm small unit. USB 3.0 high-speed interface

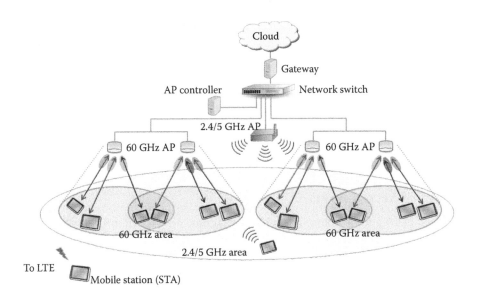

FIGURE 11.13

System architecture example. (From [29], copyright © 2017 IEICE.)

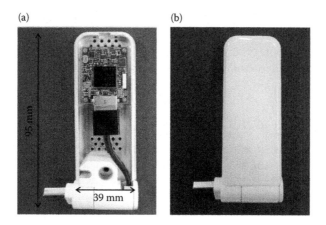

FIGURE 11.14
USB dongle prototype. (a) Internal unit and (b) exterior. (From [29], copyright © 2017 IEICE.)

TABLE 11.3
USB Dongle Specifications

Modulation	MCS 0 to MCS 9
Carrier frequency	58.32, 60.48, 62.64 GHz
Transmit power	+10 dBm EIRP max.
No. of antenna elements	4Tx/4Rx
Antenna polarization	Circular polarization
Coverage	120°
Interface	USB 3.0
Total power consumption	1 W max.

can be plugged into the USB port of mobile devices such as a laptop PC and tablet. Table 11.3 summarizes the prototype's specifications. The dongle supports up to MCS 9 (2.5 Gb/s at PHY rate) while consuming only 1 W total power. It achieves about +10 dBm maximum effective isotropic radiated power (EIRP).

11.4.3   Access point

Figure 11.15 shows a simplified block diagram of the AP prototype. Since the beam steering range of the RF module is limited to about 120°, the AP utilizes three RF modules to realize 360° area coverage. The 10 Gb/s Ethernet (10GbE) is adopted as a high-speed external interface to handle multi Gb/s system throughput. To extend communication distance, the external low noise amplifiers (LNAs) and the PAs are added in the RF signal path between antennas and the RFIC. The LNA and the PA employ the same gallium arsenide monolithic microwave integrated circuit (GaAs MMIC) with 4

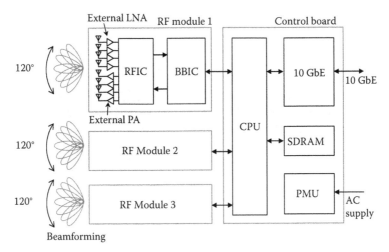

FIGURE 11.15
Block diagram of AP prototype. (From [29], copyright © 2017 IEICE.)

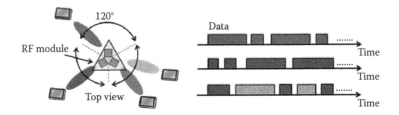

FIGURE 11.16
Operation principle of spatial sharing and time sharing.

dB noise figure and 10 dB nominal gain. The AP can establish up to three concurrent basic service sets (BSSs) by using either the same frequency channel or different frequency channels. As illustrated in Figure 11.16, by taking advantage of antenna directivity, the AP can use the same frequency channel for up to three concurrent links. Each RF module can handle up to four STAs, therefore, accommodating a maximum of 12 users by the single AP.

Figure 11.17 shows the photos of the RF module. Similar to the one used in the STA, it offers a 7-step beam steering capability. Figure 11.18a shows the measured output power for different beam patterns in the azimuth direction, and Figure 11.18b shows the one in the elevation angle. Since each RF module provides relatively wide beamwidth (more than 30° of half-power beam width), the AP prototype utilizes beam4 as a quasi-omni mode that can detect the incoming MCS 0 signals with sufficient coverage, i.e. more than 120° per one RF module. Due to the additional gain provided by the external PAs, the RF module achieves +20 dBm maximum EIRP. Figure 11.19 shows the photos

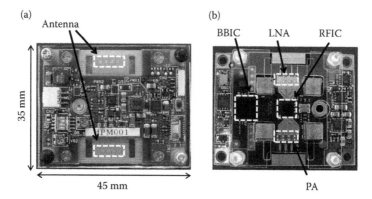

FIGURE 11.17
RF module for AP. (a) Top view and (b) bottom view. (From [29], copyright © 2017 IEICE.)

TABLE 11.4
AP Specifications

Modulation	MCS 0 to MCS 9
Carrier frequency	58.32, 60.48, 62.64 GHz
Transmit power	+20 dBm EIRP max.
No. of antenna elements	4Tx/4Rx
Antenna polarization	Circular polarization
Coverage	360° (= 120° × 3)
Interface	IEEE 802.3an (10GBase-T)
CPU	TI AM5K2E04 (Cortex-A15/4core/1.4 GHz)
SDRAM	DDR3L-1600 4 GB
Total power consumption	42 W max.

of the AP prototype. Three RF modules and the control board are housed in the 80 × 150 × 150 mm unit. Table 11.4 summarizes the AP's specifications.

11.4.4   Multiuser Gigabit/s wireless system

To evaluate system performance in a real environment, the prototype system was set up in Narita International Airport, as shown in Figure 11.20. The provided 4K tablets were equipped with the USB dongles, as shown in Figure 11.21 and used to download high-resolution video contents (2 GB maximum) from the content server. The system behavior was monitored through background system logging. Even though IEEE 802.11ad/WiGig supports both TDMA and CSMA, in this setup only the CSMA/collision avoidance (CSMA/CA) was chosen as the access method for simplicity.

Figure 11.22 shows the prototype network system. Three APs are wired with 10 Gb/s Ethernet cables. The local content server bundles four 10 Gb/s

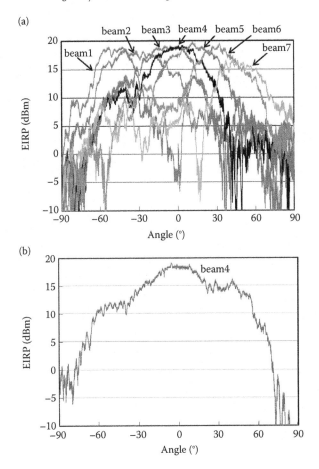

FIGURE 11.18
Measured output power (at 58.32 GHz). (a) Azimuth direction and (b) elevation direction. (From [29], copyright © 2017 IEICE.)

Ethernet cables to achieve 40 Gb/s maximum throughput. The APC manages beamforming control and handover between RF modules and multiple APs. This is done by collecting the Rx signal quality (SNR of the Rx signal) for each beam direction periodically, and selecting the optimum mmWave link based on the beam routing table that includes the sector index (i.e. the beam direction index), MCS and the Rx signal quality. Three APs were installed at a 2.5 m height, providing 10 × 5 m area coverage. To minimize interference among multiple APs and STAs, different frequency channels were allocated for each of the three RF modules within the APs. As STAs, nine 4K tablets were placed on counter desks at a 1.0 m height. They are equipped with IEEE 802.11ad/WiGig USB dongle prototypes to establish ultra-high

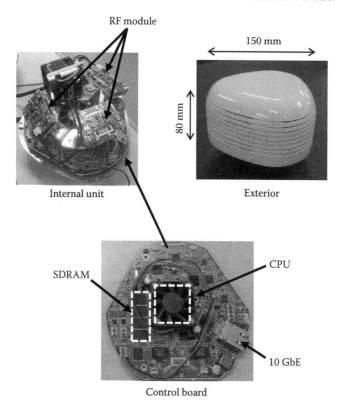

FIGURE 11.19
AP prototype. (From [29], copyright © 2017 IEICE.)

speed 60 GHz wireless links with APs. The STA achieves 1.7 Gb/s maximum throughput, enabling to download a compressed 2 h high-resolution video content (2 GB) within 10 s. Figure 11.23 shows the measured throughput of the content server, which shows 3.7 Gb/s maximum throughput. This is because the content download takes less than 15 s in most cases, which minimized the probability of packet collisions, providing a multiuser Gb/s wireless access experience. During the nine-day open period, 816 participants joined the experimental demonstration, and 99.3% of the positive feedbacks (high expectation for practical realization) were obtained.

11.5   Evolution of mmWave Technologies

This section gives the next-generation 60 GHz standardization activities as well as future perspectives in mmWave technologies.

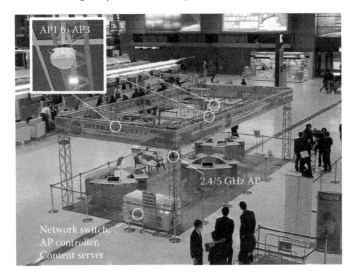

FIGURE 11.20
Experimental demonstration at Narita International Airport. (From [29], copyright © 2017 IEICE.)

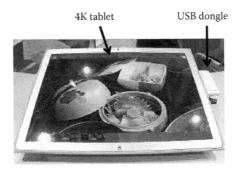

FIGURE 11.21
4K tablet with USB dongle prototype. (From [29], copyright © 2017 IEICE.)

11.5.1 IEEE 802.11ay: next-generation 60 GHz wireless

To develop an enhancement to IEEE 802.11ad/WiGig, Task Group of IEEE 802.11ay was established in May 2015. The project authorization request (PAR) includes backward compatibility and coexistence with legacy IEEE 802.11ad, maximum throughput of at least 20 Gb/s, while maintaining or improving the power efficiency per station [31]. The typical use cases are listed in Table 11.5, which require advanced technologies for enhanced throughput such as channel bonding and multiple-input and multiple-output (MIMO). IEEE 802.11ay releases the draft (D1.0) in 2017, targeting its practical realization around 2020.

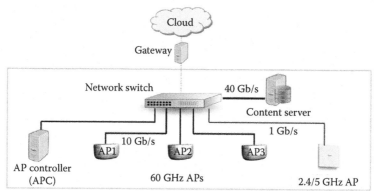

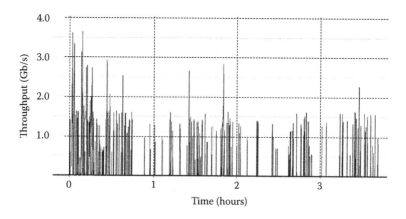

FIGURE 11.22

Prototype network system. (From [29], copyright © 2017 IEICE.)

FIGURE 11.23

Measured content server throughput in nominal operation. (From [29], copyright © 2017 IEICE.)

11.5.2 mmWave technologies toward 5G and beyond

The mmWave wireless is also expected as a key enabler for realizing efficient and high-capacity network required in future 5G networks. The signal characteristics unique to the mmWave frequency band, such as large path loss and high directivity, are suitable to establish densely distributed small cell networks to achieve multi-Gb/s throughput while minimizing interference among multiple APs and STAs. The drawback of large path loss as well as signal blockage due to obstacles can be alleviated by employing a heterogeneous network architecture, which allocates a lower frequency band as a control plane for robust connectivity and mobility management.

TABLE 11.5

IEEE 802.11ay Use Cases [30]

Use case
1 Ultra short-range communications
2 8K UHD wireless transfer at smart home
3 Augmented reality/virtual reality headsets and other high-end wearables
4 Data center 11ay interrack connectivity
5 Video/mass-data distribution/video on demand system
6 Mobile offloading and multiband operation
7 Mobile fronthauling
8 Wireless backhauling

On the other hand, as the number of mmWave devices increases, the backhaul congestion will become the throughput bottleneck. Therefore, to enable the full benefit of its high data rate, it is required to combine mmWave technology with the mobile edge computing/multi access edge computing (MEC), which brings computation and storage support at the edge of the network to avoid backhaul congestion [32]. Figure 11.24 shows a configuration example of future mmWave networks, where the edge cloud locates at the edge of the network, and it consists of an MEC server and a local storage. The mmWave small cells are deployed in densely populated area. To lower installation cost, it may be possible to use the mmWave backhaul whose communication distance can be extended to more than 150 m using a highly directional antenna [33]. Based on the users' context information (e.g. location, traffic and types of applications), the traffic and users' requests are forecast, and the users'

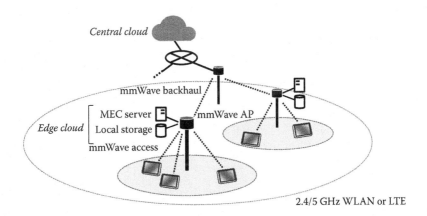

FIGURE 11.24

System architecture example composed of mmWave access and mobile edge cloud.

data are moved to the target local storage proactively. Then, the MEC server executes application computations with minimum latency.

Combining the mmWave technology and MEC, it will be made possible to realize ultra-high speed and low latency applications required for the future 5G networks, including mobile augmented reality applications, dynamic map update for automated driving, to name a few [32, 34]. The technologies required for the integration of MEC and mmWave in 5G are currently being investigated by 5G-MiEdge [35, 36], an EU-Japan cofunded research project started in July 2016.

11.6 Conclusion

This chapter has presented the design and experimental evaluation of 60 GHz wireless system based on IEEE 802.11ad/WiGig. The developed prototype system utilizes multiple APs that combine three RF modules with beamforming technology to achieve wide area coverage. As a proof of concept, the prototype system was set up at Narita International Airport and multiuser Gb/s wireless access was successfully demonstrated. The design and measurement results presented in this chapter will contribute toward developing future mmWave radio access technologies suitable for 5G networks.

Acknowledgment

Part of this work was supported by "The research and development project for expansion of radio spectrum resources" of The Ministry of Internal Affairs and Communications, Japan (MIC). Part of this work also has been done under a project named "Millimeter-wave Edge Cloud as an Enabler for 5G Ecosystem (5G-MiEdge)," supported by the European Commission Horizon 2020 and MIC. The authors would like to thank the members of Panasonic Corporation for their valuable advice and technical support.

Bibliography

[1] ARIB 2020 and Beyond Ad Hoc Group White Paper, "Mobile communications systems for 2020 and beyond, v1.0.0," Oct. 2014.

[2] ITU-R Administrative Circular CA/226, "Results of the first session of the conference preparatory meeting for WRC-19 (CPM19-1)," Dec. 2015.

[3] IEEE Std. 802.11ad, "Part 11: Wireless LAN medium access control (MAC) and physical layer (PHY) specifications, amendment 3: Enhancements for very high throughput in the 60 GHz band," Dec. 2012.

[4] Wireless Gigabit Alliance, "WiGig MAC and PHY specifications, version 1.3," Jun. 2013.

[5] C. Dehos, J. L. González, A. D. Domenico, D. Kténas, and L. Dussopt, "Millimeter-wave access and backhauling: The solution to the exponential data traffic increase in 5G mobile communications systems?" *IEEE Commun. Mag.*, vol. 52, no. 9, pp. 88–95, Sep. 2014.

[6] K. Sakaguchi, G. K. Tran, H. Shimodaira et al., "Millimeter-wave evolution for 5G cellular networks," *IEICE Trans. Electron.*, vol. E98-B, no. 3, pp. 388–402, Mar. 2015.

[7] K. Sakaguchi, E. M. Mohamed, H. Kusano et al., "Millimeter-wave wireless LAN and its extension toward 5G heterogeneous networks, (accepted for publication)," *IEICE Trans. Electron.*, vol. E98-B, no. 10, Oct. 2015.

[8] K. Okada, K. Kondou, M. Miyahara et al., "A full 4-channel 6.3 Gb/s 60 GHz direct-conversion transceiver with low-power analog and digital baseband circuitry," in *IEEE International Solid-State Circuits Conference (ISSCC) Digest of Technical Papers*, San Francisco, CA, Feb. 19–23, 2012, pp. 218–219.

[9] T. Tsukizawa, N. Shirakata, T. Morita et al., "A fully integrated 60 GHz CMOS transceiver chipset based on WiGig/IEEE802.11ad with built-in self calibration for mobile applications," in *IEEE International Solid-State Circuits Conference (ISSCC) Digest of Technical Papers*, San Francisco, CA, Feb. 17–21, 2013, pp. 230–231.

[10] M. Boers, I. Vassiliou, S. Sarkar et al., "A 16TX/16RX 60 GHz 802.11ad chipset with single coaxial interface and polarization diversity," in *IEEE International Solid-State Circuits Conference (ISSCC) Digest of Technical Papers*, San Francisco, CA, Feb. 9–13, 2014, pp. 344–345.

[11] A. M. Niknejad and H. Hashemi, *mm-Wave Silicon Technology, 60 GHz and Beyond*. Springer, New York, 2008.

[12] N. Guo, R. C. Qiu, S. S. Mo, and K. Takahashi, "60-GHz millimeter-Wave radio: Principle, technology, and new results," *Hindawi Publishing Corporation EURASIP J. Wirel. Commun. Netw.*, vol. 2007, no. 1, pp. 19–26, Jan. 2007.

[13] S. Emami, R. F. Wiser, E. Ali et al., "A 60 GHz CMOS phased-array transceiver pair for multi-Gb/s wireless communications," in *IEEE International Solid-State Circuits Conference (ISSCC) Digest of Technical Papers*, San Francisco, CA, Feb. 20–24, 2011, pp. 164–165.

[14] IEEE Std. 802.11, "Part 11: Wireless lan medium access control (MAC) and physical layer (PHY) specifications," Dec. 2016.

[15] B. Razavi, *RF Microelectronics, 2nd ed.* Prentice Hall, 2011.

[16] A. Siligaris, O. Richard, B. Martineau et al., "A 65 nm CMOS fully integrated transceiver module for 60 GHz wireless HD applications," in *IEEE International Solid-State Circuits Conference (ISSCC) Digest of Technical Papers*, San Francisco, CA, Feb. 20–24, 2011, pp. 162–163.

[17] T. Mitomo, Y. Tsutsumi, H. Hoshino et al., "A 2 Gb/s-throughput CMOS transceiver chipset with in-package antenna for 60 GHz short-range wireless communication," in *IEEE International Solid-State Circuits Conference (ISSCC) Digest of Technical Papers*, San Francisco, CA, Feb. 19–23, 2012, pp. 266–267.

[18] M. Tanomura, Y. Hamada, S. Kishimoto et al., "TX and RX front-ends for 60 GHz band in 90 nm standard bulk CMOS," in *IEEE International Solid-State Circuits Conference (ISSCC) Digest of Technical Papers*, San Francisco, CA, Feb. 3–7, 2008, pp. 558–559.

[19] C. Marcu, D. Chowdhury, C. Thakkar et al., "A 90 nm CMOS low-power 60 GHz transceiver with integrated baseband circuitry," in *IEEE International Solid-State Circuits Conference (ISSCC) Digest of Technical Papers*, San Francisco, CA, Feb. 8–12, 2009, pp. 314–315.

[20] K. Okada, K. Matsushita, K. Bunsen et al., "A 60 GHz 16QAM/8PSK/ QPSK/BPSK direct-conversion transceiver for IEEE 802.15.3c," in *IEEE International Solid-State Circuits Conference (ISSCC) Digest of Technical Papers*, San Francisco, CA, Feb. 20–24, 2011, pp. 160–161.

[21] V. Vidojkovic, G. Mangraviti, K. Khalaf et al., "Low-power 57-to-66 GHz transceiver in 40 nm LP CMOS with -17 dB EVM at 7 Gb/s," in *IEEE International Solid-State Circuits Conference (ISSCC) Digest of Technical Papers*, San Francisco, CA, Feb. 19–23, 2012, pp. 268–269.

[22] K. Takinami, J. Sato, T. Shima et al., "A 60 GHz CMOS transceiver IC for a short-range wireless system with amplitude/phase imbalance cancellation technique," *IEICE Trans. Electron.*, vol. E95-C, no. 10, pp. 1598–1609, Oct. 2012.

[23] B. Razavi, Z. Soe, A. Tham et al., "A low-power 60-GHz CMOS transceiver for WiGig applications," in *Proceedings of IEEE Symposium on VLSI Circuits (VLSIC)*, Kyoto, Japan, June 11–14, 2013, pp. 300–301.

[24] A. Yehezkely and O. Sasson, "Single transmission line for connecting radio frequency modules in an electronic device," U.S. Patent 0 307 695 A1, Dec. 6, 2012.

[25] E. Cohen, M. Ruberto, M. Cohen, O. Degani, S. Ravid, and D. Ritter, "A CMOS bidirectional 32-element phased-array transceiver at 60 GHz with LTCC antenna," *IEEE Trans. Microw. Theory Tech.*, vol. 61, no. 3, pp. 1359–1375, Mar. 2013.

[26] K. Takinami, H. Motozuka, T. Urushihara et al., "A 60 GHz hybrid analog/digital beamforming transceiver with interference suppression for multiuser Gigabit/s radio access," *IEICE Trans. Electron.*, vol. E99-C, no. 7, pp. 856–865, Jul. 2016.

[27] N. Saito, T. Tsukizawa, N. Shirakata et al., "A fully integrated 60-GHz CMOS transceiver chipset based on WiGig/IEEE 802.11ad with built-in self calibration for mobile usage," *IEEE J. Solid-State Circuits*, vol. 48, no. 12, pp. 3146–3159, Dec. 2013.

[28] T. Nakatani, T. Shima, and J. Sato, "Small and low-loss quadrature hybrid and T/R local signal selection switch for 60 GHz direct conversion transceivers," in *Proceedings of IEEE Topical Meetings on Silicon Monolithic Integrated Circuits in RF Systems (SiRF)*, Austin, TX, Jan. 21–23, 2013, pp. 3–5.

[29] K. Takinami, N. Shirakata, M. Kobayashi et al., "Design and experimental evaluation of 60 GHz multiuser Gigabit/s small cell radio access based on IEEE 802.11ad/WiGig," *IEICE Trans. Commun.*, vol. E100-C, no. 7, pp. 1075–1085, Jul. 2017.

[30] IEEE802.11-15/0625r3, "IEEE 802.11 TGay use cases," Sep. 2015.

[31] IEEE802.11-14/1151r8, "802.11 NG60 SG proposed PAR," Mar. 2015.

[32] K. Sakaguchi, T. Haustein, S. Barbarossa et al., "Where, when, and how mmWave is used in 5G and beyond," *IEICE Trans. Electron.*, vol. E100-C, no. 10, Oct. 2017.

[33] A. Maltsev, A. Sadri, A. Pudeyev, and I. Bolotin, "Highly directional steerable antennas: High-gain antennas supporting user mobility or beam switching for reconfigurable backhauling," *IEEE Veh. Technol. Mag.*, vol. 11, no. 1, pp. 32–39, Mar. 2016.

[34] K. Takahashi, H. Yomo, T. Matsuoka et al., "Evolution of millimeter-wave multi-antenna systems in the IoT era," *IEICE Trans. Electron.*, vol. E100-C, no. 10, pp. 809–817, Oct. 2017.

[35] "5G-MiEdge: Millimeter-wave edge cloud as an enabler for 5G ecosystem." [Online]. Available: https://5g-miedge.eu/.

[36] V. Frascolla, F. Miatton, G. K. Tran et al., "5G-MiEdge: Design, standardization and deployment of 5G phase 2 technologies," in *Proceedings of IEEE Conference on Standards for Communications & Networking (CSCN)*, Helsinki, Finland, Sep. 18–20, 2017.

12

Adaptive and Efficient Integrated Power Management Structures for Inductive Power Delivery

Hesam Sadeghi Gougheri and Mehdi Kiani

Pennsylvania State University

CONTENTS

12.1 Introduction

Wireless power transmission (WPT) is gaining more momentum for powering electronic devices without any direct electrical contact. Inductive coupling,

(a)   (b)   (c)   (d)

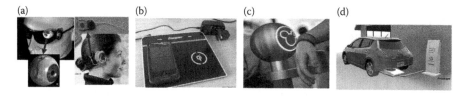

FIGURE 12.1
Some inductive coupling applications. (a) Biomedical applications, (b) charging mobile electronics, (c) RFID, and (d) charging electric cars [10–14].

coupled-mode magnetic resonance, capacitive coupling, and ultrasound are among conventional WPT techniques that have widely been investigated so far [1–9]. Inductive coupling is known as the most suitable method for short-range WPT due to its high power transmission efficiency (PTE) and safety. As shown in Figure 12.1, inductive coupling currently covers a wide range of applications with different power requirement from nanowatts in some wireless sensors and radiofrequency identification (RFID) tags to milliwatts in implantable medical devices (IMDs), watts in mobile electronics, and kilowatts in electric vehicles [10–14].

Wireless IMDs are good examples of where short-range WPTs using inductive coupling can be very effective to remove transcutaneous wires or bulky batteries with limited lifetime. IMDs can treat a wide range of ailments and disabilities from bradycardia [15] and chronic back pain to epilepsy [16] and deafness [17], and have the potential to alleviate more challenging types of disabilities such as blindness [18], paralysis [19], and loss of limbs [20]. Another category of IMDs that can highly benefit from inductive WPT is sensory prosthetic devices, which interface with the central nervous system to restore a sensory function, such as hearing or vision. These devices can stimulate neural tissue by means of tens to hundreds of stimulating channels [21]. Another well-known example for inductive WPT is an RFID that employs a pair of coils for interrogating (powering and communicating) RFID tags [10].

A key requirement in all of the aforementioned applications is to wirelessly deliver sufficient power to the receiver with high PTE in the worst-case conditions of powering distance, alignment, orientation, and loading variations. A typical inductive WPT system includes a power driver, an inductive link, and power management. Therefore, high performance (sufficient power delivery with optimal PTE) can be achieved if the power management can tolerate a wide range of input AC voltages (proportional to coils' distance, alignment, and orientation variations) as well as loading conditions. High PTE can lead to reduced tissue exposure to the RF magnetic field in IMDs, minimal heat dissipation within the coils, smaller size of the external energy source, and lower potential interference with nearby electronics [21–24].

12.2 Inductive Power Transmission Concept and Theory

Inductive coupling principle is based on Faraday's induction law, which states that when the total magnetic flux through a conductive loop varies with time, a current is induced in the loop itself [25]. Therefore, a primary loop generates the varying magnetic field, which concatenates with the secondary loop, resulting in an induced current in secondary loop as shown in Figure 12.2.

For the design of an inductive WPT link, a number of parameters directly associated with the magnetic coupling described earlier must be considered, such as self and mutual inductances. In the following, basic models to approximate self and mutual inductances for common coil configurations and their role in the design of the WPT link will be discussed. More details about theory of inductive coupling can be found in a number of other sources [26–29].

12.2.1 Inductive coupling

Self-inductance is the ratio of the magnetic flux generated in an area enclosed by a conductor loop to the current passing through the loop. Under the condition of $r/R \ll 1$, where r and R are the radii of the wire and the circular loop, self-inductance can be approximated by

$$L(R,r) \approx \mu_0 R \left(\ln \left(\frac{8R}{r} \right) - 2 \right),
\qquad (12.1)$$

where μ_0 is the permeability of the free space [25].

For the case of circular coils with N turns, if the coil length, d, is much smaller than R, the self-inductance is approximately equal to $N^2 L$, where L

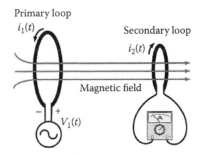

FIGURE 12.2
The principle of inductive coupling between two wire loops. The time-variant voltage across the primary coil, $V_1(t)$, generates a time-variant magnetic field, leading to induced current of $i_2(t)$ in the secondary loop.

is the self-inductance of a single-turn loop derived in Eq. (12.1). Whereas, for the case of planar spiral coils having N turns with different radii $R_i(i = 1, 2, \ldots N)$, the total self-inductance should be calculated from

$$L = \sum_{i=1}^{N} L(R_i, r) + \sum_{i=1}^{i=N} \sum_{j=1}^{j=N} M_{ij}(R_i, R_j, d_r = 0)(1 - \alpha_{i,j}), \qquad (12.2)$$

where $\alpha_{i,j} = 1$ if $i = j$, and $\alpha_{i,j} = 0$ otherwise [25].

M_{ij} is the mutual inductance between two conductor loops, which depends on the proportion of the magnetic flux generated by one loop that passes through the other loop. Therefore, it highly depends on coils' geometries, relative orientation, and magnetic properties of the medium. In a simplified case for two perfectly aligned parallel coaxial circular coils, separated by relative distance d_{12} in the air with the radius and number of turns of the first coil to be R_1 and N_1, and those of the second coil to be, R_2 and N_2, then,

$$M_{12}(R_1, R_2, d_{12}) = \frac{\pi \mu_0 N_1 R_1^2 N_2 R_2^2}{2\sqrt{\left(R_1^2 + d_{12}^2\right)^3}}. \qquad (12.3)$$

To more accurately calculate self and mutual inductances of coils with various geometries, one should either use tabulated parameterized equations [30, 31] or coil analysis software such as FastHenry or High Frequency Structure Simulator (HFSS).

In addition to distance and geometry, coils' alignment has a significant effect on their mutual inductance [32, 33]. For example, if one of the coils is tilted by an angle θ, their mutual inductance reduces by a factor of $\cos(\theta)$

$$M_{12}(d_{12}, \theta) = M_{12}(d_{12}, 0)\cos(\theta). \qquad (12.4)$$

We can normalize the mutual inductance between two coils to get a qualitative sense of how strongly they are coupled and compare the coupling between different pairs of coils. The coupling coefficient, k_{12}, between two coils with self-inductance, L_1 and L_2, is defined as

$$k_{12} = \frac{M_{12}}{\sqrt{L_1 \times L_2}} \qquad 0 \le k_{12} \le 1 \qquad (12.5)$$

Assuming $R_2 < R_1$, the coupling coefficient in this case can be approximated by [10]

$$k_{12}(d_{12}) = \frac{R_1^2 R_2^2}{\sqrt{R_1 R_2}\left(\sqrt{\left(R_1^2 + d_{12}^2\right)}\right)^3}. \qquad (12.6)$$

As a result of Eq. (12.4), k_{12} also depends on coils orientation and alignment,

$$k_{12}(d_{12}, \theta) = k_{12}(d_{12})\cos(\theta). \qquad (12.7)$$

According to Eq. (12.7), parallel and perfectly aligned coaxial coils would provide maximum k_{12}.

12.2.2 Power transmission efficiency of resonant inductive links

Figure 12.3 shows a simplified schematic diagram of a resonant inductive power transmission link. Within the transmitter side (Tx), a power source, modeled by an AC voltage source of V_s and a resistance of R_s, provides a time-variant current of $i_1(t)$ in the primary coil (L_1) that generates time-varying magnetic field in the secondary coil (L_2) due to the mutual coupling of M_{12} between L_1 and L_2. Within the secondary side (Rx), induced $i_2(t)$ in L_2 is delivered to the load (R_L). In Figure 12.3, R_1 and R_2 represent the loss in L_1 and L_2, respectively. It is well known that to achieve high PTE both L_1 and L_2 should operate at resonance by adding C_1 and C_2 resonance capacitors in Tx and Rx, respectively [10]. A detailed discussion of series vs. parallel connection of resonance capacitors has been provided [27].

For a sinusoidal input of V_s at the power carrier frequency of $\omega_p = 2\pi f_p$, the load voltage across the $L_2 C_2$-tank in parallel with R_L, i.e., V_L can be found from

$$V_L(j\omega) = j\omega M_{12} \cdot i_1 + j\omega L_2 \cdot i_2 + i_2 R_2. \tag{12.8}$$

Substituting i_2 with $-V_L/(R_L \| C_2)$ gives V_L as a function of i_1 and other circuit parameters as

$$V_L(j\omega) = \frac{j\omega M_{12} \cdot i_1}{1 + (j\omega L_2 + R_2)\left(\frac{1}{R_L} + j\omega C_2\right)}. \tag{12.9}$$

The power delivered to the load (PDL) can be found from

$$PDL = \frac{V_L{}^2}{2R_L}. \tag{12.10}$$

Therefore, to increase PDL for a constant R_L, $|V_L|$ in Eq. (12.9) should be maximized. To significantly increase $|V_L|$, C_2 and L_2 should resonate at the power carrier frequency (f_p) [28],

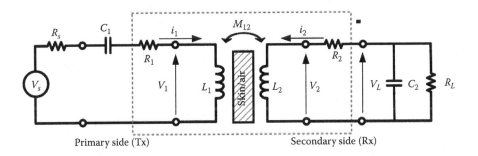

FIGURE 12.3

Schematic diagram of an inductive power transmission link with series and parallel resonance within the Tx and Rx sides, respectively.

$$f_{res} = \frac{1}{2\pi\sqrt{L_2 \cdot C_2}} = f_p. \qquad (12.11)$$

Inductive link PTE, which is key in efficient and safe WPT, is defined as the ratio of the PDL to the delivered power to the primary loop in Figure 12.3 and can be found from the reflected load theory [28]. In inductive links, the Rx effect on Tx can be modeled with a reflected impedance. To calculate the reflected impedance, Rx side at resonance can be modeled with only a parallel resistance equals to $R_P = R_{P2}\|R_L$, where $R_{P2} = Q_2^2 R_2$ is the equivalent parallel resistance of the $L_2 C_2$-tank. The quality factors of L_1 and L_2 coils are defined as $Q_1 = \omega_p L_1/R_1$ and $Q_2 = \omega_p L_2/R_2$, respectively [28]. Due to the $L_1 - L_2$ mutual coupling, the Rx side impedance can be reflected to the Tx side with a reflected resistance, R_{ref}, and capacitance, C_{ref}, as shown in Figure 12.4a.

$$R_{ref} = k_{12}^2(L_1/L_2)R_P = k_{12}^2\omega_p L_1 Q_{2L},$$
$$C_{ref} = (L_2/L_1)(C_2/k_{12}^2) = 1/(\omega_p^2 L_1 k_{12}^2). \qquad (12.12)$$

$Q_{2L} = R_P/\omega L_2$ is referred to as the loaded quality factor of L_2 [28]. In Figure 12.4a, it can be seen that C_{ref} resonates out with $k_{12}^2 L_1$ at ω_p, leaving behind only a resistance, R_{ref}, in the Tx side, as shown in Figure 12.4b. Since $L_1 - L_2$ are loosely coupled in most WPT applications that involve a large powering distance (k_{12} is very small), $(1 - k_{12}^2)L_1$ can safely be approximated to L_1.

In the simplified equivalent circuit in Figure 12.4b, L_1 and C_1 also resonate out at ω_p. Therefore, the input power provided by the source is divided between R_1 and R_{ref}. The consumed power by R_1 is wasted as heat in the Tx coil and the power delivered to R_{ref} represents the portion of the input power delivered to the Rx side. Within Rx, this delivered power is divided between R_2 as heat and R_L as the load power. Therefore, the PTE of the inductive link in Figure 12.3 can be found from

$$\eta_{12} = \frac{R_{ref}}{R_1 + R_{ref}}\frac{R_{P2}}{R_{P2} + R_L} = \frac{k_{12}^2 Q_1 Q_{2L}}{1 + k_{12}^2 Q_1 Q_{2L}} \cdot \frac{Q_{2L}}{Q_L}, \qquad (12.13)$$

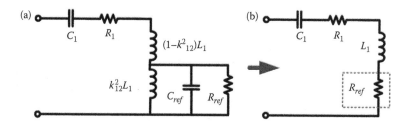

FIGURE 12.4
(a) Equivalent circuit diagram of the inductive link in Figure 12.3 with the effect of Rx shown on the Tx side. (b) C_{ref} resonates out with $k_{12}^2 L_1$ at the power carrier frequency, leaving behind R_{ref} as the only effect of Rx on the Tx side.

where $Q_{2L} = Q_2Q_L/(Q_2+Q_L)$ and $Q_L = R_L/\omega_p L_2$ are often referred to as the load quality factor [28]. According to Eq. (12.13), to achieve high PTE, k_{12}, Q_1, and Q_2 need to be maximized. However, for a given set of Q_1, Q_2 and k_{12} values, there is an optimal load, $R_{L,PTE} = \omega_p L_2 Q_{L,PTE}$, which can maximize the PTE at that particular arrangement. $R_{L,PTE}$ can be found by taking the derivative of PTE in Eq. (12.13) with respect to Q_L, leading to

$$R_{L,PTE} = \omega_p L_2 Q_{L,PTE} = \frac{Q_2}{(1 + k_{12}^2 Q_1 Q_2)^{1/2}}. \qquad (12.14)$$

12.3 Integrated Power Managements for Inductive Power Delivery

Figure 12.5 shows the generic block diagram of a conventional inductive WPT system. In the Tx side, an efficient power amplifier (PA) drives the Tx L_1C_1-tank at the resonance frequency of $\omega = 2\pi f = 1/(L_1C_1)^{1/2} = 1/(L_2C_2)^{1/2}$. Inside Rx, the AC voltage across the Rx L_2C_2-tank, V_R, is first converted to DC using an AC–DC converter, which often is a voltage rectifier. Then, a voltage regulator provides a constant voltage of V_L across R_L using a filtering capacitor (C_L).

There are four key parameters in inductive WPT system design: (1) delivered power to R_L (P_L or PDL), (2) overall PTE (η_{ov}) defined as P_L/P_S, where P_S is the PA input power, (3) power conversion efficiency (PCE) of the power management defined as P_L/P_R, where P_R is the power management input power, and (4) voltage conversion efficiency (VCE) within the Rx side defined as $V_L/V_{R,peak}$, where $V_{R,peak}$ is the amplitude of V_R in steady state. Taking into account the PA, inductive link, and power management losses, η_{ov} can be written as

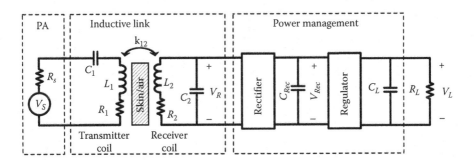

FIGURE 12.5
Generic block diagram of a conventional inductive WPT system.

$$\eta_{ov} = \eta_{PA} \times \eta_{IL} \times \eta_{PM}, \tag{12.15}$$

where η_{PA}, η_{IL}, and η_{PM} are the power efficiencies of the PA, inductive link, and power management, respectively.

In the design of inductive WPT systems, achieving high PTE and sufficient PDL should always be considered. But, maximizing PCE or VCE depends on $V_{R,peak}$. When $V_{R,peak}$ is larger than the required V_L, which is the case when the powering distance is relatively small and coils are well aligned, high PCE is more desirable to maximize η_{ov}, and VCE < 1 V/V is quite acceptable. However, for $V_{R,peak} < V_L$ with large d and/or misaligned coils, VCE > 1 V/V is paramount to achieve the required V_L. Therefore, for most WPT applications that involve powering distance and coils' orientation variations and misalignments, the power management should be smart enough to sense $V_{R,peak}$ and decide to whether maximize VCE or not.

12.3.1 Voltage-mode integrated power managements

12.3.1.1 Passive and active voltage rectifier and regulator

To achieve high power efficiency and small size, integrated power managements (IPMs) have become very popular over the past decade, particularly for small power levels, which is the focus of this chapter. Conventional IPMs, such as voltage rectifiers, utilize Rx LC-tank as a voltage source and rectify the AC input to extract the received power [33–43]. Therefore, such IPMs operate in voltage mode (VM). Passive voltage rectifiers with diode-connected transistors are among the first class of such IPMs [33, 34]. However, they suffer from poor PCE and VCE due to the large voltage drop across diode-connected transistors (large threshold voltage). Several threshold-voltage cancelation techniques have been proposed to reduce the forward-voltage drop of passive rectifiers [36, 37].

Active rectifiers with synchronous active switches can further reduce the forward-voltage drop and improve PCE. Figure 12.6 shows the block diagram of a full-wave active rectifier, in which two transistor switches (M_1 and M_2) are driven with high-speed comparators (A_1 and A_2), forming an active switch. When $V_R > 0$, M_4 is on and A_2 compares V_R with V_L. If $V_R > V_L$, A_2 outputs low to turn M_2 on and charge C_L. Similarly, when $V_R < 0$, M_3 is on and A_1 controls M_1 to charge C_L when $| V_R |> V_L$. Since M_1 and M_2 are used in the triode region with small drain-source voltage, active rectifiers can achieve much higher PCE compared with their passive counterparts.

Switching timing is key in active rectifiers to achieve maximum PCE, which makes it challenging to operate efficiently at high frequencies (>20 MHz). If switches are turned on too late, the desired forward current is reduced, resulting in poor PCE. If switches are turned off too late, a reverse current flows from C_L to the Rx LC-tank, resulting in poor PCE again. To achieve optimal switching timing, various comparator structures have been proposed. In [38], high-speed comparators with intentional offsets have been employed

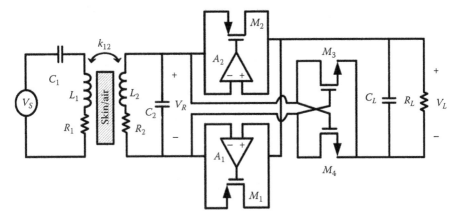

FIGURE 12.6
Schematic diagram of a full-wave active rectifier [38].

to compensate for the comparator delay during falling and rising transitions. Similarly, in Ref. [42] a sampling-based calibration technique with automatic circuit-delay compensation has been proposed to achieve optimal switching timing. An active rectifier with cross-coupled latched comparators has also been proposed in Ref. [41] to reduce the reverse current. Although active rectifiers can achieve high PCE, they still suffer from poor VCE of < 1.

As shown in Figure 12.5, conventional IPMs employ a two-step AC–DC conversion by adding a regulator after the rectifier. However, these structures further suffer from power loss in the regulator (low PCE). To mitigate this problem, combined rectification–regulation techniques, the concept of which is shown in Figure 12.7, have recently been proposed [44–47]. Unlike conventional active rectifiers with optimal on-time windows, these structures adjust the on-time duration of the rectifier to control the amount of forward current to simultaneously rectify V_R and self-regulate V_L with high PCE. However, these structures still suffer from low VCE < 1 as well as possible large voltage peaks across V_R, which can damage the IPM.

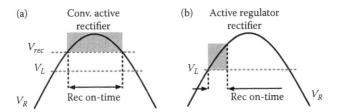

FIGURE 12.7
Simplified waveforms for (a) conventional two-step voltage rectification and regulation and (b) combined rectifier-regulator.

12.3.1.2 Passive and active voltage doubler

To enhance VCE, voltage doublers (see Figure 12.8), and in general voltage multipliers, have been proposed [48–53]. These structures utilize multiple capacitors and diodes (or diode-connected transistors) to boost the input voltage. Similar to passive rectifiers, passive voltage doublers suffer from large voltage drop across diodes. To mitigate this problem, active switches can be substituted for diode-connected transistors as shown in Figure 12.8. In the active voltage doubler, when $V_{R2} < 0$, A_2 outputs high to turn M_2 on in the triode region with a low dropout voltage of V_{DS,M_2}, and C_D is charged to $V_{R,peak} - V_{DS,M_2}$. Then, when $V_{R2} > V_L$, A_2 and A_1 output low to turn M_2 off and M_1 on in the triode region with a low dropout voltage of $V_{DS,M1}$, charging the load. Therefore, after several power carrier cycles, V_L reaches up to $2V_{R,peak} - V_{DS,M2} - V_{DS,M1}$. Ideally VCE can be close to two. However, commonly reported VCEs in the literature are < 1.7 due to the considerable drain-source dropout voltages, particularly at high power levels [48–50]. Similar to active rectifiers, switching timing is also key in active voltage doublers.

12.3.1.3 Reconfigurable active voltage doubler/rectifier

To take advantage of both high PCE in active rectifiers and high VCE in active doublers, a reconfigurable voltage doubler/rectifier has recently been proposed for applications that involve large $V_{R,peak}$ variations [54–56]. As shown in Figure 12.9, this IPM structure includes three main blocks: active voltage rectifier, active voltage doubler, and mode selection (MS). If $V_{R,peak}$ is larger than the required V_L considering the active switch voltage drop, the mode-selection block enables the active rectifier. Otherwise, it employs the voltage doubler to still charge C_L with smaller $V_{R,peak}$. As shown in [55], to save some chip area, voltage rectifier and doubler can share several transistors as active switches.

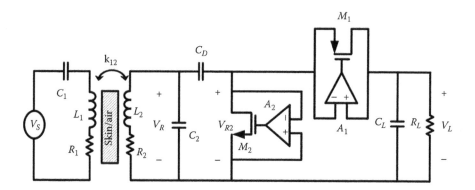

FIGURE 12.8
Schematic diagram of an active voltage doubler to achieve high VCE [48].

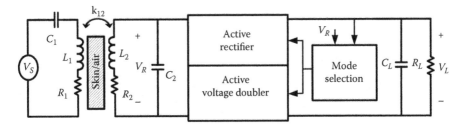

FIGURE 12.9
Schematic diagram of a reconfigurable active voltage doubler/rectifier [55].

12.3.2 Current-mode integrated power management

Unlike the VM structure, which employs the parallel-resonance Rx LC-tank as a voltage source, a current mode (CM) structure utilizes the series-resonance Rx LC-tank as a current source [57–64]. Conventionally, CM structures have been used for small R_L to achieve a high quality factor within Rx and consequently increase PTE. However, several CM IPM structures have recently been proposed that can significantly improve VCE and also provide load matching within Rx.

Figure 12.10 shows the conventional CM structure using a full-wave active rectifier [57]. When V_R and i_R are positive for a half cycle, M_2 and M_3 act as active switches with minimal resistance to charge C_L to a positive V_L. For the negative cycle with $i_R < 0$, active switches, M_1 and M_2, are turned on to again charge C_L to a positive V_L. It should be noted that R_L should be small enough (suitable for high power applications) to minimize its loading on the series-connected L_2C_2-tank. Although this structure can achieve high PCE for small R_L, the VCE is still less than one (Figure 12.10).

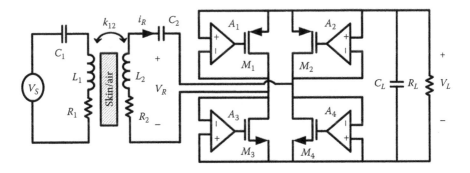

FIGURE 12.10
Schematic diagram of a current-mode full-wave active rectifier with series Rx LC-tank [57].

Recently, several switching IPM structures, operating in CM, have been proposed to improve PCE and VCE, as well as to achieve dynamic on-chip load matching [59–64]. As it can be seen in Eqs. (12.13) and (12.14), the inductive link PTE highly depends on R_L, which is often given by the application and can change during the operation (e.g., charging a battery). To achieve the optimal $R_{L,PTE}$ in Eq. (12.14), multicoil (three- and four-coil) inductive links and off-chip matching circuits have been used in the past to transform any given R_L to $R_{L,PTE}$ [28, 58]. But these solutions add to the size and cost of the system, and they still suffer from load mismatch with dynamic R_L changes during the operation.

A technique, called Q-modulation (Figure 12.11a), has recently been proposed to dynamically transform R_L to $R_{L,PTE}$ during the operation [59, 60]. As shown in Figure 12.11b, in Q-modulation the current of the Rx coil (i_{L2}) is sensed, and a switch (SW) shorts the series-connected Rx LC-tank at the zero-crossing times of i_{L2} for the duration of T_{on} in every half power carrier cycle ($T_p = 1/f_p$). By closing SW during Φ_1, which shorts R_L, a high-Q L_2C_2-tank is formed to store the maximum energy that is being delivered by the Tx side. During Φ_2, SW is opened and the L_2C_2-tank is connected to R_L to deliver its stored energy. Thus, the equivalent Q_{2L} and the amount of transferred power to the load can be modulated by controlling the duty cycle of SW (i.e., $D = 2T_{on}/T_p$).

The equivalent quality factor ($Q_{2L,eq}$) in Q-modulation can be found by calculating the ratio of the stored energy inside the L_2C_2-tank to the average power dissipation within the Rx side in the steady state

$$Q_{2L,eq} = \frac{\omega_p L_2}{R_2 + R_{SW}[D - \sin(2\pi D)/2\pi] + R_L[1 - D + \sin(2\pi D)/2\pi]}, \tag{12.16}$$

where R_{SW} is the switch resistance [60]. According to Eq. (12.16), when D increases from 0% to 100%, $Q_{2L,eq}$ can be adjusted from $\omega_p L_2/(R_2+R_L)$ to

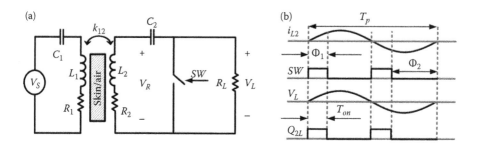

FIGURE 12.11

(a) A Q-modulation technique for dynamic transformation of R_L during operation. (b) Key switching waveforms to control $Q_{2L,eq}$ by the adjustment of $D = 2T_{on}/T_p$ [60].

$\omega_p L_2/(R_2+R_{SW})$. Therefore, D can be controlled dynamically during the operation to achieve optimal load quality factor $(Q_{L,PTE})$ in the presence of dynamic R_L variations.

Although the Q-modulation technique can increase V_R, V_L is still smaller than V_R particularly for large R_L, resulting in a small VCE < 1. The Q-modulation technique is suitable for applications that involve high power consumption, i.e., small R_L in the range of hundreds of ohms and below. In such conditions, Q-modulation helps to match small R_L to the larger equivalent resistance of the Rx LC-tank [60]. However, in applications that involve small power consumption, such as RFID and some IMDs, R_L is in the order of several kiloohms and above, in which the Q-modulation technique is not applicable [60].

To achieve VCE > 1 and also perform load matching for large R_L, recently a multicycle switching CM-resonant power delivery (CRPD) technique has been presented [63, 64]. In CRPD, only a single switch (SW) is added to the Rx side. Figure 12.12a and b shows the simplified circuit schematic and key operational waveforms of the CRPD technique, respectively. CRPD operation can be divided into three regions. In region (I), which is shown as $t_0 < t < t_1$ in Figure 12.12b, SW is closed for several power carrier cycles $(T_p = 1/f_p)$. Therefore, the high-Q L_2C_2-tank stores the energy provided by the inductive link. In region (II), which is shown as $t_1 < t < t_2$ in Figure 12.12b, at the peak of i_{L2}, i.e., $i_{L2,max}$, where the voltage across C_2 is zero and all the energy is stored in L_2, SW is opened to deliver all L_2 energy to the load for less than $0.25T_p$. At $t = t_2$, i_{L2} reaches zero and the switching state remains open in region (III), which is shown as $t_2 < t < t_3$, for $t_3 - t_1 = 0.25T_p$. At $t = t_3$ the next power cycle starts, SW is closed again, and i_{L2} starts building up.

The switching timing is key in CRPD to achieve the optimal performance, because the L_2C_2-tank should have enough time of several T_ps to store high amount of energy, and the energy transfer to the load needs to be started at $i_{L2,max}$ by opening SW for $T_{off} = 0.25T_p$. While the onset of switching SW is fixed in CRPD, its switching frequency, i.e., $f_{sw} = 1/T_{sw}$, is a degree of freedom that has been provided by CRPD, compared with conventional inductive links.

At $t = t_1$, since V_R is zero and the only path for discharging L_2 is the rectifier, V_R suddenly increases to $> V_D + V_L$, where V_D represents the forward voltage of the rectifier. Therefore, the steady-state peak voltage of V_R in region (I), i.e., $V_{R,peak}$, does not need to be higher than V_L, because L_2C_2-tank is being effectively used as a current source, leading to high VCE. Since the L_2C_2-tank is in series with $C_L \| R_L$ in regions (II) and (III) and C_L is much larger than C_2, L_2 is always at resonance in the CRPD technique.

Figure 12.13 shows V_L and PTE of a CRPD-based inductive link vs. f_{sw} for the R_L of 100 $k\Omega$ and $|Vs|$ of 0.39 V at the powering distance of 7 cm [64]. It can be seen that at the optimal f_{sw} of 50 kHz, maximum V_L and PTE of 3.1 V and 5.3% have been achieved, respectively. For f_{sw} much greater than 50 kHz, the L_2C_2-tank cannot store maximum energy and, therefore,

i_{L2} and V_R in Figure 12.12b are small, resulting in low V_L values. For f_{sw} much smaller than 50 kHz, L_2C_2-tank is shorted for a long period, and more energy is wasted into the L_2C_2-tank ($R_2 + R_{sw}$), resulting in low V_L again. It is worth noting that, for the same conditions, the conventional inductive

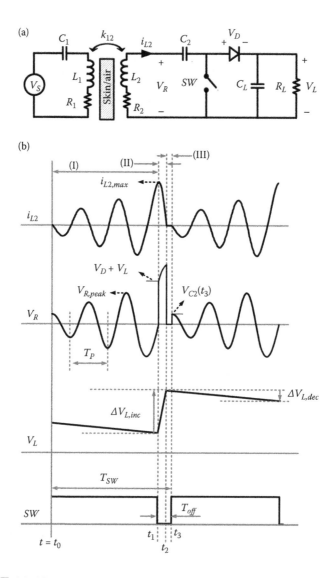

FIGURE 12.12
(a) Simplified circuit schematic of CM-resonant power deliver (CRPD) technique to achieve VCE > 1 and load matching for large R_L with only adding a single switch (SW). (b) Key operational waveform of the CRPD technique [64].

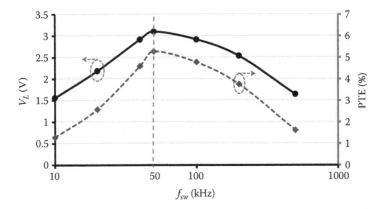

FIGURE 12.13
Measured V_L and PTE of the CRPD-based inductive link vs. f_{sw} for $R_L = 100$ kΩ [64].

link has achieved V_L and PTE of 0.95 V and 0.45%, respectively. Therefore, the CRPD-based inductive link has increased V_L and PTE by $\sim$ 3.3 and 11.8 times compared with the conventional inductive link, respectively.

Figure 12.14 shows PTE vs. R_L for both CRPD-based and conventional inductive links as well as the optimal f_{sw} to maximize V_L at each R_L, respectively [64]. It can be seen that for small R_L, f_{sw} should be increased to provide required high output power. For large R_L, optimal f_{sw} is decreased, because output load does not need high power, and therefore, L_2 has more time to store more energy and then deliver it to the output. The CRPD-based link achieved

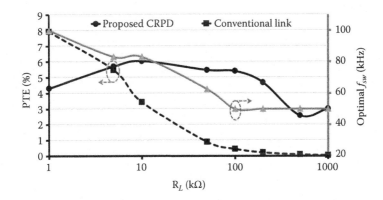

FIGURE 12.14
Measured PTE and optimal f_{sw} of the CRPD-based inductive link vs. R_L, compared with a conventional half-wave rectifier [64].

higher V_L and consequently PTE for $R_L \geq 10$ kΩ with the optimal f_{sw}s of 50–100 kHz. However, the conventional link was superior for $R_L < 5$ kΩ, at which the equivalent resistance of parallel-connected L_2C_2-tank was matched to R_L. Nonetheless, the CRPD-based link achieves higher V_L and PTE for a wide range of R_L partly due to impedance matching. As shown in Figure 12.14, the CRPD technique is mostly suitable for applications that either involve low-power consumption in the Rx side (large R_L), such as RFID and some IMDs, or require a duty-cycled high-power and high-voltage Rx, in which a large capacitor (C_L) is often charged through the inductive link and then discharged on a small R_L.

12.3.3 Self-regulated reconfigurable voltage/current-mode integrated power management

As discussed earlier, maximizing PCE or VCE of an inductive WPT system highly depends on the V_R amplitude. Therefore, neither VM-only IPMs with high PCE nor CM-only IPMs with high VCE are optimal in WPT applications with powering distance, alignment, orientation, and loading variations. To address this issue, a reconfigurable voltage/current-mode integrated power management (VCIPM) has recently been proposed with the capability of adaptively switching between VM and CM structures based on the V_R amplitude [65, 66].

Figure 12.15 shows the block diagram and die micrograph of a prototype VCIPM chip, which has been designed at the f_p of 1 MHz to regulate V_L at $V_{DD} = 3.2$ V. The VCIPM chip operates in either VM or CM based on the V_R amplitude using M_2 and $M_1 - M_3$ transistors, respectively, and performs rectification, regulation, and overvoltage protection (OVP) all in one step with a single off-chip capacitor (C_L). In VCIPM chip, V_R amplitude

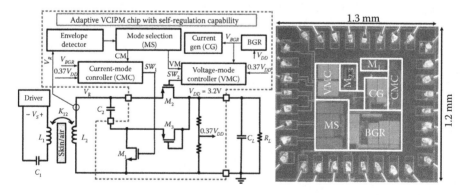

FIGURE 12.15
Block diagram of the self-regulated reconfigurable voltage/current-mode integrated power management (VCIPM) chip that can operate in either VM or CM based on the V_R amplitude [66].

is first detected by a passive envelope detector. Then, an MS block determines whether VCIPM chip should operate in VM (if $V_R > 3.2$ V) or CM (if $V_R \leq 3.2$ V) by enabling voltage-mode controller (VMC) or current-mode controller (CMC) blocks, respectively. If VMC is enabled, M_1 is turned on by setting $SW_1 = 3.2$ V, and M_2 is controlled by SW_2 to form a half-wave active rectifier. In VM, diode-connected M_3 is always off, because its source-gate voltage is negative. If CMC is enabled, M_2 is turned off by setting $SW_2 = 3.2$ V, and M_1 is controlled by SW_1. Self-regulation will also be achieved in VMC and CMC by adjusting SW_2 and SW_1 pulses, respectively. A bandgap reference (BGR) provides a constant 1.2 V, from which a reference bias current of 60 nA is generated by a current generator.

12.3.3.1 Voltage/current-mode self-regulation

As shown in Figure 12.7, conventional IPMs in VM control on-time duration (T_{on}) of active switches to regulate V_L at the desired level. This technique suffers from serious problems, particularly in cases where R_L or V_R are very large, because they require ultrasmall T_{on} (in range of several ns), the generation of which consumes extra power. Moreover, large V_R values in these techniques can damage the IPM.

The VCIPM chip performs self-regulation in both VM and CM. In VM, intentional reverse current from C_L to the Rx LC-tank (see Figure 12.15) is utilized for simultaneous voltage regulation and OVP. Figure 12.16 shows key waveforms of the VCIPM chip for voltage regulation in VM (and CM). For $t_0' < t < t_1'$, SW_2 is high (M_2: off), and L_2 and C_2 resonate since SW_1 is always high (M_1: on) in VM. For $t_1' < t < t_2'$, SW_2 becomes low (M_2: on) to provide a low-resistance path for i_{L2} to charge the load. For self-regulation and OVP, M_2 remains on intentionally for longer periods of time (T_d) to allow reverse current to flow from C_L to the Rx LC-tank. In other words, C_L detunes the Rx LC-tank and sends back excessive charge to the tank for regulation and OVP. Indeed, T_d is a new degree of freedom provided by this technique to adjust V_L at the desired level.

During CM operation, as shown in Figure 12.16, SW_2 is always high (M_2: off) and SW_1 is also high (M_1: on) for several cycles for L_2 and C_2 to resonate. Then, at the time zero crossings of V_R (maximum i_{L2}), SW_1 becomes low to turn M_1 off. This results in a sudden jump in V_R to $V_L + V_{GS3}$ (where V_{GS3} is gate-source voltage of M_3) to turn M_3 on and provide a path for i_{L2} to charge C_L. The switching frequency ($1/T_{sw}$ in Figure 12.16) is the degree of freedom in CM to adjust V_L at the desired level. Moreover, T_{sw1} can be used for impedance matching as discussed in detail in [64].

12.3.3.2 VCIPM chip circuits

Figure 12.17a and b shows the circuit diagrams and key operational waveforms of VMC and CMC, respectively. In VMC, a regulation amplifier (Reg_Amp in Figure 12.17a), controlling the bias current (I_{bias}) of the active rectifier

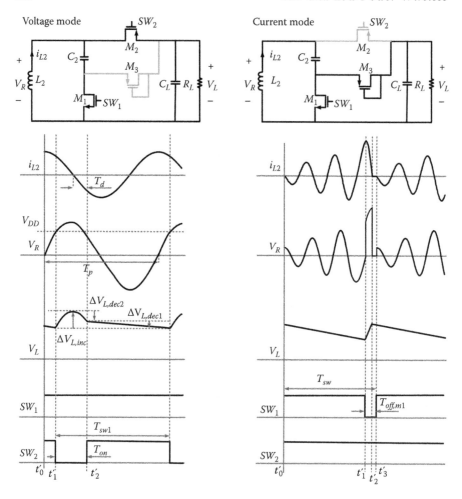

FIGURE 12.16
Key waveforms for self-regulation and OVP in VCIPM chip during VM and CM operation [66].

comparator (VM_Comp), amplifies the difference between V_L and required $V_{DD} = 3.2$ V by comparing $0.37 \times V_L$ with $V_{BGR} = 1.2$ V. If $V_L < 3.2$ V, this amplifier outputs low and I_{bias} is maximized. Therefore, VM_Comp operates at its maximum speed with intentional offsets so that it can maximize the forward current and minimize the reverse current to achieve the highest PCE as well as to quickly charge C_L and increase V_L. When V_L surpasses 3.2 V, Reg_Amp reduces I_{bias}, slowing down VM_Comp in turn-off, that allows reverse current from C_L to the L_2C_2-tank by increasing the width of SW_2 pulses (T_d in Figure 12.16), as clearly seen in Figure 12.17a inset waveforms.

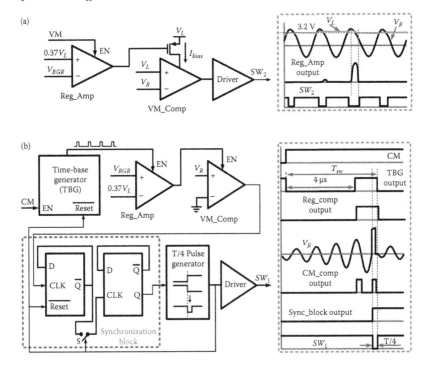

FIGURE 12.17
Schematic circuit diagrams and key waveforms of (a) VMC and (b) CMC blocks in VCIPM to generate proper SW_2 and SW_1 signals, respectively [66].

In CMC as shown in Figure 12.17b, a time-base generator (TBG), whenever it is reset, outputs high after 4 μs to enable a regulation comparator (Reg_Comp) that compares $0.37 \times V_L$ with $V_{BGR} = 1.2$ V. If $V_L < 3.2$ V, the CM comparator (CM_Comp) with an intentional offset of 170 mV is enabled by Reg_Comp to detect the time zero-crossings of V_R, where i_{L2} reaches its maximum, with the help of a synchronization block and consequently generates a sharp SW_1 pulse to charge C_L through M_3. The synchronization block includes two cascaded D-flip-flops that count two pulses to generate a transition, which is then converted to a short pulse (active low) with the width of $T_p/4$ by a pulse-generator block. The pulse-generator output controls M_1 with a driver (SW_1 pulses) and also resets D-flip-flops and TBG for the same process to be repeated. The inset in Figure 12.17b shows how a synchronization block can eliminate false CM_Comp pulses, which are not at the time zero crossings of V_R. The intentional offset in CM_Comp compensates for the circuit delays in CMC path, ensuring M_1 switching occurs at i_{L2} peaks. If $V_L > 3.2$ V, CM_Comp is disabled and, therefore, SW_1 remains high and C_L is not charged. It can be seen that f_{sw} is automatically adjusted to regulate V_L at 3.2 V. Since the synchronization block requires ~2 clock cycles (~ 2μs)

for synchronization and then resting TBG, the TBG delay is set to 4 μs to
achieve the maximum f_{sw} of 166.6 kHz.

12.3.3.3   VCIPM chip measurement results

Figure 12.18a and b shows the measured V_L and V_R waveforms in VM at $R_L =$
100 kΩ when the Tx voltage (V_s in Figure 12.15) has been increased from 11
to 15 V peak-to-peak, demonstrating that despite the V_s increase, the VCIPM
chip has adaptively adjusted the width of SW_2 pulses and consequently the
amount of reverse currents to regulate V_L at 3.2 V. For lower $V_s = 11\ V_{p-p}$,
since R_L of 100 kΩ is very large, demanding ultralow power in each cycle,
and the incoming power is moderate, the chip has generated large reverse
currents every $\sim$ 125 μs whenever V_L has exceeded 3.2 V, which has resulted
in a sudden voltage drop in V_R for several cycles to slightly reduce V_L below
3.2 V. Since the Rx LC-tank has received more power at $V_s = 15\ V_{p-p}$, the
VCIPM chip has employed the reverse current more frequently, seen as sudden
decreases in V_R, to regulate V_L. Since $V_{R,peak}$ is higher than 3.3 V, the chip
has automatically operated in VM. It should also be noted that, thanks to the
reverse-current regulation, V_R amplitude has been maintained fairly constant
despite the V_s increase.

Figure 12.19a and b shows the measured V_L, V_R, and V_s waveforms in
CM at $R_L = 100$ kΩ when V_s has been increased from 4 V_{p-p} to 9 V_{p-p},
demonstrating that for $V_s = 4\ V_{p-p}$, (1) since $V_{R,peak}$ is 1.2 V in the steady
state without switching ($< V_{DD} = 3.2$ V), the VCIPM chip has automatically
operated in CM, and (2) V_R has jumped from 1.2 to $\sim$ 5 V by turning M_1
off with proper SW_1 pulses to charge C_L to 3.2 V. Despite the V_s increase
to 9 V_{p-p}, in which $V_{R,peak}$ has increased to 2.9 V (still below 3.3 V), the
VCIPM chip has remained in the CM configuration and adaptively adjusted
f_{sw} to regulate V_L at 3.2 V. At the lower V_s of 4 V_{p-p}, resulting in less power

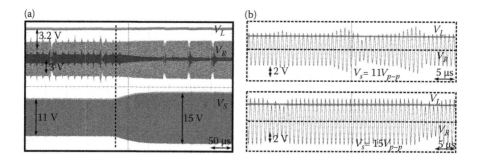

FIGURE 12.18
(a) Measured V_L and V_R waveforms in VM when the Tx voltage (V_s in Figure
12.15) has been increased from 11 to 15 V_{p-p} at $R_L = 100$ kΩ. (b) Zoomed
waveforms for V_L and V_R, demonstrating how reverse current has regulated
V_L at 3.2 V despite V_s variations [66].

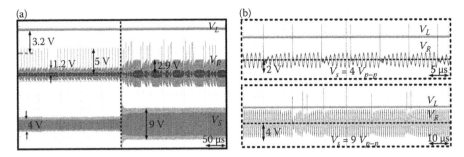

FIGURE 12.19
(a) Measured V_L and V_R waveforms in CM when V_s is increased from 4 V_{p-p} to 9 V_{p-p} at $R_L = 100k\Omega$. (b) Zoomed waveforms for V_L and V_R demonstrating how changes in f_{sw} has regulated V_L at 3.2 V despite V_s variations [66].

delivered to Rx, the chip has generated SW_1 pulses at the highest f_{sw} of 166.6 kHz to more frequently charge C_L. In contrast, at higher V_s of 9 V_{p-p} with increased received power, f_{sw} is automatically decreased to charge C_L less frequently and regulate V_L at 3.2 V. It should be noted that the VCIPM chip has achieved a high VCE of 2.7 V/V at $V_s = 4$ V_{p-p}. Nonetheless, the maximum measured VCE in VCIPM chip is 4.1 V/V at $f_{sw} = 166.6$ kHz, $R_L = 100$ kω, $V_L = 3.2$ V, and steady-state $V_{R,peak}$ of 0.78 V.

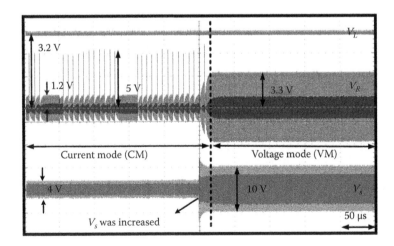

FIGURE 12.20
Measured V_L, V_R, and V_s waveforms when V_s is manually increased from 4 V_{p-p} to 10 V_{p-p}, resulting in the automatic reconfiguration of the VCIPM chip from CM to VM based on the V_R amplitude (1.2 V vs. 3.35 V) to regulate V_L at 3.2 V [66].

Figure 12.20 shows an automatic reconfiguration of the VCIPM chip from CM to VM when V_s is suddenly increased from 4 V_{p-p} to 10 V_{p-p} in measurements with $R_L = 100$ kΩ. At lower $V_s = 4V_{p-p}$, the steady-state $V_{R,peak}$ is 1.2 V, and therefore, the chip has operated in CM to regulate V_L at 3.2 V by large $V_{R,peak}$ of $\sim$ 5 V, i.e., operating with high VCE of 2.7 V/V. As V_s is increased to 10 V_{p-p}, $V_{R,peak}$ has gradually been increased to 3.35 V after $\sim$ 15μs (higher than required V_{DD} of 3.2 V), in which the VCIPM chip has automatically changed its configuration to VM. Figure 12.20 clearly shows that V_L remained constant at 3.2 V for a drastic change in V_R amplitude.

Figure 12.21a–c compares measured V_L between a conventional VM structure and the VCIPM chip vs. the powering distance (d), Rx coil's orientation (ϕ), and misalignment at $R_L = 100$ kΩ and fixed input power of 145 mW, respectively. Figure 12.21a shows that for generating a constant V_L of 3.2 V at the same V_s, the VCIPM chip can extend d from 6 to 13.5 cm for 125%, because at $d > 6$ cm the VCIPM chip has employed the CM configuration with high VCE to provide $V_L = 3.2$ V. As shown in Figure 12.21b, the VCIPM chip can extend ϕ for 150% from 30° to 75° thanks to the CM operation. Finally, the VCIPM chip has improved robustness against Rx coil misalignment at $d = 6$ cm for 500% from 1 to 6 cm.

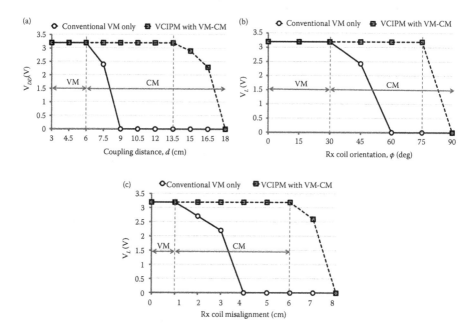

FIGURE 12.21
Measured V_L vs. (a) coupling distance, d, (b) Rx coil orientation, ϕ, and (c) Rx coil misalignment for conventional VM only and VCIPM chip at $R_L = 100$ kΩ and fixed input power of 145 mW. The VCIPM chip can extend robustness against d, ϕ, and misalignment for 125%, 150%, and 500%, respectively [66].

12.4 Conclusion

In this chapter, fundamental principles, design parameters, and different power-management structures for inductive WPT have been described. Fundamental equations governing self- and mutual-inductance between magnetically coupled coils have been reviewed. Also important geometric and circuit parameters that can affect the PTE of an inductive link have been specified. The inductive link PTE and PDL were formulated utilizing reflected load theory. It can be concluded that both PTE and PDL highly depend on the coupling between the coils, the primary and secondary coil quality factors, as well as the load resistance. To achieve the optimal PTE, load matching with multicoil links, discrete matching circuits, and switching power management structures have been discussed.

A review on different passive and active power management structures, operating in VM or CM, has been provided. The theory behind the Q-modulation technique for dynamic load-matching during operation has been discussed. The theory, concept, chip implementation, and measurement results of a reconfigurable VCIPM with self-regulation have also been provided. The VCIPM chip has achieved high VCE and PCE by operating in current and voltage modes, respectively. The VCIPM chip has also increased the inductive link PTE for large R_L values by dynamic load transformation in the CM configuration. By adjusting the reverse current in VM and f_{sw} in CM, regulation and OVP have been achieved along with rectification, eliminating the need for two off-chip capacitors. The VCIPM chip has achieved a maximum VCE of 4.1 V/V and extended the powering range by 125%.

Bibliography

[1] M. Baker, and R. Sarpeshkar, "Feedback analysis and design of RF power links for low-power bionic systems," *IEEE Trans. Biomed. Circuits Syst.*, vol. 1, pp. 28–38, Mar. 2007.

[2] G. Kendir, W. Liu, G. Wang, M. Sivaprakasam, R. Bashirullah, M. Humayun, and J. Weiland, "An optimal design methodology for inductive power link with class-E amplifier," *IEEE Trans. Circuits Syst. I*, vol. 52, pp. 857–866, May 2005.

[3] H. Sadeghi Gougheri, and M. Kiani, "Optimal frequency for powering millimeter-sized biomedical implants inside an inductively-powered homecage," *Proc. 38th Int. IEEE EMBS Conf.*, Orlando, FL, pp. 4804–4807, Aug. 2016.

[4] H. Haus, and W. Huang, "Coupled-mode theory," *Proc. IEEE*, vol. 79, pp. 1505–1518, Oct. 1991.

[5] A. Kurs, A. Karalis, R. Moffatt, J. D. Joannopoulos, P. Fisher, and M. Soljacic, "Wireless power transfer via strongly coupled magnetic resonances," *Sci. Expr.*, vol. 317, pp. 83–86, Jul. 2007.

[6] A. Sample, D. Meyer, and J. Smith, "Analysis, experimental results, and range adaptation of magnetically coupled resonators for wireless power transfer," *IEEE Trans. Ind. Electron.*, vol. 58, no. 2, pp. 544–554, Feb. 2011.

[7] S. Ozeria, D. Shmilovitza, S. Singera, and C. Wang, "Ultrasonic transcutaneous energy transfer using a continuous wave 650 kHz Gaussian shaded transmitter," *Ultrasonics*, vol. 50, pp. 666–674, Jun. 2010.

[8] M. Meng, and M. Kiani, "Design and optimization of ultrasonic wireless power transmission links for millimeter-sized biomedical implants," *IEEE Trans. Biomed. Circuits Syst.*, vol. 11, pp. 98–107, Feb. 2017.

[9] R. Erfani, F. Marefat, A. M. Sodaghar, and P. Mohseni, "Modeling and experimental validation of a capacitive link for wireless power transfer to biomedical implants," *IEEE Trans. Circuits Syst. II*, 2017.

[10] K. Finkenzeller, *RFID-Handbook*, 2nd ed. Hoboken, NJ: Wiley, 2003.

[11] D. Zhou, and E. Greenbaum, *Implantable Neural Prostheses 1*, New York: Springer, 2009.

[12] Near field communication (NFC) forum. Available: http://www.nfc-forum.org [accessed 19.02.14].

[13] S. Hui, and W. Ho, "A new generation of universal contactless battery charging platform for portable consumer electronic equipment," *IEEE Trans. Power Electron.*, vol. 20, pp. 620–627, 2005.

[14] C. Wang, O. Stielau, and G. Covic, "Design considerations for a contactless electric vehicle battery charger," *IEEE Trans. Ind. Electron.*, vol. 52, pp. 1308–1314, Oct. 2005.

[15] R. Allan, "Medtronic sets the pace with implantable electronics," *Electron. Design*, vol. 51, pp. 52–56, Oct. 2003.

[16] R. Fisher, "Direct brain stimulation is an effective therapy for epilepsy," *Neurology*, vol. 77, pp. 1220–1221, Sep. 2011.

[17] F. Zeng, S. Rebscher, W. Harrison, X. Sun, and H. Feng, "Cochlear implants: system design, integration, and evaluation," *IEEE Reviews Biomed. Eng.*, vol. 1, pp. 115–142, Nov. 2008.

[18] J. Weiland, and M. Humayun, "Visual prosthesis," *Proc. IEEE*, vol. 96, pp. 1076–1084, Jul. 2008.

[19] A. Schwartz, T. Cui, D. Weber, and D. Moran, "Brain-controlled interfaces: movement restoration with neural prosthetics," *Neuron*, vol. 52, pp. 205–220, Oct. 2006.

[20] T. Kuiken, L. Miller, R. Lipschutz, B. Lock, K. Stubblefield, P. Marasco, P. Zhou, and G. Dumanian, "Targeted reinnervation for enhanced prosthetic arm function in a woman with a proximal amputation: a case study," *Lancet*, vol. 369, pp. 371–380, Feb. 2007.

[21] L. Theogarajan, "Strategies for restoring vision to the blind: current and emerging technologies," *Neurosci. Lett.*, vol. 519, pp. 129–133, Jun. 2012.

[22] G. Lazzi, "Thermal effects of bioimplants," *IEEE Eng. Med. Biol. Mag.*, vol. 24, no. 5, pp. 75–81, Sep./Oct. 2005.

[23] IEEE Standard for Safety Levels With Respect to Human Exposure to Radio Frequency Electromagnetic Fields, 3 kHz to 300 GHz, IEEE Std. C95.1, 1999 Edition, pp. 1–83, 1999.

[24] Federal Communication Commission, Wireless Medical Telemetry. Available: http://www.wireless.fcc.gov/services/index.htm?job=service_home &id=wireless_medical_telemetry.

[25] M. Sadiku, *Elements of Electromagnetics*, 4th ed., New York, Oxford University Press, 2007.

[26] M. Kiani, and M. Ghovanloo, "The circuit theory behind coupled-mode magnetic resonance based wireless power transmission," *IEEE Trans. Circuits Syst.-I*, vol. 59, pp. 2065–2074, Sep. 2012.

[27] M. Kiani, and M. Ghovanloo, "A figure-of-merit for designing high performance inductive power transmission links," *IEEE Trans. Ind. Electron.*, vol. 60, pp. 5292–5305, Nov. 2013.

[28] M. Kiani, U. Jow, and M. Ghovanloo, "Design and optimization of a 3-coil inductive link for efficient wireless power transmission," *IEEE Trans. Biomed. Circuits Syst.*, vol. 5, pp. 579–591, Dec. 2011.

[29] U. Jow, and M. Ghovanloo, "Design and optimization of printed spiral coils for efficient transcutaneous inductive power transmission," *IEEE Trans. Biomed. Circuits Syst.*, vol. 1, pp. 193–202, Sep. 2007.

[30] F. Grover, *Inductance Calculations Working Formulas and Tables*, New York: D. Van Nostrand Company, 1946.

[31] F. Terman, *Radio Engineers Handbook*, New York: McGraw-Hill, 1943.

[32] M. Soma, D. C. Galbraith, and R. L. White, "Radio-frequency coils in implantable devices: misalignment analysis and design procedure," *IEEE Trans. Biomed. Eng.*, vol. 34, pp. 276–282, Apr. 1987.

[33] H. Sadeghi Gougheri, and M. Kiani, "Optimal wireless receiver structure for omnidirectional inductive power transmission to biomedical implants," *Proc. 38th Int. IEEE EMBS Conf.*, Orlando, FL, pp. 1975–1978, Aug. 2016.

[34] J. Ham, and R. Puers, "A power and data front-end IC for biomedical monitoring systems," *Sens. Actuators A*, vol. 147, pp. 641–648, Oct. 2008.

[35] C. Sauer, M. Stanacevic, G. Cauwenberghs, and N. Thakor, "Power harvesting and telemetry in CMOS for implanted devices," *IEEE Trans. Circuits Syst. I*, vol. 52, no. 12, pp. 2605–2613, Dec. 2005.

[36] F. Mounaim, and M. Sawan, "Integrated high-voltage inductive power and data-recovery front end dedicated to implantable devices," *IEEE Trans. Biomed. Circuits Syst.*, vol. 5, no. 3, pp. 283–291, Jun. 2011.

[37] J. Yoo, L. Yan, S. Lee, Y. Kim, H. Kim, B. Kim, and H.-J. Yoo, "A 5.2 mW self-configured wearable body sensor network controller and a 12 W wirelessly powered sensor for a continuous health monitoring system," *IEEE J. Solid State Circuits*, vol. 45, no. 1, pp. 178–188, Jan. 2010.

[38] H. Lee, and M. Ghovanloo, "An integrated power-efficient active rectifier with offset-controlled high speed comparators for inductively powered applications," *IEEE Trans. Circuits Syst. I, Regul. Pap.*, vol. 58, pp. 1749–1760, Aug. 2011.

[39] Y. Lu, and W.-H. Ki, "A 13.56 MHz CMOS active rectifier with switched offset and compensated biasing for biomedical wireless power transfer systems," *IEEE Trans. Biomed. Circuits Syst.*, vol. 8, no. 3, pp. 334–344, Jun. 2014.

[40] S. Guo, and H. Lee, "An efficiency-enhanced CMOS rectifier with unbalanced-biased comparators for transcutaneous-powered high-current implants," *IEEE J. Solid State Circuits*, vol. 44, no. 6, pp. 1796–1804, Jun. 2009.

[41] H. Cha, W. T. Park, and M. K. Je, "A CMOS rectifier with a cross-coupled latched comparator for wireless power transfer in biomedical applications," *IEEE Trans. Circuits Syst. II: Express Briefs*, vol. 59, no. 7 pp. 409–413, Jul. 2012.

[42] C. Huang, T. Kawajiri, and H. Ishikuro, "A near-optimum 13.56 MHz CMOS active rectifier with circuit-delay real-time calibrations for high-current biomedical implants," *IEEE J. Solid State Circuits*, vol. 51, no. 8, pp. 1797–1809, Aug. 2016.

[43] L. Cheng, W.-H. Ki, Y. Lu, and T.-S. Yim, "Adaptive on/off delay compensated active rectifiers for wireless power transfer systems," *IEEE J. Solid State Circuits*, vol. 51, no. 3, pp. 712–723, Mar. 2016.

[44] H. M. Lee, H. Park, and M. Ghovanloo, "A power-efficient wireless system with adaptive supply control for deep brain stimulation," *IEEE J. Solid State Circuits*, vol. 48, no. 9, pp. 2203–2216, Sep. 2013.

[45] E. Lee, "A timing controlled AC-DC converter for biomedical implants," *IEEE Int. Solid State Circuits Conf. (ISSCC) Dig. Tech. Pap.*, San Francisco, CA, pp. 128–129, Feb. 2010.

[46] C. Kim, S. Ha, J. Park, A. Akinin, P. P. Mercier, and G. Cauwenberghs, "A 144MHz integrated resonant regulating rectifier with hybrid pulse modulation," *Proc. Symp. VLSI Circuits*, Kyoto, Japan, pp. C284–C285, Jun. 2015.

[47] X. Li, C. Y. Tsui, and W. H. Ki, "A 13.56 MHz wireless power transfer system with reconfigurable resonant regulating rectifier and wireless power control for implantable medical devices," *IEEE J. Solid State Circuits*, vol. 50, no. 4, pp. 978–989, Apr. 2015.

[48] H. Lee, and M. Ghovanloo, "A high frequency active voltage doubler in standard CMOS using offset-controlled comparators for inductive power transmission," *IEEE Trans. Biomed. Circuits Syst.*, vol. 7, pp. 213–224, Jun. 2012.

[49] E. Lee, "A voltage doubling passive rectifier/regulator circuit for biomedical implants," *IEEE Custom Integrated Circuits Conf. (CICC)*, San Jose, CA, pp. 1–4, Sep. 2015.

[50] C. Wu, X. Qian, M. Cheng, Y. Liang, and W. Chen, "A 13.56 MHz 40 mW CMOS high-efficiency inductive link power supply utilizing on-chip delay-compensated voltage doubler rectifier and multiple LDOs for implantable medical devices," *IEEE J. Solid State Circuits*, vol. 49, no. 11, pp. 2397–2407, Nov. 2014.

[51] Z. Hameed, and K. Moez, "Hybrid forward and backward threshold-compensated RF-DC power converter for RF energy harvesting," *IEEE J. Emerging Sel. Topics Circuits Syst.*, vol. 4, pp. 335–343, Jul. 2014.

[52] Z. Chen, P. Yang, G. Zhou, J. Xu, and Z. Chen, "Variable duty cycle control for quadratic boost PFC converter," *IEEE Trans. Ind. Electron.*, vol. 63, pp. 4222–4232, Jul. 2017.

[53] J. Dias, and T. Lazzarin, "A family of voltage-multiplier unidirectional single-phase hybrid boost PFC rectifiers," *IEEE Trans. Ind. Electron.*, 2017.

[54] Y. Lu, X. Li, W. H. Ki, C. Y. Tsui, and C. Yue, "A 13.56 MHz fully integrated 1X/2X active rectifier with compensated bias current for inductively powered devices," *IEEE Int. Solid State Circuits Conf. (ISSCC) Dig. Tech. Pap.*, San Francisco, CA, pp. 66–67, Feb. 2013.

[55] H. Lee, and M. Ghovanloo, "An adaptive reconfigurable active voltage doubler/rectifier for extended-range inductive power transmission," *IEEE Int. Solid State Circuits Conf. (ISSCC) Dig. Tech. Pap.*, San Francisco, CA, pp. 286–288, Feb. 2012.

[56] C. Kim, J. Park, A. Akinin, S. Ha, R. Kubendran, H. Wang, P. P. Mercier, and G. Cauwenberghs, "A fully integrated 144 MHz wireless-power-receiver-on-chip with an adaptive buck-boost regulating rectifier and low-loss h-tree signal distribution," *Proc. Symp. VLSI Circuits*, Honolulu, HI, pp. C94–C95, Jun. 2016.

[57] L. Cheng, W. Ki, and C. Tsui, "A 6.78MHz single-stage wireless power receiver using a 3-mode reconfigurable resonant regulating rectifier," *IEEE J. Solid State Circuits*, vol. 52, no. 5, pp. 1412–1423, May 2017.

[58] R. F. Xue, K. W. Cheng, and M. Je, "High-efficiency wireless power transfer for biomedical implants by optimal resonant load transformation," *IEEE Trans. Circuits Syst. I*, vol. 60, no. 4, pp. 867–874, Aug. 2013.

[59] M. Kiani, B. Lee, P. Yeon, and M. Ghovanloo, "A power-management ASIC with Q-modulation capability for efficient inductive power transmission," *IEEE Int. Solid State Circuits Conf. (ISSCC) Dig. Tech. Pap.*, San Francisco, CA, pp. 226–227, Feb. 2015.

[60] M. Kiani, B. Lee, P. Yeon, and M. Ghovanloo, "A Q-modulation technique for efficient inductive power transmission," *IEEE J. Solid State Circuits*, vol. 50, pp. 2839–2848, Jul. 2015.

[61] B. Lee, P. Yeon, and M. Ghovanloo, "A multi-cycle Q-modulation for dynamic optimization of inductive links," *IEEE Trans. on Ind. Electron.*, vol. 63, pp. 5091–5100, Apr. 2016.

[62] M. Choi, T. Jang, J. Jeong, S. Jeong, D. Blaauw, and D. Sylvester, "A current-mode wireless power receiver with optimal resonant cycle tracking for implantable systems," *IEEE Int. Solid State Circuits Conf. (ISSCC) Dig. Tech. Pap.*, San Francisco, CA, pp. 372–373, Feb. 2016.

[63] H. Sadeghi Gougheri, and M. Kiani, "A resonant voltage multiplier for long-range inductive power transmission," *Texas Symp. Wireless Microwave Circuits Sys. (WMCS)*, Waco, TX, Mar. 2016.

[64] H. Sadeghi Gougheri, and M. Kiani, "Current-based resonant power delivery with multi-cycle switching for extended-range inductive power transmission," *IEEE Trans. Circuits Syst. I*, vol. 63, pp. 1543–1552, Sep. 2016.

[65] H. Sadeghi Gougheri, and M. Kiani, "An adaptive reconfigurable voltage/current-mode power management with self-regulation for extended-range inductive power transmission," *IEEE Int. Solid State Circuits Conf. (ISSCC) Dig. Tech. Pap.*, San Francisco, CA, pp. 374–375, Feb. 2017.

[66] H. Sadeghi Gougheri, and M. Kiani, "Self-regulated reconfigurable voltage/current-mode power management with self-regulation for extended-range inductive power transmission," *IEEE J. Solid State Circuits*, vol. 52, pp. 3056–3070, Nov. 2017.

Index

Page numbers followed by f indicate figures; those followed by t indicate tables.